AF333829

Electronic Structure Calculations
on Fullerenes and Their Derivatives

TOPICS IN PHYSICAL CHEMISTRY
*A Series of Advanced Textbooks and Monographs*

Series Editor, Donald G. Truhlar

F. Iachello and R. D. Levine, *Algebraic Theory of Molecules*

P. Bernath, *Spectra of Atoms and Molecules*

J. Simons and J. Nichols, *Quantum Mechanics in Chemistry*

J. Cioslowski, *Electronic Structure Calculations on Fullerenes and Their Derivatives*

# Electronic Structure Calculations on Fullerenes and Their Derivatives

JERZY CIOSLOWSKI

Professor of Theoretical Chemistry,
Florida State University

New York    Oxford
OXFORD UNIVERSITY PRESS
1995

Oxford University Press

Oxford   New York
Athens   Auckland   Bangkok   Bombay
Calcutta   Cape Town   Dar es Salaam   Delhi
Florence   Hong Kong   Istanbul   Karachi
Kuala Lumpur   Madras   Madrid   Melbourne
Mexico City   Nairobi   Paris   Singapore
Taipei   Tokyo   Toronto

and associated companies in
Berlin   Ibadan



Published by Oxford University Press, Inc.,
200 Madison Avenue, New York, New York 10016

Oxford is a registered trademark of Oxford University Press, Inc.



Library of Congress Cataloging-in-Publication Data
Cioslowski, Jerzy.
Electronic structure calculations on fullerenes
and their derivatives / Jerzy Cioslowski.
p.  cm.   (Topics  in physical chemistry series)
Includes bibliographical references and index.
ISBN 0-19-508806-9
1. Fullerenes.  I. Title.  II. Series.
QD181.C1C495  1995  546'.681422—dc20  94-43158

9  8  7  6  5  4  3  2  1

Printed in the United States of America
on acid-free paper

# Preface

The explosive growth of fullerene research during the last five years has resulted in an abundance of conferences, talks, and scientific publications. Benefiting from recent advances in quantum chemistry, a substantial portion of this research has dealt with electronic structures of fullerenes and related systems, such as fullerene derivatives, graphitic microtubules, and polymeric allotropes of carbon. As it is always the case with a rapidly changing scientific field, one finds it rather difficult to present all these developments in a concise manner without running the risk of almost instantaneous obsolescence. Such a risk notwithstanding, in this book I attempt to provide the reader with a self-contained, up-to-date account of the recent progress in quantum-chemical calculations on fullerenes and their derivatives.

The intended reading audience for this monograph is the broad community of chemists and physicists. I would like to see this book become a fullerene research primer for graduate students who are novices in the field of fullerenes. I also hope that experienced researchers will find it useful as a handbook from which diverse data on fullerenes can be conveniently extracted with minimal effort. Accordingly, the book has been organized around chapters that deal with individual classes of molecules. In addition, for the benefit of readers with a limited previous exposure to quantum chemistry, Chapter 2 is devoted to a short overview of electronic structure methods that are currently employed in fullerene research.

This book is the result of collective effort. My former post-docs Drs. Eugene D. Fleischmann and Asiri Nanayakkara have carried out some fullerene research mentioned herein. Dr. Stacey T. Mixon contributed to the early stages of this book project by assisting in the creation of the fullerene database. Mrs. Elena Kenderova has expertly performed most of the TeX formatting and has also been instrumental in the tedious task of finding typos and other bugs. Mr. Dick Roche has drawn all figures and Mrs. Marietta Peytcheva has provided invaluable secretarial support. My graduate students Dr. Matt Challacombe and Mr. Boris B. Stefanov have spent a lot of time and effort proofreading the text, while Mr. Martin Martinov has formatted most of the equations.

Over the past several years I have enjoyed many enlightening discussions on fullerenes with Profs. Helmut Schwarz, Gustavo E. Scuseria, Walter Thiel,

Martin Saunders, Drs. Krishnan Raghavachari, Rodney S. Ruoff, Brett I. Dunlap, John D. Mintmire, and Paul A. Cahill. This project would never have seen completion without the initial gentle prompting from Prof. Harry M. Walborsky ("... you got to do it ... so you will know how unpleasant it is ... and will never agree to write another book ...") and the never-faltering love and support of my wife Dr. Qing-Xiang Sang, to whom this book is dedicated.

J. C.

Tallahassee, FL
October 1994

# Contents

Electronic Structure Calculations
on Fullerenes and Their Derivatives

# Chapter 1

# Preliminaries and Historical Overview

*Fullerenes* are clusters composed of an even number of tricoordinated carbon atoms that are located at vertices of polyhedra with only pentagonal and hexagonal faces [1–3]. Topological considerations (Chapter 7) lead to the conclusion that fullerene cages possess exactly 12 five-membered rings. Six-membered rings are lacking in the smallest fullerene, which is derived from a dodecahedron and has 20 carbon atoms. For every even $N \geq 24$, at least one $C_N$ fullerene structure with both pentagons and hexagons is possible [4].

Among a myriad of fullerenes, only a few have been isolated and experimentally characterized. All of these stable species satisfy the *isolated pentagon rule (IPR)* that forbids the presence of abutting pentagons in their structures [1, 2, 5–7]. This very important rule makes it possible to focus fullerene research on a limited number of isomers. It can be justified on both steric and electronic grounds, as fused pentagons both increase the local strain and, forming a cycle of length eight, have a $\pi$-electron destabilizing effect on the total energy. The smallest fullerene that conforms to the isolated pentagon rule is the celebrated $C_{60}$ buckminsterfullerene (Chapter 3). The next smallest IPR fullerene structure is the $C_{70}$ species (Chapter 4). Cages composed of more than 70 carbon atoms give rise to multiple IPR isomers (Chapters 5 and 6). These isomers are often interrelated through the so-called Stone–Wales transformations in which a $C_2$ fragment is rotated by 90° [8].

As is often the case, theoretical research has predated experimental investigations of fullerenes. The Hückel molecular orbitals of the $C_{60}$ fullerene were calculated as early as in 1981 [9], and several papers dealing with simple electronic structure calculations on $C_{60}$ were cited in a 1984 review on hypothetical allotropes of carbon [10]. The possibility of experimental observation of buckminsterfullerene was first brought up in late 1985 in order to explain the unusual prominence of the $C_{60}$ peak in mass spectra of carbon vapors generated by laser ablation of graphite [11]. That initial report has triggered a flurry of theoretical activity, including the first MNDO calculations on $C_{60}$ [12] and the first *ab initio* HF/STO-3G study [13], both of which appeared in 1986. However, the icosahedral structure of the $C_{60}$ cluster has remained a mere speculation for the following five years as the isolation of buckminsterfullerene has eluded researchers until 1990. In that year macroscopic quantities of both $C_{60}$ and $C_{70}$ were obtained for the first time [14]. The history of this exciting discovery has been recently reviewed [15]. Higher fullerenes, including $C_{76}$ [16–18], $C_{78}$ [18–20], $C_{82}$ [18, 20], and $C_{84}$ [16, 18, 20, 21], were isolated and characterized in 1991 and 1992.

In a parallel development, a suggestion that metals such as lanthanum could be encapsulated in fullerene cages [22] has resulted in the emergence of an entirely new branch of chemistry. In 1987, the first MNDO calculations on the $Li^+$ cation located at the center of the $C_{60}$ fullerene were published [23], followed in 1988 by an LDA investigation of an analogous structure involving the La atom [24]. In 1991, the first *ab initio* study of atoms and ions encaged in $C_{60}$ was published and the name *endohedral complexes* was coined for these species [25]. Further calculations immediately followed [26, 27]. The existence of endohedral complexes involving noble gas atoms as guests was unequivocally proven by collisional insertion experiments carried out in the same year [28]. In 1993, macroscopic amounts of these complexes were obtained [29], and the $^3$He NMR spectrum of the $He@C_{60}$ species was recorded for the first time [30]. That year also witnessed the isolation of pure $La@C_{82}$ [31].

Several review articles on fullerenes and their derivatives are available. Among them, the 1989 review on carbon clusters [32] provides some perspective on the early fullerene research. Properties of the $C_{60}$ fullerene were reviewed in 1991 [33] and those of higher fullerenes in 1992 [34]. Endohedral metallofullerenes were the subject of a short review published in 1993 [35]. Early calculations on fullerenes and their derivatives were summarized in two chapters of a book that appeared in 1993 [36, 37]. Electronic structures of endohedral complexes were reviewed in 1992 [38].

The discovery of fullerenes has prompted several of the largest electronic structure calculations ever reported. The $C_{960}$ species has been one of the biggest molecules optimized with the MM3 method [39]. The largest MNDO calculations have been carried out for the $C_{540}$ fullerene [40]. In these calculations, which involved 2160 basis functions, each total energy evaluation required 2404 seconds of CPU time on a single-processor CRAY Y-MP supercomputer, whereas each energy gradient computation took 31 seconds. The total amount of CPU time needed to complete the geometry optimization was 12.3 hours. The $C_{60}F_{60}$ molecule has been the subject of the two largest *ab initio* Hartree–Fock calculations. In the first case, a dzP basis set resulted in 1800 molecular orbitals [41], whereas the second HF/6-31G*(5d) calculation employed 1680 basis functions [42]. The largest MP2 calculations have been carried out for the $C_{60}$ fullerene [43]. At the MP2/TZP level of theory (corresponding to 1140 basis functions), total energies at several geometries have been computed with a semidirect algorithm at a cost of 13 hours of the CRAY X-MP CPU time per point [43].

The need for large-scale calculations on fullerenes has provided incentives for the development of new electronic structure algorithms. Among these, the direct approach to GIAO CPHF computations of the NMR chemical shifts is particularly worth mentioning. The first calculations of this kind were carried out for the $C_{60}$ fullerene in 1992 [44]. The largest GIAO CPHF study reported to date has been that on the isomers of $C_{84}$, where each calculation involving 1596 basis functions has required between 4 and 7 days on the HP 735 workstation [45]. Direct RPA and CIS algorithms for the computation of the singlet and triplet excitation energies have also been developed and tested on the $C_{60}$ fullerene [46].

# References


1. H. W. Kroto, *The Stability of the Fullerenes $C_n$, with $n = 24$, 28, 32, 36, 50, 60, and 70*, Nature **329**, 529 (1987).
2. T. G. Schmalz, W. A. Seitz, D. J. Klein, and G. E. Hite, *Elemental Carbon Cages*, J. Am. Chem. Soc. **110**, 1113 (1988).
3. D. J. Klein, W. A. Seitz, and T. G. Schmalz, *Icosahedral Symmetry Carbon Cage Molecules*, Nature **323**, 703 (1986).
4. P. W. Fowler, J. E. Cremona, and J. I. Steer, *Systematics of Bonding in Non-icosahedral Carbon Clusters*, Theor. Chim. Acta **73**, 1 (1988).
5. T. G. Schmalz, W. A. Seitz, D. J. Klein, and G. E. Hite, *$C_{60}$ Carbon Cages*, Chem. Phys. Lett. **130**, 203 (1986).
6. H. W. Kroto and K. McKay, *The Formation of Quasi-icosahedral Spiral Shell Carbon Particles*, Nature **331**, 328 (1988).
7. D. E. Manolopoulos, *Proposal of a Chiral Structure for the Fullerene $C_{76}$*, J. Chem. Soc. Faraday Trans. **87**, 2861 (1991).
8. A. J. Stone and D. J. Wales, *Theoretical Studies of Icosahedral $C_{60}$ and Some Related Species*, Chem. Phys. Lett. **128**, 501 (1986).
9. R. A. Davidson, *Spectral Analysis of Graphs by Cyclic Automorphism Subgroups*, Theor. Chim. Acta **58**, 193 (1981).
10. I. V. Stankevich, M. V. Nikerov, and D. A. Bochvar, *The Structural Chemistry of Crystalline Carbon: Geometry, Stability, and Electronic Spectrum*, Russ. Chem. Rev. **53**, 640 (1984).
11. H. W. Kroto, J. R. Heath, S. C. O'Brien, R. F. Curl, and R. E. Smalley, *$C_{60}$: Buckminsterfullerene*, Nature **318**, 162 (1985).
12. M. D. Newton and R. E. Stanton, *Stability of Buckminsterfullerene and Related Carbon Clusters*, J. Am. Chem. Soc. **108**, 2469 (1986).
13. R. L. Disch and J. M. Schulman, *On Symmetrical Clusters of Carbon Atoms: $C_{60}$*, Chem. Phys. Lett. **125**, 465 (1986).

14. W. Krätschmer, L. D. Lamb, K. Fostiropoulos, and D. R. Huffman, *Solid $C_{60}$: A New Form of Carbon*, Nature **347**, 354 (1990).

15. D. R. Huffman and W. Krätschmer, *Solid $C_{60}$ — How We Found It*, Mat. Res. Soc. Symp. Proc. **206**, 601 (1991).

16. F. Diederich, R. Ettl, Y. Rubin, R. L. Whetten, R. Beck, M. Alvarez, S. Anz, D. Sensharma, F. Wudl, K. C. Khemani, and A. Koch, *The Higher Fullerenes: Isolation and Characterization of $C_{76}$, $C_{84}$, $C_{90}$, $C_{94}$, and $C_{70}O$, an Oxide of $D_{5h}$-$C_{70}$*, Science **252**, 548 (1991).

17. R. Ettl, I. Chao, F. Diederich, and R. L. Whetten, *Isolation of $C_{76}$, a Chiral ($D_2$) Allotrope of Carbon*, Nature **353**, 149 (1991).

18. K. Kikuchi, N. Nakahara, T. Wakabayashi, M. Honda, H. Matsumiya, T. Moriwaki, S. Suzuki, H. Shiromaru, K. Saito, K. Yamauchi, I. Ikemoto, and Y. Achiba, *Isolation and Identification of Fullerene Family: $C_{76}$, $C_{78}$, $C_{82}$, $C_{84}$, $C_{90}$, and $C_{96}$*, Chem. Phys. Lett. **188**, 177 (1992).

19. F. Diederich, R. L. Whetten, C. Thilgen, R. Ettl, I. Chao, and M. M. Alvarez, *Fullerene Isomerism: Isolation of $C_{2v}$-$C_{78}$ and $D_3$-$C_{78}$*, Science **254**, 1768 (1991).

20. K. Kikuchi, N. Nakahara, T. Wakabayashi, S. Suzuki, H. Shiromaru, Y. Miyake, K. Saito, I. Ikemoto, M. Kainosho, and Y. Achiba, *NMR Characterization of Isomers of $C_{78}$, $C_{82}$, and $C_{84}$ Fullerenes*, Nature **357**, 142 (1992).

21. D. E. Manolopoulos, P. W. Fowler, R. Taylor, H. W. Kroto, and D. R. M. Walton, *An End to the Search for the Ground State of $C_{84}$?*, J. Chem. Soc. Faraday Trans. **88**, 3117 (1992).

22. J. R. Heath, S. C. O'Brien, Q. Zhang, Y. Liu, R. F. Curl, H. W. Kroto, F. K. Tittel, and R. E. Smalley, *Lanthanum Complexes of Spheroidal Carbon Shells*, J. Am. Chem. Soc. **107**, 7779 (1985).

23. M. L. McKee and W. C. Herndon, *Calculated Properties of $C_{60}$ Isomers and Fragments*, J. Mol. Struct. (Theochem) **153**, 75 (1987).

24. A. Rosén and B. Wästberg, *First-Principle Calculations of the Ionization Potentials and Electron Affinities of the Spheroidal Molecules $C_{60}$ and $LaC_{60}$*, J. Am. Chem. Soc. **110**, 8701 (1988).

25. J. Cioslowski and E. D. Fleischmann, *Endohedral Complexes: Atoms and Ions Inside the $C_{60}$ Cage*, J. Chem. Phys. **94**, 3730 (1991).

26. A. H. H. Chang, W. C. Ermler, and R. M. Pitzer, *The Ground and Excited States of $C_{60}M$ and $C_{60}M^+$ (M=O, F, K, Ca, Mn, Cs, Ba, La, Eu, U)*, J. Chem. Phys. **94**, 5004 (1991).

27. J. Cioslowski, *Endohedral Chemistry: Electronic Structures of Molecules Trapped Inside the $C_{60}$ Cage*, J. Am. Chem. Soc. **113**, 4139 (1991).

28. T. Weiske, D. K. Böhme, J. Hrušák, W. Krätschmer, and H. Schwarz, *Endohedral Cluster Compounds: Inclusion of Helium within $C_{60}^{\bullet\oplus}$ and $C_{70}^{\bullet\oplus}$ through Collision Experiments*, Angew. Chem. Int. Ed. Engl. **30**, 884 (1991).

29. M. Saunders, H. A. Jiménez-Vázquez, R. J. Cross, S. Mroczkowski, M. L. Gross, D. E. Giblin, and R. J. Poreda, *Incorporation of Helium, Neon, Argon, Krypton, and Xenon into Fullerenes Using High Pressure*, J. Am. Chem. Soc. **116**, 2193 (1994).

30. M. Saunders, H. A. Jiménez-Vázquez, R. J. Cross, S. Mroczkowski, D. I. Freedberg, and F. A. L. Anet, *Probing the Interior of Fullerenes by $^3He$ NMR Spectroscopy of Endohedral $^3He@C_{60}$ and $^3He@C_{70}$*, Nature **367**, 256 (1994).

31. K. Kikuchi, S. Suzuki, Y. Nakao, N. Nakahara, T. Wakabayashi, H. Shiromaru, K. Saito, I. Ikemoto, and Y. Achiba, *Isolation and Characterization of the Metallofullerene $LaC_{82}$*, Chem. Phys. Lett. **216**, 67 (1993).

32. W. Weltner, Jr., and R. J. Van Zee, *Carbon Molecules, Ions, and Clusters*, Chem. Rev. **89**, 1713 (1989).

33. H. W. Kroto, A. W. Allaf, and S. P. Balm, *$C_{60}$: Buckminsterfullerene*, Chem. Rev. **91**, 1213 (1991).

34. F. Diederich and R. L. Whetten, *Beyond $C_{60}$: The Higher Fullerenes*, Acc. Chem. Res. **25**, 119 (1992).

35. D. S. Bethune, R. D. Johnson, J. R. Salem, M. S. de Vries, and C. S. Yannoni, *Atoms in Carbon Cages: The Structure and Properties of Endohedral Fullerenes*, Nature **366**, 123 (1993).

36. G. E. Scuseria, *Ab Initio Theoretical Predictions of Fullerenes*, in *Buckminsterfullerenes* (W. E. Billups and M. A. Ciufolini, eds.), VCH Publishers, New York, 1993, Ch. 5, p. 103.

37. C. T. White, J. W. Mintmire, R. C. Mowrey, D. W. Brenner, D. H. Robertson, J. A. Harrison, and B. I. Dunlap, *Predicting Properties of Fullerenes and their Derivatives*, in *Buckminsterfullerenes* (W. E. Billups and M. A. Ciufolini, eds.), VCH Publishers, New York, 1993, Ch. 6, p. 125.

38. J. Cioslowski, *Ab Initio Electronic Structure Calculations on Endohedral Complexes of the $C_{60}$ Cluster*, in *Spectroscopic and Computational Studies of Supramolecular Systems* (J. E. D. Davies, ed.), Kluwer Academic Publishers, Dordrecht, 1992, Ch. 10, p. 269.

39. M. Yoshida and E. Ōsawa, *Molecular Mechanics Calculations of Giant- and Hyperfullerenes with Eicosahedral Symmetry*, Full. Sci. Tech. **1**, 55 (1993).

40. D. Bakowies and W. Thiel, *MNDO Study of Large Carbon Clusters*, J. Am. Chem. Soc. **113**, 3704 (1991).

41. G. E. Scuseria, *Ab Initio Theoretical Predictions of the Equilibrium Geometries of $C_{60}$, $C_{60}H_{60}$ and $C_{60}F_{60}$*, Chem. Phys. Lett. **176**, 423 (1991).

42. J. Cioslowski, *Electronic Structures of the Icosahedral $C_{60}H_{60}$ and $C_{60}F_{60}$ Molecules*, Chem. Phys. Lett. **181**, 68 (1991).

43. M. Häser, J. Almlöf, and G. E. Scuseria, *The Equilibrium Geometry of $C_{60}$ as Predicted by Second-Order (MP2) Perturbation Theory*, Chem. Phys. Lett. **181**, 497 (1991).

44. M. Häser, R. Ahlrichs, H. P. Baron, P. Weis, and H. Horn, *Direct Computation of Second-Order SCF Properties of Large Molecules on Workstation Computers with an Application to Large Carbon Clusters*, Theor. Chim. Acta **83**, 455 (1992).

45. U. Schneider, S. Richard, M. M. Kappes, and R. Ahlrichs, *Ab Initio $^{13}C$ NMR Shifts of Several $C_{84}$ Isomers*, Chem. Phys. Lett. **210**, 165 (1993).

46. H. Weiss, R. Ahlrichs, and M. Häser, *A Direct Algorithm for Self-Consistent-Field Linear Response Theory and Application to $C_{60}$: Excitation Energies, Oscillator Strengths, and Frequency-Dependent Polarizabilities*, J. Chem. Phys. **99**, 1262 (1993).

# Chapter 2

# Theoretical Tools of Fullerene Research

Diverse theoretical techniques with varied degrees of sophistication are employed in research on electronic structures of fullerenes and their derivatives. Some of these methods give rise to complicated, time-consuming calculations that can only be carried out with commercial suites of programs such as GAUSSIAN 92 [1] and TURBOMOLE [2], whereas the others do not even require a calculator. From the researcher's point of view, the tradeoff between the cost of calculations and the reduced accuracy of the computed data is the single most important consideration in making the decision of which method should be used. For this reason, limitations of individual theoretical approaches have to be well understood before actual calculations can be performed with confidence.

## 2.1 The Born–Oppenheimer Approximation and the Electronic Hamiltonian

Within the nonrelativistic quantum mechanics, the stationary wavefunctions $\Psi$ of a chemical system composed of $N$ electrons and $M$ nuclei are the solutions of the *time-independent Schrödinger equation* [3],

$$\hat{H}\Psi = E\Psi , \tag{2.1}$$

with the *Hamiltonian operator* $\hat{H}$ given by

$$\hat{H} = -\frac{1}{2}\sum_{I=1}^{M} m_I^{-1}\vec{\nabla}_I^2 - \frac{1}{2}\sum_{i=1}^{N}\vec{\nabla}_i^2 + \frac{1}{2}\sum_{I\neq J=1}^{M} Z_I Z_J \left|\vec{R}_I - \vec{R}_J\right|^{-1}$$

$$- \sum_{I=1}^{M}\sum_{i=1}^{N} Z_I \left|\vec{R}_I - \vec{r}_i\right|^{-1} + \frac{1}{2}\sum_{i\neq j=1}^{N} \left|\vec{r}_i - \vec{r}_j\right|^{-1} . \tag{2.2}$$

In Eq. (2.2), which is written in atomic units, $m_I$, $Z_I$, and $\vec{R}_I$ are, respectively, the mass, the atomic number, and the instantaneous position of the nucleus $I$, whereas $\vec{r}_i$ denotes the instantaneous position of the $i$th electron. The five terms in Eq. (2.2) describe, respectively, the kinetic energy of nuclei, the kinetic energy of electrons, the repulsion between nuclei, the attraction between electrons and nuclei, and the electron–electron repulsion.

The *Born–Oppenheimer approximation* [4] takes advantage of the fact that nuclei are much heavier than electrons and thus can be treated as a source of an average potential to which the electrons are subjected. In other words, the total wavefunction $\Psi$ that depends on both $\{\vec{R}_I\}$ and $\{\vec{r}_i\}$ can be approximately written as a product of the electronic and nuclear wavefunctions. The *electronic wavefunction* $\Psi_e$ is an eigenfunction of the *electronic Hamiltonian*,

$$\hat{H}_e(\vec{R}_I) = -\frac{1}{2}\sum_{i=1}^{N}\vec{\nabla}_i^2 - \sum_{I=1}^{M}\sum_{i=1}^{N}Z_I\,|\vec{R}_I - \vec{r}_i|^{-1}$$
$$+ \frac{1}{2}\sum_{i\neq j=1}^{N}|\vec{r}_i - \vec{r}_j|^{-1}\,, \tag{2.3}$$

which depends parametrically on the nuclear coordinates. The respective eigenvalue $E_e(\{\vec{R}_I\})$, known as *electronic energy*, enters the nuclear Hamiltonian as a part of the potential,

$$\hat{H}_n = -\frac{1}{2}\sum_{I=1}^{M}m_I^{-1}\vec{\nabla}_I^2 + \frac{1}{2}\sum_{I\neq J=1}^{M}Z_I Z_J\,|\vec{R}_I - \vec{R}_J|^{-1}$$
$$+ E_e(\vec{R}_I)\,. \tag{2.4}$$

The ground-state eigenfunction of $\hat{H}_n$ is the nuclear wavefunction $\Psi_n$, which corresponds to the eigenvalue $E_n$.

The advantages stemming from the use of the Born–Oppenheimer approximation extend beyond simplification of the wavefunction. In fact, many chemical concepts, such as molecular geometries, potential energy hypersurfaces, transition states, and reaction paths, which are often taken for granted, can be defined only within the Born–Oppenheimer formalism. In particular, the *potential energy hypersurface* is given by the function

$$E_{BO}(\vec{R}_I) = E_e(\vec{R}_I) + \frac{1}{2}\sum_{I\neq J=1}^{M}Z_I Z_J\,|\vec{R}_I - \vec{R}_J|^{-1}\,. \tag{2.5}$$

The sets of the equilibrium $\{\vec{R}_I\}$ at which $E_{BO}$ attains its minima correspond closely to experimental geometries (determined by X-ray or electron diffraction) of individual chemical species. The corresponding values of $E_{BO}$ constitute good approximations to the total energies of these species. The accuracy of the total energy can be further improved by adding to $E_{BO}$ the so-called *zero-point energy (ZPE)*,

$$E_{ZPE} = E_n - \frac{1}{2} \sum_{I \neq J = 1}^{M} Z_I Z_J \left| \vec{R}_I - \vec{R}_J \right|^{-1} , \qquad (2.6)$$

evaluated at the equilibrium $\{\vec{R}_I\}$.

The Born–Oppenheimer factorization of the total wavefunction into the electronic and nuclear components is one of the least severe approximations of quantum chemistry. It works extremely well in most instances, except for the systems with degenerate or almost degenerate electronic energy levels. In the former case, a *Jahn–Teller distortion* ensues, breaking molecular symmetry in such a way that the degeneracy is lifted [5]. In the latter case, *symmetry lowering* is often observed.


## 2.2 The One-Electron Reduced Density Matrix and the Electron Density

In addition to the parametric dependence on the nuclear coordinates and the explicit dependence on the positions of electrons, full specification of the electronic wavefunction $\Psi_e$ requires inclusion of the electron spin [3]. The $z$ component of the electron spin $m_z$ can assume only two values, namely $-1/2$ ("spin down") and $1/2$ ("spin up"). Electrons with $m_z = -1/2$ are assigned the $\beta(m_z)$ spin function, whereas those with $m_z = 1/2$ are described by the $\alpha(m_z)$ function. Thus, for a system with $N$ electrons there are $3N$ Cartesian (spatial) and $N$ spin coordinates, for a total of $4N$ electronic degrees of freedom. The Cartesian coordinates $\{\vec{r}_i\}$ are often combined with their spin counterparts $\{m_{zi}\}$ to yield the set $\{\vec{x}_i\}$.

Approximations $\tilde{\Psi}_e$ to the electronic wavefunction $\Psi_e$ are usually obtained by minimizing the Rayleigh quotient,

$$\tilde{E}_e = \frac{\langle \tilde{\Psi}_e | \hat{H}_e | \tilde{\Psi}_e \rangle}{\langle \tilde{\Psi}_e | \tilde{\Psi}_e \rangle} . \qquad (2.7)$$

In Eq. (2.7) and in the following, the parametric dependences of $\tilde{\Psi}_e$, $\hat{H}_e$, $\tilde{E}_e$, and so on, on the nuclear coordinates $\{\vec{R}_I\}$ are omitted for the sake of clarity. By virtue of the *variational principle*, $\tilde{E}_e$ is an upper bound to the true energy $E_e$, provided the trial wavefunction $\tilde{\Psi}_e$ satisfies proper boundary conditions (*i.e.* vanishes at infinity). In addition, $\tilde{\Psi}_e$ must conform to the Pauli principle that implies the antisymmetry property,

$$\tilde{\Psi}_e \left( \vec{x}_1, ..., \vec{x}_i, ..., \vec{x}_j, ..., \vec{x}_N \right) = -\tilde{\Psi}_e \left( \vec{x}_1, ..., \vec{x}_j, ..., \vec{x}_i, ..., \vec{x}_N \right) , \qquad (2.8)$$

for all pairs of indices $(i, j)$.

In actual electronic structure calculations, it is often more convenient to work with the *reduced density matrices* [6]. The *one-particle reduced density matrix* (also known as the 1-*matrix*) is defined as

$$\Gamma\left(\vec{x}_1', \vec{x}_1\right) = \int \cdots \int \Psi_e^*\left(\vec{x}_1', \vec{x}_2, ..., \vec{x}_N\right) \Psi_e\left(\vec{x}_1, \vec{x}_2, ..., \vec{x}_N\right) d\vec{x}_2 \, \cdots \, d\vec{x}_N \,, \quad (2.9)$$

where the integration signs stand for integrations over spatial coordinates and summations over the respective spin coordinates. The 1-matrix can be written as an infinite sum,

$$\Gamma\left(\vec{x}_1', \, \vec{x}_1\right) = \sum_i n_i \, \psi_i^*\left(\vec{x}_1'\right) \psi_i\left(\vec{x}_1\right) \,, \quad (2.10)$$

which involves one-particle functions $\{\psi_i\}$ called *natural spinorbitals*. The natural spinorbitals form an orthonormal set of functions. The corresponding *occupation numbers* (or simply *occupancies*) $\{n_i\}$ sum up to $N$ and obey the inequalities $0 \leq n_i \leq 1$ for all $i$. Natural spinorbitals describing electrons with $m_z = 1/2$ and $-1/2$ are given by the $\phi_i(\vec{r}_1)\,\alpha(m_z)$ and $\phi_i(\vec{r}_1)\,\beta(m_z)$ products, respectively. The one-particle functions $\phi_i(\vec{r}_1)$ are known as *natural orbitals*.

The *electron density* $\rho(\vec{r})$ is given by

$$\rho\left(\vec{r}\right) = \sum_i n_i \, \phi_i^*\left(\vec{r}\right) \phi_i\left(\vec{r}\right) \,, \quad (2.11)$$

and satisfies several important conditions. In particular, $\rho(\vec{r})$ possesses cusps at $\vec{r}$ coinciding with any of the nuclear positions [7] and decays exponentially at large distances [8]. The electron density is the primary quantity of the *density functional theory* (Section 2.15).

## 2.3 *Ab Initio* Calculations and Basis Sets

Methods of quantum chemistry that follow the general philosophy of constructing approximate electronic wavefunctions $\tilde{\Psi}_e$ that minimize the corresponding energies $\tilde{E}_e$, Eq. (2.7), are called *ab initio* (meaning "from the beginning") approaches [9]. The overwhelming majority of the contemporary *ab initio* electronic structure calculations is based upon expansion of one-particle functions, such as the natural orbitals, in terms of *basis functions*. Although many different forms of basis functions are possible in principle, only the nuclei-centered *Cartesian Gaussian-type functions (CGTFs)*,

$$\phi_\nu\left(\vec{r}\right) = (x - X_I)^{l_x}(y - Y_I)^{l_y}(z - Z_I)^{l_z} \exp\left[-\zeta(\vec{r} - \vec{R}_I)^2\right] \,, \quad (2.12)$$

are used in *ab initio* calculations on large molecules such as fullerenes. In Eq. (2.12), $(l_x, l_y, l_z)$ is the angular momentum vector, $\zeta$ is the CGTF exponent, $\vec{r} \equiv (x, y, z)$, and $\vec{R}_I \equiv (X_I, Y_I, Z_I)$. Individual functions $\phi_\nu$, also known as *primitives*, are usually multiplied by fixed coefficients and combined into *contractions*. These contractions are the actual basis functions of *ab initio* methods. Sets of contractions are called *basis sets*.

In general, the accuracy of computed wavefunctions improves with the increasing number of basis functions. However, this improvement comes at the cost of increased computational effort. Over the last 30 years, vast experience with many commonly used basis sets, denoted by acronyms such as *STO-3G*, *DZP*, and 6-311++$G^{**}$, has been accumulated. These basis sets consist of a predetermined number of primitives with optimized exponents, combined into a predetermined number of contractions with optimized coefficients. The exponents and the contraction coefficients vary from one type of atom to another. Depending on the number of contractions, basis sets fall into different categories, including the *minimal* (such as *STO-3G* and *SZ*), *split-valence* (such as 4-31*G* and 6-31*G*), *double-zeta (DZ)*, *triple-zeta (TZ)*, *polarized* (such as 6-31$G^*$ and 6-311$G^{**}$), and *diffuse* (such as 6-311++G$^{**}$) ones. A comprehensive review that explains this terminology is available [10]. It also provides some advice on which basis sets should be used in different kinds of electronic structure calculations.

As mentioned above, the primitives are usually fixed for a given type of atom, meaning that they are independent of external fields. However, the requirement of gauge invariance in calculations of magnetic properties such as susceptibilities and shielding tensors dictates the use of *"gauge-including atomic orbitals" (GIAOs)*, which are field-dependent basis functions [11]. When ordinary basis functions are used instead of GIAOs, the computed magnetic properties depend on the choice of the coordinate system, which renders them practically useless.

## 2.4 The Hartree–Fock Approximation

Constraining occupation numbers of natural spinorbitals to either zero or one leads to the *Hartree–Fock (HF) approximation* [12]. This approximation is also known as the *self-consistent field (SCF) theory*. The $N$ natural spinorbitals with occupancies of one are called *occupied Hartree–Fock spinorbitals*. The other natural spinorbitals that are unoccupied are referred to as *virtual spinorbitals*. A general class of trial wavefunctions compatible with the above constraint is given by the *Slater determinants*,

$$\tilde{\Psi}_e(\vec{x}_1, ..., \vec{x}_N) = (N!)^{-1/2} \begin{vmatrix} \psi_1(\vec{x}_1) & ... & \psi_N(\vec{x}_1) \\ ... & ... & ... \\ \psi_1(\vec{x}_N) & ... & \psi_N(\vec{x}_N) \end{vmatrix}, \qquad (2.13)$$

built from the orthonormal set of occupied spinorbitals. One should note that the antisymmetry requirement, Eq. (2.8), is automatically satisfied by Slater determinants.

The variational energy, Eq. (2.7), obtained from the determinant (2.13) is called the *Hartree–Fock energy* $E_{HF}$ and depends solely on the occupied spinorbitals,

$$E_{HF} = \sum_i \langle \psi_i | \hat{h} | \psi_i \rangle + \frac{1}{2} \sum_{ij} \left( \langle ij|ij \rangle - \langle ij|ji \rangle \right) , \qquad (2.14)$$

where

$$\hat{h} = -\frac{1}{2}\, \vec{\nabla}^2 - \sum_{I=1}^{M} Z_I\, |\vec{R}_I - \vec{r}|^{-1} , \qquad (2.15)$$

is the one-electron *core Hamiltonian* and

$$\langle ij|kl \rangle = \iint \psi_i^*\left(\vec{x}_1\right) \psi_j^*\left(\vec{x}_2\right) \psi_k\left(\vec{x}_1\right) \psi_l\left(\vec{x}_2\right) d\vec{x}_1 d\vec{x}_2 , \qquad (2.16)$$

is the *electron repulsion integral.* In actual Hartree–Fock calculations, the spinorbitals $\{\psi_i\}$ are expressed in terms of $K$ basis functions $\{\psi_\nu\}$ (which are products of $\{\phi_\nu\}$ and the appropriate spin functions $\alpha$ or $\beta$),

$$\psi_i = \sum_{\nu=1}^{K} C_{\nu i}\, \psi_\nu , \qquad\qquad i = 1, \ ..., \ K , \qquad (2.17)$$

where $K \geq N$. The coefficients $\{C_{\nu i}\}$ are determined by extremizing $E_{HF}$ while maintaining orthonormality among $\{\psi_i\}$. This optimization results in a set of nonlinear equations,

$$\sum_\nu \left[ h_{\mu\nu} + \sum_{\lambda\sigma} P_{\lambda\sigma}\Big( \langle \mu\sigma|\nu\lambda \rangle - \langle \mu\sigma|\lambda\nu \rangle \Big) \right] C_{\nu i} = \epsilon_i \sum_\nu S_{\mu\nu}\, C_{\nu i} , \qquad (2.18)$$

which are satisfied for $1 \leq \mu, i \leq K$. The matrix elements of the core Hamiltonian,

$$h_{\mu\nu} = \langle \psi_\mu | \hat{h} | \psi_\nu \rangle , \qquad (2.19)$$

and the *overlap integrals*,

$$S_{\mu\nu} = \langle \psi_\mu | \psi_\nu \rangle , \qquad (2.20)$$

are collectively known as the *one-electron integrals.* In Eq. (2.18),

$$P_{\mu\nu} = \sum_{i \,\epsilon\, occ} C_{\nu i}^*\, C_{\mu i} , \qquad (2.21)$$

are the elements of the *density matrix*. The sum that appears in Eq. (2.21) is restricted to the occupied spinorbitals. The decision of which of the $K$ spinorbitals should be occupied is commonly based on their orbital energies $\{\epsilon_i\}$ (*the Aufbau principle*). Such an approach usually, but not always, produces the lowest $E_{HF}$ possible.

According to *Koopmans' theorem*, negative orbital energies of the occupied and virtual spinorbitals approximate ionization potentials (*IPs*) and electron affinities (*EAs*), respectively [13]. The assumption behind this theorem is that the spinorbitals of the parent system do not change upon electron detachment or attachment. Another way of computing *IPs* and *EAs*, called the $\Delta SCF$ *method*, is to carry out separate HF calculations for the parent system and the corresponding ions. The differences between the respective *IPs* or *EAs* obtained with these two methods are called the *orbital relaxation energies*. Due to a fortuitous cancellation of errors, Koopmans' theorem often yields *IPs* that are more accurate than those calculated with the $\Delta$SCF approach.

There are two variants of the Hartree–Fock approximation. In the *restricted Hartree–Fock (RHF) method*, the same orbitals $\{\phi_i\}$ are multiplied by the spin functions $\alpha$ and $\beta$, furnishing spinorbitals $\{\psi_i\}$ that possess pairwise identical spatial parts. In systems where the number of electrons with spins down does not equal that of electrons with spins up, some of the spinorbitals do not have opposite-spin counterparts. In such a case, one deals with the *restricted open-shell Hartree–Fock (ROHF) approximation*. In the *unrestricted Hartree–Fock (UHF) method*, the spatial parts of the $\alpha$ and $\beta$ spinorbitals are different. Eqs. (2.18) are often called the *Roothaan–Hall* [14, 15] and *Pople–Nesbet equations* [16] when used in the context of RHF and UHF calculations, respectively.

In general, because of the additional variational freedom, $E_{HF}$ obtained within the UHF formalism is lower than its RHF counterpart. However, RHF (or ROHF) wavefunctions are preferable to the UHF ones. This is so because the UHF wavefunctions are usually not eigenfunctions of the total spin operator $\hat{S}^2$. For this reason, the corresponding expectation values $\langle \tilde{\Psi}_e | \hat{S}^2 | \tilde{\Psi}_e \rangle$ differ significantly from those expected from the multiplicity of the electronic state. This phenomenon is known as *spin contamination*. For systems with no unpaired electrons (the *closed-shell systems*), the UHF energy is often the same as the RHF one. In such a case, the RHF wavefunction is called *triplet stable*. There are other possible instabilities of the Hartree–Fock wavefunctions, indicating that the HF energy is a saddle point rather than a minimum with respect to unitary transformations among spinorbitals [17]. Tests for these instabilities are well known and involve quantities that are closely related to the *random-phase approximation (RPA)*, which is commonly used in the calculation of excitation energies. The same quantities enter the *coupled-perturbed Hartree–Fock (CPHF) equations* that obtain when the second-order response properties, such as electric polarizabilities, are calculated within the Hartree–Fock approximation [18].

The *canonical Hartree–Fock spinorbitals* that are solutions of Eq. (2.18) are highly delocalized. However, since the Hartree–Fock energy is invariant to unitary transformations among occupied spinorbitals with the same spin part, it is possible to construct orbitals that are localized on atoms and bonds [19]. Such *localized molecular orbitals (LMOs)* lend themselves easily to chemical interpretation.

For reasons spelled out in Section 2.6, the standard enthalpies of formations of molecules cannot be accurately calculated from atomization energies obtained within the *ab initio* Hartree–Fock formalism. On the other hand, the use of the so-called *atom* [20] and *group equivalents* [21] makes it possible to calculate standard enthalpies of formation $\Delta H_f^0$ with average errors of only 1–3 kcal/mol. Specialized sets of group equivalents for converting the HF energies of benzenoid hydrocarbons obtained with the STO-3G, 3-21G, and 6-31G* basis sets to the corresponding values of $\Delta H_f^0$ are available [22]. For a hydrocarbon $C_n H_m$, the standard enthalpy of formation is given by

$$\Delta H_f^0 = E_{HF} - mE_{=\mathrm{CH}-} - (n - m)E_{=\mathrm{C}<} , \qquad (2.22)$$

where $E_{=\mathrm{CH}-}$ and $E_{=\mathrm{C}<}$ are the equivalents for the =CH– and =C< fragments, respectively. Standard enthalpies of formation computed from Eq. (2.22) differ from the respective experimental figures by no more than 1 kcal/mol.

## 2.5   The Conventional and Direct Self-Consistent Field Procedures

Since the equations of the Hartree–Fock approximation are nonlinear, they have to be solved in an iterative manner. In each iteration, the density matrix elements $\{P_{\mu\nu}\}$ are combined with two-electron integrals and added to the core Hamiltonian matrix elements $\{h_{\mu\nu}\}$ to form the left-hand side of Eq. (2.18). After the overlap integrals $\{S_{\mu\nu}\}$ are taken into account, the orbital energies $\{\epsilon_i\}$ and coefficients $\{C_{\nu i}\}$ are calculated. The choice is made which orbitals are occupied and which are virtual, and the new density matrix elements are computed from Eq. (2.21). The entire calculation is iterated until the orbital energies and coefficients change between iterations by less than predetermined thresholds. In practice, the aforedescribed iterative process is slightly more complicated, as various forms of convergence accelerators are employed. One such accelerator is the frequently used *direct inversion of iterative space (DIIS) algorithm* [23].

From the foregoing discussion it is clear that in every cycle of SCF calculations both one-electron and two-electron integrals are input. For a basis set consisting of $K$ real-valued functions, there are a total of $K(K+1)$ distinct overlap and core Hamiltonian integrals, and about $K^4/8$ distinct two-electron integrals. The relatively small number of one-electron integrals means that they can be calculated once and then stored in the core memory. On the other hand, two-electron integrals are usually too numerous to be stored in the main memory. One possible solution of this problem is to calculate the integrals once, dump them to a disk, and then read them back in each iteration. This approach is known as the *conventional SCF procedure*. An alternative to the conventional procedure is offered by the *direct SCF algorithm*, in which the integrals are recalculated on demand in each iteration [24].

For many years it has been believed that the quartic dependence of the number of two-electron integrals on the basis set size would make *ab initio* electronic structure calculations on large molecules prohibitively expensive. However, the number of integrals with significant contributions to the computed electronic properties grows only like $K^2$, as most of the $K^4/8$ integrals are very small. This observation, which was first made in 1973 [25], has opened the avenue for large-scale quantum-chemical calculations.

Contemporary SCF procedures assess batches of two-electron integrals for their contributions to the Hartree–Fock energy. In this prescreening process, which involves computationally cheap upper bounds [26, 27], the unimportant integrals are eliminated from further calculations, dramatically reducing the overall cost of calculations. Efficient evaluation of the remaining integrals is possible thanks to sophisticated algorithms that employ recursive formulas [28]. Additional savings are obtained by an incremental assembly of the Fock matrix [24, 29] and the use of symmetry blocking [30–32]. With all these improvements, direct SCF calculations that would otherwise require disk storages available only in a few largest computer centers are now carried out routinely on small workstations. Further details of the direct SCF method are discussed in a recent review [33].

A similar methodology can be applied to the CPHF and RPA formalisms (Section 2.4). Direct and semidirect algorithms have been developed for the purpose of large-scale CPHF [34] and RPA calculations [35]. They have been successfully tested on fullerenes.


## 2.6 Approximate Electron Correlation Methods

Even at the limit of a complete basis set, electronic properties calculated within the Hartree–Fock approximation differ from their exact counterparts. In particular, the exact electronic energy is lower than $E_{HF}$; the difference amounting to the *correlation energy $E_{corr}$*. Formalisms allowing practical calculations of exact electronic wavefunctions and the corresponding exact correlation energies for systems of chemical interest are presently unknown. However, several approximate electron correlation methods have been proposed. Among these, the *Møller–Plesset second-order perturbation theory (MP2)* [36, 37] is presently the only approach feasible for large-scale *ab initio* calculations of the ground-state properties.

The MP2 approximation to the correlation energy reads [12]

$$E_{corr} = \sum_{a<b} \sum_{r<s} \frac{|\langle ab \| rs \rangle|^2}{\epsilon_a + \epsilon_b - \epsilon_r - \epsilon_s} \,, \qquad (2.23)$$

where

$$\langle ab \| rs \rangle = \langle ab|rs \rangle - \langle ab|sr \rangle \,, \qquad (2.24)$$

is the *antisymmetrized electron repulsion integral* involving the canonical Hartree–Fock spinorbitals $\psi_a$, $\psi_b$, $\psi_r$, and $\psi_s$. In Eq. (2.23), the sums run over all pairs of occupied spinorbitals $(a, b)$ and all pairs of virtual spinorbitals $(r, s)$. Computation of the MP2 correlation energy requires a *four-index transformation* of the two-electron integrals over basis functions. These integrals, which are also used by the SCF procedure (Sections 2.4 and 2.5), can be either read from a disk or calculated on the fly. In the latter case, the intermediate quantities of the four-index transformation can be either stored on a disk (the *semidirect MP2 procedure*) [2, 38] or recalculated several times (the *direct MP2 procedure*) [38, 39]. The direct MP2 calculations are more expensive but, since they do not require any disk storage, can be carried out even for very large systems.

Molecular geometries optimized within the MP2 approximation are usually more accurate than those computed with the HF approach. In particular, application of the MP2 method eliminates the overestimation of bond alternation in benzenoid systems such as fullerenes. This improvement in accuracy is attributable to the fact that the HF method overestimates binding, predicting bonds that are too strong and too short. Therefore, it comes as no surprise that the computed MP2 vibrational frequencies are lower than their HF counterparts. Another deficiency of the HF approximation that is alleviated by the MP2 method is the exaggerated ionicity of bonds.

Both the HF and the MP2 approximations are unsuitable for a description of bond dissociation. For this reason, atomization energies calculated with these methods are often very inaccurate. However, excellent predictions of energy differences between molecules possessing the same numbers of analogous bonds are often obtained even with the Hartree–Fock method. The same is true about the standard enthalpies of *isodesmic reactions*, that is chemical transformations in which the numbers of bonds of each formal type are conserved [9, 40].

The *configuration interaction (CI)* formalism is the other electron correlation method often used in electronic structure calculations on fullerenes [12]. In its simplest version, called *CIS*, the CI electronic wavefunction is written as a linear combination of Slater determinants, each of them constructed by replacing one occupied spinorbital of the ground-state HF wavefunction by its virtual counterpart. The linear combination coefficients are obtained by minimizing the corresponding electronic energy through diagonalization of an appropriate Hamiltonian matrix. Although the ground-state energy is unaffected by such a procedure (thanks to *Brillouin's theorem*), energies of the excited states are. The CIS approximation and its extensions (including *CISD*) are often used in conjunction with semiempirical methods such as CNDO/S (Section 2.12) and INDO/S (Section 2.13) to calculate approximate electron excitation energies.

The *coupled-cluster singles and doubles (CCSD) approximation* [12, 41] has been used extensively in calculations on small carbon clusters (Chapter 7). The CCSD approach yields electronic energies and other properties that are more accurate than those obtained with the MP2 method. Moreover, the CCSD formalism is capable of correctly describing the dissociation of single bonds. The nonlinearity of CCSD equations implies that they have to be solved iteratively. Since direct CCSD procedures are not available at present and the calculations are computationally demanding, applications of this method are limited to small and medium-size molecules.

The *level of theory* is defined for a given electronic structure calculation as a combination of the quantum-chemical method and the basis set used. For example, the HF/STO-3G level of theory denotes a Hartree–Fock calculation performed with the STO-3G basis set. When the level of theory used in the calculation of electronic properties differs from that employed in geometry optimization, both levels are listed and separated by a double slash. For example, CCSD/DZP//MP2/DZ stands for a CCSD calculation employing the DZP basis set carried out at the geometry optimized at the MP2/DZ level of theory.

## 2.7 The Topological Hückel Hamiltonian and Its Properties

The highly sophisticated electron correlation methods discussed in Section 2.6, aim at quantitative prediction of electronic properties. The *Hückel molecular orbitals (HMO) formalism* [42–44] embodies an entirely different philosophy of quick and crude electronic structure calculations. In the HMO method, only the $\pi$ electrons of conjugated unsaturated hydrocarbons are taken into account. The $\pi$-electron orbital energy levels are obtained by diagonalization of the *Hückel matrix* **H**, which is given by

$$\mathbf{H} = \beta \mathbf{A} + \alpha \mathbf{1} \ , \tag{2.25}$$

where the empirical parameters $\alpha$ and $\beta$, which are both negative, are called the *Coulomb* and *resonance integrals*, respectively. Exact values of these parameters are unimportant because all electronic properties calculated within the Hückel approximation satisfy straightforward scaling relationships with respect to both $\alpha$ and $\beta$. The *adjacency matrix* **A** has a particularly simple structure,

$$A_{IJ} = \begin{cases} 1 & \text{if carbon atoms } I \text{ and } J \text{ are linked through a bond} \\ 0 & \text{otherwise} \end{cases} \ . \tag{2.26}$$

Although its exact physical meaning is unclear, $\mathbf{H}$ can be regarded as an effective Fock matrix, expressed in terms of orthogonalized $p_z$ orbitals centered at carbon atoms, that includes only the nearest-neighbor interactions.

The usefulness of the HMO formalism stems from the fact that the adjacency matrix lends itself easily to mathematical analysis. Since explicit diagonalization of $\mathbf{A}$ is often possible, exact Hückel orbital energies and coefficients can be calculated in many cases. Even more importantly, the general properties of $\mathbf{A}$ are well understood thanks to advances in the *chemical graph theory* [45]. The chemical graph theory disregards relative positions of atoms in the Cartesian space, focusing instead on the topology of bonds. In *molecular graphs*, the hydrogen atoms of conjugated hydrocarbons are discarded, the carbon atoms are replaced by graph vertices, and graph edges are put in place of the carbon–carbon bonds. The graphs can be freely deformed, provided their vertex connectivities are left unaltered. In particular, fullerene cages can be "squashed" onto a plane, furnishing the so-called *Schlegel diagrams*.

The *total $\pi$-electron energy* is obtained in the Hückel approach by adding the energies of all occupied orbitals,

$$E_\pi = \sum_{i \,\epsilon\, \text{occ}} n_i \, \epsilon_i \;, \tag{2.27}$$

and the *$\pi$-electron bond orders* are calculated as

$$P_{IJ} = \sum_{i \,\epsilon\, \text{occ}} n_i \, C_{Ii}^* \, C_{Ji} \;. \tag{2.28}$$

In Eqs. (2.27) and (2.28), $C_{Ii}$ is the $I$th component of the $i$th orbital and $n_i$ is its occupancy (either 1 or 2). The $\pi$-electron bond orders correlate well with the corresponding bond lengths (Section 2.8). Since the $\sigma$-electron energies are approximately proportional to their $\pi$-electron counterparts [46], the relative values of $E_\pi$ usually follow thermodynamic stabilities of conjugated hydrocarbons.

The influence of various structural motifs present in molecular graphs on the total $\pi$-electron energy is well understood. For example, according to the celebrated Hückel $4N + 2$ rule [42], $(4N + 2)$-membered rings confer aromaticity upon monocyclic systems, whereas $4N$-membered rings bring in antiaromaticity. A generalization of this rule is provided by the *conjugated circuits theory* [47, 48], which assigns energy increments to circuits, that is, even-membered rings of alternating single and double bonds. The destabilizing influence of circuits with length eight furnishes one possible explanation for the isolated pentagon rule (Chapter 1).

The Hückel *delocalization energy* is defined as the difference between the total $\pi$-electron energy of a given system (possessing $N$ carbon atoms) and that of $N/2$ isolated ethylene molecules,

$$E_{del} = E_\pi - 2N \;. \tag{2.29}$$

Delocalization energies of linear polyenes do not equal zero, which implies that values of $E_{del}$ should *not* be used for the purpose of assessing aromaticity. On the other hand, the *Dewar resonance energy (DRE)*,

$$E_{DRE} = E_\pi - a\,M_{\text{C}-\text{C}} - b\,M_{\text{C}=\text{C}} \; , \tag{2.30}$$

where $M_{\text{C}-\text{C}}$ and $M_{\text{C}=\text{C}}$ are, respectively, the numbers of formal single and double carbon–carbon bonds, is a proper measure of aromaticity [49]. The constants $a$ and $b$ that enter Eq. (2.30) are obtained by fitting the energies of acyclic polyenes. Calculations of the *Hess–Schaad resonance energy (HSRE)* [50, 51] and the *topological resonance energy (TRE)* [52] involve energies of fully localized reference structures obtained from either bond increments or the so-called acyclic polynomial. Dividing the values of HSRE and TRE by $N$ leads to resonance energies per $\pi$ electron, called *HSREPE* and *TREPE*, that are widely used as aromaticity measures.

## 2.8　The Variable-$\beta$ and Curvature-Corrected HMO Methods

The lack of dependence of resonance integrals on both bond lengths and relative orientations of the $p_z$ orbitals is the major shortcoming of the simple (topological) Hückel approach. The former deficiency is rectified in the *variable-$\beta$ HMO (or self-consistent $\beta$-HMO) method*, in which the off-diagonal elements of the Hückel matrix are given as

$$H_{IJ} = \beta_{IJ}\,A_{IJ} \; , \tag{2.31}$$

with the resonance integral $\beta_{IJ}$ dependent on the corresponding distance $R_{IJ}$ between nuclei $I$ and $J$ [53]. The dependence of $\beta_{IJ}$ on $R_{IJ}$ is usually assumed to be exponential. The distances $\{R_{IJ}\}$ are calculated for pairs of adjacent nuclei from an expression involving the respective $\pi$-electron bond orders,

$$R_{IJ} = a - b\,P_{IJ} \; , \tag{2.32}$$

where $a$ and $b$ are positive constants.

Application of either the simple or the variable-$\beta$ HMO method to three-dimensional systems, such as fullerenes, raises questions about its validity in situations where the $p_z$ orbitals of carbon atoms are no longer parallel. The extent of orbital misalignment can be assessed with the help of the *$\pi$-orbital axis vector (POAV) method* [54, 55]. In the POAV approach, the hybrid $\sigma$ orbitals of each tricoordinated carbon atom are assumed to extend in the directions of three bonds. The requirement that the $p_z$-like $\pi$ orbital be orthogonal to these $\sigma$ hybrids yields its direction in space. Once the mutual orientation of two adjacent $\pi$ orbitals is determined from this simple geometrical principle, the corresponding resonance integral can be scaled down by a product of the cosines of two tilt angles and the cosine of the twist angle. This *curvature-corrected HMO* (or *3D HMO) method* is clearly superior to its uncorrected congener [56–58]. In particular, when applied to fullerenes, it yields energies that correlate well with those obtained with more sophisticated approaches (such as MNDO). In contrast, the energies furnished by topological HMO calculations do not follow their all-electron counterparts at all.

## 2.9  The PPP, Hubbard, and Gutzwiller Approximations

The *Pariser–Parr–Pople (PPP) method* [59–61] is based upon two theoretical assumptions, namely, the *σ-π separability* and the *zero differential overlap (ZDO) approximation*. The concept of $\sigma$-$\pi$ separability, which is also employed in the HMO approach (Sections 2.7 and 2.8), allows one to construct a $\pi$-electron Hamiltonian, which is then expressed in the basis of $p_z$ orbitals. Further simplification is provided by the ZDO approximation [62],

$$\psi_\nu^* \, \psi_\mu \approx 0 \,, \qquad \text{for } \nu \neq \mu \,, \qquad (2.33)$$

which is applied to the electron repulsion integrals. Under these assumptions, the Roothaan–Hall equations of the RHF method become

$$\sum_J F_{IJ} \, C_{Ji} = \epsilon_i \, C_{Ii} \,, \qquad (2.34)$$

where the matrix $\mathbf{F}$ has the elements

$$F_{IJ} = \begin{cases} U_I + \frac{1}{2}P_{II} \, \gamma_{II} + \sum_{K \neq I} (P_{KK} - Z_K) \, \gamma_{IK} & \text{for } I = J \\[2ex] \beta_{IJ} \, A_{IJ} - \frac{1}{2} \, P_{IJ} \, \gamma_{IJ} & \text{for } I \neq J \end{cases} \,. \qquad (2.35)$$

In Eq. (2.35), $\{A_{IJ}\}$ and $\{P_{IJ}\}$ are the elements of the adjacency and the $\pi$-electron bond-order matrices, Eqs. (2.26) and (2.28), respectively. Because of the explicit dependence of $\mathbf{F}$ on $\mathbf{P}$, the PPP equations have to be solved iteratively. Since PPP is a *semiempirical method*, the quantities $U$, $\beta$, and $\gamma$ that enter Eq. (2.35) are estimated from both atomic data and empirical formulas. In particular, the resonance integrals $\{\beta_{IJ}\}$ are either set proportional to the corresponding overlaps of the $p_z$ orbitals or related to $\{R_{IJ}\}$. One should note that the former procedure is preferred, as it automatically corrects resonance integrals for the curvature effects (Section 2.8). The *core integrals* $\{U_I\}$ are approximated by the energy of the $p_z$ orbital in a carbon atom. The *core charges* $\{Z_K\}$ are equal to one for carbons. The electron repulsion integrals,

$$\gamma_{IJ} \equiv \langle IJ | IJ \rangle \,, \qquad (2.36)$$

are calculated from empirical expressions, such as the *Ohno–Klopman–Sabelli formula* [63–66],

$$\gamma_{IJ} = \left(\bar{\gamma}_{IJ}^{-2} + R_{IJ}^2\right)^{-1/2} , \qquad (2.37)$$

or the *Mataga–Nishimoto formula* [67],

$$\gamma_{IJ} = \left[2\left(\gamma_{II} + \gamma_{JJ}\right)^{-1} + R_{IJ}\right]^{-1} , \qquad (2.38)$$

where $\bar{\gamma}_{IJ}$ is a harmonic average of $\gamma_{II}$ and $\gamma_{JJ}$. The one-center repulsion integrals $\{\gamma_{II}\}$ are approximated by differences between the $\pi$-electron ionization potentials and electron affinities of the corresponding atoms.

At present, the PPP method is often used in conjunction with the configuration interaction formalism involving singly excited determinants (CIS, Section 2.6) to calculate the $\pi \to \pi^*$ singlet and triplet excitation energies for planar conjugated hydrocarbons and heterocyclic compounds. In addition, with the help of Eq. (2.32), the ground-state bond lengths can be obtained within the PPP approach from the computed $\pi$-electron bond orders. The PPP method employing curvature-corrected resonance integrals (Section 2.8) has also been applied successfully to three-dimensional systems including fullerenes.

The *Hubbard Hamiltonian* [68], which is frequently used in the interpretation of band structures of solids, obtains from its PPP counterpart by setting all electron repulsion integrals $\{\gamma_{IJ}\}$ with $I \neq J$ to zero. In other words, only the one-center repulsion integrals are retained in the Hubbard formalism. These integrals are equated to a parameter $U$, which is called the *on-site electron repulsion*. Moreover, all resonance integrals are set equal to a single parameter $W$, known as the *hopping integral*. Sometimes the first-neighbor two-electron integrals are retained as well and set to $V$, the *nearest-neighbor repulsion*. Unlike the parameters of the PPP Hamiltonian, $U$, $W$, and $V$ do not refer specifically to the $p_z$ orbitals. Instead, they are empirical parameters that describe interactions between *sites* in solids. Each site, which can be a single atom, an ion, or an entire molecule, is presumed to contribute one electron to the system under consideration. The advantages of using the Hubbard Hamiltonian stem from the fact that it can be exactly diagonalized in many cases.

The *Gutzwiller method* [69, 70] is closely related to both the PPP and the Hubbard formalisms. It begins with a Slater determinant constructed from Hückel orbitals, which is then projected onto a complete set of determinants built from individual $p_z$ orbitals. The ground-state wavefunction is given as a linear combination of these determinants, with the coefficients expressed in terms of a single parameter, which is determined by minimization of the corresponding electronic energy. In this respect, the Gutzwiller approximation resembles a constrained configuration interaction formalism.

## 2.10 The QCFF/PI Method and Other Approaches of Molecular Mechanics

Although both the HMO and PPP methods are capable of predicting bond lengths in conjugated hydrocarbons with reasonable accuracy, full geometry optimizations that include bond and dihedral angles are possible only when the $\sigma$-electron interactions are incorporated into the total energy expressions. The simplest approach to augmentation of the total energy with these interactions is provided by the formalism of *molecular mechanics* in which molecules are treated as collections of atoms connected by flexible bonds. The energies associated with distortions of these bonds are approximated by simple, transferable functions, forms of which are inspired by mechanical considerations.

In the *quantum consistent force field (QCFF/PI) method* [71], the total energy is expressed as a sum of the $\pi$-electron energy calculated within the PPP approximation and the $\sigma$ contribution,

$$E_\sigma = E_{sat} + E_{conj} + E_{sat/conj} \, . \tag{2.39}$$

The $E_{sat}$ term, which describes the $sp^3$-hybridized carbon atoms and the hydrogens, consists of contributions due to bond stretching, bond angle distortion, rotation around the C–C bond, and interactions between remote nuclei (the nonbonded interactions). The $E_{conj}$ and $E_{sat/conj}$ terms, which encompass contributions from the $sp^2$-hybridized carbons and the mixed $sp^2$-$sp^3$ interactions, respectively, have forms similar to that of $E_{sat}$. The parameters that enter the aforementioned contributions are obtained by fitting the calculated standard enthalpies of formation, geometries, and vibrational frequencies of several hydrocarbons to the experimental data. The resulting parameterization has been found to yield accurate geometries and energies of both the ground and the excited states of a large number of hydrocarbons. Lattice parameters of molecular crystals composed of hydrocarbons are also well reproduced.

A similar approach has been used in the *MMP2 method* [72]. Further refinement has been achieved by combining the *MM3 force field* [73–76] with the PPP formalism. The resulting method is capable of predicting molecular properties very accurately [77], as reflected in the typical errors that are less than 1 kcal/mol for the standard enthalpies of formation, 0.02 Å for bond lengths, and 50 cm$^{-1}$ for vibrational frequencies. In particular, thanks to explicit inclusion of the van der Waals interactions, the computed energies and geometries of systems involving weakly interacting fragments (such as, for example, the cyclopentadiene rings in ferrocene) surpass in accuracy those obtained with *ab initio* techniques.

In fullerene research, one often deals with systems, such as molecular crystals with large unit cells, for which quantum-mechanical calculations are excessively expensive if not simply impossible. For these systems, one often resorts to the use of *atom-atom pair potentials*. This approach, in which the total energy is given as a sum of atom–atom interaction energies, is usually too crude for optimization of molecular geometries. However, the atom–atom pair potentials, including the 6-12 *Lennard-Jones potential*,

$$E_{IJ}(R_{IJ}) = E_0 \left[ \left( \frac{R_0}{R_{IJ}} \right)^{12} - \left( \frac{R_0}{R_{IJ}} \right)^{6} \right] \tag{2.40}$$

where $E_0$ and $R_0$ are adjustable parameters, work fairly well for weakly interacting systems. For this reason, they are widely used in the prediction of packing in molecular crystals. Potentials designed specifically for carbon atoms, such as the *Tersoff potential* [78], are also available.

## 2.11 The EHT Method and the Tight-Binding Hamiltonian

A straightforward generalization of the HMO method to all valence electrons is provided by the *extended Hückel theory (EHT)* [79]. In this approach, an effective Hamiltonian matrix is constructed within the basis of all valence orbitals of a given molecule. Thus, EHT is an example of the *all-valence-electron semiempirical method*. Within the EHT formalism, orbital energies and coefficients are obtained by solving the equations [compare Eq. (2.18)]

$$\sum_{\nu} h_{\mu\nu}\, C_{\nu i} = \epsilon_i \sum_{\nu} S_{\mu\nu}\, C_{\nu i}\ . \tag{2.41}$$

The set of basis functions used in the EHT method consists of normalized *Slater-type orbitals (STOs)*,

$$\phi_\nu(\vec{r}) = N_{nml}\, |\vec{r} - \vec{R}_I|^n \exp\left( -\zeta |\vec{r} - \vec{R}_I| \right) Y_l^m(\theta, \phi)\ , \tag{2.42}$$

where $N_{nml}$ is a normalization constant, $n$ is called the *effective quantum number*, $\zeta$ is the orbital exponent, $Y_l^m$ is the spherical harmonic, and $(\theta, \phi)$ are spherical coordinates corresponding to the vector $\vec{r} - \vec{R}_I$. The STOs are centered at individual nuclei and describe atomic valence electrons.

The matrix elements of the effective Hamiltonian $\{h_{\mu\nu}\}$ are usually approximated with the *Wolfsberg-Helmholz formula* [80],

$$h_{\mu\nu} = \frac{K}{2} \left( h_{\mu\mu} + h_{\nu\nu} \right) S_{\mu\nu}\ , \tag{2.43}$$

where the constant $K$ varies between 1.0 and 3.0, and the diagonal elements $\{h_{\mu\mu}\}$ are related to respective ionization potentials. The total energy in the

EHT method is given by Eq. (2.27), that is, it is obtained by summing the energies of all occupied orbitals, as neither the electron–electron repulsion nor the repulsion between nuclei is included. The proportionality between the total energy and the sum of orbital energies in the Hartree–Fock theory [81] is often used to justify such an approach despite the fact that the proportionality constant equals *ca.* 1.55 rather than 1.0. However, this proportionality holds only at equilibrium geometries, meaning that the EHT energy minima do not coincide with those of the exact total energy [82]. For this reason, the EHT method produces optimized geometries that are usually highly inaccurate and occasionally bizarre. As the accuracy of the predicted orbital energy levels is also very poor, EHT calculations are rarely carried out nowadays.

For very large systems with little or no symmetry, even EHT calculations are too expensive. In the *tight-binding Hamiltonian approximation*, the EHT matrix elements between basis functions residing on atoms that are not directly bonded are set to zero. Such an approach offers substantial savings in computational effort and is capable of producing surprisingly accurate results, provided the remaining matrix elements are properly parameterized. A set of tight-binding Hamiltonian parameters for carbon has been published [83] and employed with encouraging results in a variety of calculations on fullerenes. A similar approximation has been used in conjunction with the LDA method (Section 2.16) [84].

## 2.12  The CNDO/1, CNDO/2, and CNDO/S Methods

Application of the ZDO approximation to all valence orbitals leads to a variety of semiempirical methods [85]. The number of electron repulsion integrals required by these methods scales only with the square of the molecular size and is far smaller than that arising in the corresponding *ab initio* calculations carried out even with a minimal basis set. The ensuing reduction of computational effort has been the major driving force behind the development of the ZDO-type approaches. However, the widespread popularity that these methods once enjoyed has begun to wane with the advent of direct SCF procedures employing efficient integral thresholding (Section 2.5). Although the approximations used in all-valence-electron semiempirical methods have been severely criticized [86], attempts to justify them have also been made [87].

The most drastic variant of the ZDO approximation is known as the *complete neglect of differential overlap (CNDO)* [88]. In the CNDO formalism, the approximation (2.33) is applied to all electron repulsion integrals, including those corresponding to pairs of orbitals residing on the same nucleus. In other words, only the integrals of the type $\langle \mu\nu|\mu\nu \rangle$ are retained. The requirement of rotational invariance leads to the conclusion that, for a given pair of nuclei $(I, J)$,

$$\gamma_{IJ} = \langle \mu\nu|\mu\nu \rangle , \qquad \text{for all } \mu\epsilon I, \ \nu\epsilon J . \qquad (2.44)$$

As in the EHT method (Section 2.11), the basis set employed in ZDO-type calculations consists of the STOs describing atomic valence electrons. The molecular orbitals are constructed as linear combinations of these STOs with coefficients obtained by diagonalization of an effective Fock matrix $\mathbf{F}$. For closed-shell systems, the elements of $\mathbf{F}$ are given by

$$F_{\mu\nu} = \begin{cases} U_{\mu\mu} + \left(P_{II} - \frac{1}{2}P_{\mu\mu}\right)\gamma_{II} + \sum_{K \neq I}(P_{KK} - Z_K)\gamma_{IK}\,, & \text{for } \mu = \nu \\ \\ \beta_{\mu\nu} - \frac{1}{2}P_{\mu\nu}\gamma_{IJ}\,, & \text{for } \mu \neq \nu \end{cases} \,, \quad (2.45)$$

where

$$P_{II} = \sum_{\mu \in I} P_{\mu\mu}\,, \qquad (2.46)$$

and $Z_I$ is the charge of the $I$th core (the $I$th nucleus with its core electrons). The core integrals $\{U_{\mu\mu}\}$ are estimated from energies of individual atoms and the corresponding ions.

Two parameterizations of the CNDO method, called *CNDO/1* [89] and *CNDO/2* [90], have been published. In these approaches, the $\{\gamma_{IJ}\}$ integrals are computed exactly using the respective $s$-type STOs. The resonance integrals $\{\beta_{\mu\nu}\}$ are approximated as

$$\beta_{\mu\nu} = \frac{1}{2}\left(\beta_I^0 + \beta_J^0\right)S_{\mu\nu}\,, \qquad (2.47)$$

where $\{\beta_I^0\}$ are parameters obtained by fitting results of CNDO calculations on diatomic molecules to their *ab initio* counterparts computed with minimal basis sets. Both the CNDO/1 and CNDO/2 methods, which are now considered obsolete, yield very poor geometries and total energies.

The *CNDO for spectroscopy (CNDO/S) method* [91] is an all-valence analog of the PPP formalism (Section 2.9). In the CNDO/S approach, resonance integrals are calculated from Eq. (2.47) used within the context of local coordinate systems. However, overlaps between the locally defined $\pi$ orbitals are multiplied by $\kappa/2$ ($\kappa = 0.585$) instead of $1/2$. The electron repulsion integrals have been originally calculated from an extrapolation formula, which has been later replaced [92] by the Mataga–Nishimoto approximation, Eq. (2.38).

CNDO/S calculations are usually carried out in conjunction with a (limited) configuration interaction scheme involving singly excited determinants (CIS, Section 2.6). Reasonably accurate energies of low-lying singlet and triplet excited states are obtained from such calculations, which explains why the CNDO/S method is still in use. Unfortunately, the original CNDO/S method yields vanishing transition moments and singlet–triplet energy splittings for the $\sigma \to \pi^*$ and $n \to \pi^*$ excitations. While the former deficiency is easily remedied [92], the latter one cannot be eliminated without abandoning the CNDO approximation.

## 2.13  The INDO and MINDO Families of Methods

In the *intermediate neglect of differential overlap (INDO) method* [93], the
ZDO approximation is not applied to one-center electron repulsion integrals.
In other words, in addition to the general class of $\langle \mu\nu|\mu\nu \rangle$ integrals, the $\langle \mu\nu|\nu\mu \rangle$
integrals are retained, provided $\mu$ and $\nu$ are centered at the same nucleus.
These additional one-center integrals are estimated from experimental atomic
data. Under the INDO approximation, the elements of an effective Fock matrix
become [compare Eq. (2.45)]

$$
F_{\mu\nu} =
\begin{cases}
U_{\mu\mu} + \sum_{\sigma \in I} P_{\sigma\sigma} \left( \langle \mu\sigma|\mu\sigma \rangle - \frac{1}{2}\langle \mu\sigma|\sigma\mu \rangle \right) \\
\qquad + \sum_{K \neq I} (P_{KK} - Z_K)\gamma_{IK} , & \text{for } \mu = \nu \\[2mm]
P_{\mu\nu} \left( \frac{3}{2}\langle \mu\nu|\nu\mu \rangle - \frac{1}{2}\langle \mu\nu|\mu\nu \rangle \right) , & \text{for } \mu \neq \nu,\ I = J \\[2mm]
\beta_{\mu\nu} - \frac{1}{2}P_{\mu\nu}\gamma_{IJ} , & \text{for } \mu \neq \nu,\ I \neq J
\end{cases}
\tag{2.48}
$$

The original parameterizations, called *INDO/1* and *INDO/2* [85], are
only marginally better than their CNDO counterparts and are no longer used.
On the other hand, the spectroscopic version, known as *INDO/S* [94], has
been successfully applied to diverse chemical systems, including fullerenes.
Thanks to the inclusion of all one-center electron repulsion integrals, INDO/S
calculations yield nonvanishing singlet–triplet energy splittings for all excited
states.

Prior to the emergence of the NDDO-based methods (Section 2.14), approaches of the *modified intermediate neglect of differential overlap (MINDO)*
family have been extensively used in computational organic chemistry. In the
*MINDO/1* [95] parameterization for hydrocarbons, all unneglected two-center
integrals are calculated from empirical formulas. A parameterized expression
is also used instead of a simple Coulombic term for the core–core repulsion,
which is a semiempirical analog of the nuclear repulsion. The parameters have
been obtained by converting the total energies calculated for a training set
of molecules to the corresponding energies of formation and fitting those to
the experimental standard enthalpies of formation. The MINDO/1 method
yields reasonable energies, but is incapable of reproducing experimental bond
lengths. For this reason, it is no longer used.

The parameters of the *MINDO/2 method* [96] have been obtained using
an approach similar to that employed in the development of MINDO/1, the
only difference being inclusion of the C–C and C–H bond lengths in the fitting
procedure. Although both the standard enthalpies of formation and the C–C
bond lengths are calculated with satisfactory accuracy, the predicted C–H
bond lengths are systematically too long by *ca.* 0.1 Å. The MINDO/2 method
is now regarded as obsolete.

In the *MINDO/3 method* [97], the shortcomings of its predecessors have been partially eliminated at the cost of introducing a vast number of additional empirical parameters, which include separate adjustable exponents for the *s*- and *p*-type STOs and bond-specific proportionality constants between resonance and overlap integrals. These parameters, which are available for the H, B, C, N, O, F, Si, P, S, and Cl atoms, have been obtained from fitting of the standard enthalpies of formation, bond lengths, and bond angles. The initial test results [98–101] were encouraging and the arrival of the MINDO/3 method was heralded at one time as the beginning of a new era in quantum organic chemistry [102]. However, failures of the MINDO/3 approximation, including large errors in energies of acetylenic (such as $CH_3C{\equiv}CCH_3$) and globular [such as $C(CH_3)_4$] molecules, together with occasional large discrepancies between the predicted and experimental bond angles, have soon become apparent [103–105]. As the result, the MINDO/3 method has been largely superseded by approaches of the MNDO family (Section 2.14).

## 2.14 The PRDDO, MNDO, MNDOC, AM1, and PM3 Methods

The *neglect of diatomic differential overlap (NDDO)* constitutes one of the least severe ZDO-like approximations [88]. Within the NDDO formalism, only the electron repulsion integrals involving pairs of orbitals centered at two different nuclei are neglected. With fewer integrals set to zero, the resulting elements of the **F** matrix become much more complicated. They equal [compare Eqs. (2.45) and (2.48)]

$$F_{\mu\nu} = \begin{cases} U_{\mu\mu} + \sum_{\sigma\epsilon I} P_{\sigma\sigma}(\langle\mu\sigma|\mu\sigma\rangle - \frac{1}{2}\langle\mu\sigma|\sigma\mu\rangle) \\ \qquad + \sum_{K\neq I}(V_{\mu\mu,K} + \sum_{\lambda,\sigma\epsilon K} P_{\lambda\sigma}\langle\mu\sigma|\mu\lambda\rangle)\,, & \text{for } \mu = \nu \\ \frac{1}{2}P_{\mu\nu}(3\langle\mu\nu|\nu\mu\rangle - \langle\mu\nu|\mu\nu\rangle) \\ \qquad + \sum_{K\neq I}(V_{\mu\nu,K} + \sum_{\lambda,\sigma\epsilon K} P_{\lambda\sigma}\langle\mu\sigma|\nu\lambda\rangle)\,, & \text{for } \mu \neq \nu,\ I = J \\ \beta_{\mu\nu} - \frac{1}{2}\sum_{\lambda\epsilon I}\sum_{\sigma\epsilon J} P_{\lambda\sigma}\langle\mu\sigma|\lambda\nu\rangle\,, & \text{for } \mu \neq \nu,\ I \neq J \end{cases} \qquad (2.49)$$

In Eq. (2.49), $V_{\mu\nu,K}$ stands for the energy of attraction between the electron density distribution $\phi_\mu^* \phi_\nu$ and the $K$th core. In the CNDO and INDO approaches (Sections 2.12 and 2.13), this energy is usually approximated as

$$V_{\mu\nu,K} = -\delta_{\mu\nu}\, Z_K\, \gamma_{IK}\,. \qquad (2.50)$$

Although unparameterized versions of the NDDO approximation have been found rather inaccurate [106, 107], its heavily parameterized variant, known as the *modified neglect of differential overlap (MNDO)* [108], has remained a mainstay of semiempirical quantum chemistry for the last 15 years. The crucial innovation of the MNDO method is the treatment of two-center electron repulsion integrals. For a pair of first-row atoms, 22 such integrals are retained within the NDDO formalism, as opposed to only one integral that survives the CNDO or INDO approximation. Exact calculations of these NDDO integrals not only are time consuming but also result in electronic properties of poor accuracy. For this reason, the two-center electron repulsion integrals are efficiently computed in the MNDO method with the help of a parameterized multipole interaction model [109]. The resonance integrals, the matrix elements $\{V_{\mu\nu,K}\}$, and the core–core repulsion energy are all calculated from parametric expressions. The parameters have been obtained by fitting the standard enthalpies of formations, geometries, first ionization potentials, and dipole moments for a training set of molecules. The initial parameterization for the H, C, N, and O atoms [108] has been extended to other elements, including helium [110], beryllium [111], boron [112], fluorine [113], silicon [114, 115], phosphorus and sulfur [114], chlorine [114, 116], bromine [117], and iodine [118].

The MNDO method constitutes a substantial improvement over MINDO/3. For a set of 138 sample molecules containing the C, H, N, and O atoms, an average error in the calculated standard enthalpies of formation equals 11.0 kcal/mol for MINDO/3 and only 6.3 kcal/mol for MNDO [119]. Average errors in bond lengths, first ionization potentials, and dipole moments are reduced from 0.022 Å, 0.71 eV, and 0.49 D to 0.014 Å, 0.48 eV, and 0.30 D, respectively. The MNDO electron affinities are on average within 0.43 eV from the respective experimental values, although errors greater than 1.4 eV are often observed in systems for which the corresponding negative ions possess highly localized HOMOs [120]. Electric polarizabilities are predicted quite accurately by the MNDO approach, provided systematic atomic corrections are added to the calculated values [121].

Despite the aforedescribed improvements, the MNDO parameterization is far from optimal. Attempts to palliate the known weaknesses of MNDO, which include the failure to reproduce hydrogen bonds, energies that are too positive for sterically crowded compounds [such as $C(CH_4)_4$], and exaggerated torsional angles in molecules with steric hindrances (such as biphenyl or nitrobenzene), have resulted in a new parameterization called *Austin Model 1 (AM1)* [122]. In the AM1 method, the semiempirical formulas for the effective Fock matrix elements are the same as in MNDO (although the parameters are different), but the expressions for the core–core repulsions are more elaborate, giving rise to a substantially increased number of adjustable parameters. Although the energies of hydrogen bonds are reproduced reasonably well by the AM1 method, the overall improvement over MNDO is marginal.

The *MNDO parametric method 3 (MNDO-PM3* or simply *PM3)* stems from a massive reparameterization, in which a training set of 763 molecules has been used [123, 124]. PM3 parameters for the H, C, N, O, F, Al, Si, P, S, Cl, Br, and I atoms have been evaluated simultaneously with improved fitting techniques, resulting in a semiempirical method clearly superior to both MNDO and AM1. The average error in the standard enthalpies of formation of "normal-valence" molecules has been reduced from the MNDO and AM1 values of 13.9 and 12.7 kcal/mol, respectively, to 7.8 kcal/mol. Interestingly, despite the absence of the $d$-type STOs in the basis sets of all three methods, the PM3 standard enthalpies of formation of "hypervalent" compounds are quite satisfactory (average error equal to 13.6 kcal/mol), whereas those obtained from either MNDO or AM1 calculations are definitely not (errors equaling 75.8 and 83.1 kcal/mol, respectively). Similar reduction of error (from 0.054 and 0.050 Å to 0.036 Å) has been achieved for bond lengths. However, the accuracy of predictions for bond and torsional angles, ionization potentials, and dipole moments has not been significantly improved.

A correlated version of the MNDO formalism, called *MNDOC* has been proposed [125]. In the MNDOC approach, the total energies are calculated from MP2-like expressions. For ordinary closed-shell systems, electronic properties calculated with the MNDO and MNDOC methods are essentially the same, whereas some improvements in accuracy are observed for reactive intermediates and transition states [126]. The MNDO method is unsuitable for the calculation of excitation energies, which are predicted much too low [127]. Although the MNDOC energies of excited states are in reasonable agreement with the experimental data [128], even better results are obtained with the *local neglect of differential overlap for spectroscopy (LNDO/S) method* [129], which is closely related to the MNDO family of semiempirical approaches.

Reports on several attempts to go beyond the MNDO approximation have been published. Semiempirical NDDO calculations involving the STO-3G and 4-31G basis sets have been tried [130]. *The partial retention of diatomic differential overlap (PRDDO) approximation* has been proposed as the means for reducing the computational effort associated with *ab initio* calculations [131, 132]. The PRDDO method, which can be regarded as a forerunner of the integral thresholding techniques (Section 2.5), is rarely used nowadays.

Semiempirical approaches of quantum chemistry have played an important role in early electronic structure calculations on medium-size organic molecules. At present such calculations are the domain of *ab initio* methods, but approaches of the MNDO family still remain a valuable tool for studies on large systems, especially those possessing low molecular symmetries. Errors in electronic properties computed with *ab initio* techniques are often larger than those yielded by semiempirical calculations. However, the former errors are well-understood and systematic, whereas the latter ones tend to be highly unpredictable and occasionally disastrous. Further information on this topic can be found in Chapter 17 of ref. [3], in which the performance of various semiempirical and *ab initio* methods is assessed.

## 2.15 Density Functional Theory

According to the celebrated *Hohenberg–Kohn theorem* [133], the ground-state electron density solely determines all electronic properties of a Coulombic system. A somewhat simplified, though convincing, proof of this theorem is based on the fact that both the positions and charges of nuclei can be inferred from the cusps in $\rho(\vec{r})$ (Section 2.2). Knowledge of these quantities allows one to construct the electronic Hamiltonian, Eq. (2.3), from which all electronic properties of both the ground and the excited states can be, at least in principle, calculated. Thus, the electronic energy of the ground-state is given by [134, 135]

$$E_e[\rho] = T[\rho] + V_{ee}[\rho] + \int \rho(\vec{r}) V_n(\vec{r}) \, d\vec{r} \,, \tag{2.51}$$

where $V_n(\vec{r})$, which is equal to the negative of the electrostatic potential due to nuclei, is called *the external potential*. The *functionals* for the kinetic and electron–electron repulsion energies, $T[\rho]$ and $V_{ee}[\rho]$, respectively, are *universal*, that is, independent of $V_n(\vec{r})$. The functional $V_{ee}[\rho]$ is a sum of the *classical term* and the *exchange-correlation functional* $E_{xc}[\rho]$,

$$V_{ee}[\rho] = \frac{1}{2} \iint \rho(\vec{r}) \rho(\vec{r}') |\vec{r} - \vec{r}'|^{-1} d\vec{r} \, d\vec{r}' + E_{xc}[\rho] \,. \tag{2.52}$$

The exact ground-state energy and the corresponding ground-state electron density can be obtained by minimizing $E_e[\rho]$ with respect to $\rho(\vec{r})$ under the constraint

$$\int \rho(\vec{r}) \, d\vec{r} = N \,, \tag{2.53}$$

where $N$ is the number of electrons. Such a minimization leads to an *Euler equation*,

$$\frac{\delta\left(T[\rho] + E_{xc}[\rho]\right)}{\delta\rho} + \int \rho(\vec{r}') |\vec{r} - \vec{r}'|^{-1} d\vec{r}' + V_n(\vec{r}) = \mu \,, \tag{2.54}$$

where $\delta/\delta\rho$ denotes functional differentiation and the Lagrange multiplier $\mu$ is called the *chemical potential*. Eqs. (2.51) and (2.54) constitute the algebraic basis of the *density functional theory (DFT)*, which recently has been gaining prominence in quantum chemistry.

The exact forms of $T[\rho]$ and $E_{xc}[\rho]$ are unknown. In principle, they can be obtained with a procedure called a *constrained search* [136]. However, explicit implementation of the constrained search results in an extremely complicated system of equations that expresses the functionals in a convoluted manner [137]. For this reason, contemporary DFT calculations use approximations to both $T[\rho]$ and $E_{xc}[\rho]$. The simplest approximation of this kind is given by the *Thomas–Fermi theory* [138, 139], in which

$$T[\rho] = \frac{3}{10}\left(3\pi^2\right)^{2/3} \int \rho(\vec{r})^{5/3}\, d\vec{r}\,, \qquad E_{xc}[\rho] = 0\,. \qquad (2.55)$$

Although the Thomas–Fermi theory provides some interesting insights into properties of the electron gas, it is too crude for modern electronic structure calculations.

A practical alternative to the direct minimization, Eq. (2.54), is offered by the *Kohn–Sham (KS) formalism* [140]. The concept of a fictitious reference state of $N$ noninteracting electrons that is employed by the KS theory results in a set of HF-like equations for the occupied *Kohn–Sham spinorbitals* $\{\psi_i\}$,

$$\left\{\hat{h} + \int \rho(\vec{r}')\,|\vec{r} - \vec{r}'|^{-1} d\vec{r}' + V_{KS}(\vec{r})\right\}\psi_i = \epsilon_i\,\psi_i\,, \qquad 1 \leq i \leq N\,. \quad (2.56)$$

In Eq. (2.56), $V_{KS}(\vec{r})$ is the local effective *Kohn–Sham potential*, equal to the functional derivative

$$V_{KS}(\vec{r}) = \frac{\delta \tilde{E}_{xc}[\rho]}{\delta \rho}\,. \qquad (2.57)$$

The exchange-correlation functional $\tilde{E}_{xc}[\rho]$ differs from $E_{xc}[\rho]$ by the difference between the kinetic energy of the actual system and that of the noninteracting reference state. The one-particle equations (2.56) yield the *exact* ground-state density [compare Eq. (2.11)],

$$\rho(\vec{r}) = \sum_i \phi_i^*(\vec{r})\,\phi_i(\vec{r})\,, \qquad (2.58)$$

provided the exact functional $\tilde{E}_{xc}[\rho]$ is used. In other words, the electron correlation effects are absorbed in $\tilde{E}_{xc}[\rho]$ and its functional derivative $V_{KS}(\vec{r})$. Since both $\tilde{E}_{xc}[\rho]$ and $V_{KS}(\vec{r})$ depend on $\rho(\vec{r})$, Eqs. (2.56)–(2.58) have to be solved in an iterative manner. The iterations are avoided in approximate calculations based on the *Harris functional* [141]. Such calculations produce total energies, geometries, and vibrational frequencies that follow closely those obtained with the full KS formalism. A tight-binding variant of the Harris formalism has also been proposed [84] and employed in numerous calculations on carbon clusters.

In practice, the Kohn–Sham spinorbitals are expanded in terms of basis functions. Introduction of basis sets results in equations that involve matrix elements of the Kohn–Sham potential operator,

$$V_{\mu\nu}^{KS} = \langle\psi_\mu|\hat{V}_{KS}|\psi_\nu\rangle\,. \qquad (2.59)$$

Unlike their $\{S_{\mu\nu}\}$ and $\{h_{\mu\nu}\}$ counterparts, these matrix elements cannot be expressed in closed algebraic forms and have to be evaluated with numerical quadratures. An efficient scheme for performing these numerical integrations has been developed and used with good results [142].

## 2.16 The Local Density Approximation

With the exact form of $\tilde{E}_{xc}[\rho]$ unknown, practical implementations of the Kohn–Sham formalism resort to approximate exchange-correlation functionals. In the *local density approximation (LDA)* [143], $\tilde{E}_{xc}[\rho]$ is given by

$$\tilde{E}_{xc}[\rho] = \int \tilde{e}_{xc}\Big(\rho\left(\vec{r}\right),\vec{r}\Big)d\vec{r}\,, \tag{2.60}$$

that is, it is equal to an integral of a function that depends only on the local value of $\rho(\vec{r})$. The $X_\alpha$ *method* [144], which uses

$$\tilde{E}_{xc}[\rho] = -\frac{9}{8}\left(\frac{3}{\pi}\right)^{1/3}\alpha \int \rho\left(\vec{r}\right)^{4/3}d\vec{r} \tag{2.61}$$

for closed-shell systems, constitutes the simplest version of LDA. Eq. (2.61), in which $\alpha$ is an adjustable parameter, has its origins in the expression for the exchange energy of a homogeneous electron gas [145]. More sophisticated approaches augment the exchange energy with terms describing electron correlation. Several forms of these terms, including the Vosko–Wilk–Nusair [146], von Barth–Hedin [147], Hedin–Lundqvist [148], and Perdew–Zunger [149] functionals, are commonly used in LDA calculations.

Bond dissociation energies obtained with the LDA method are much more accurate than those afforded by Hartree–Fock calculations [143]. In many cases, including the transition metal dimers (such as the notoriously difficult $Cr_2$ molecule), where the HF method is incapable of describing bonding correctly, the LDA approach has no trouble yielding reasonably accurate results. However, LDA is well known for its general tendency to substantially overestimate bond dissociation energies. This *overbinding* is largely eliminated once *gradient-corrected functionals* are used in place of their local-density congeners.

## 2.17 Electron Density Functionals for Correlation Energy

The gradient-corrected functionals depend not only on the electron density but also on its gradient and Laplacian,

$$\tilde{E}_{xc}[\rho] = \int \tilde{e}_{xc}\Big(\rho\left(\vec{r}\right),\vec{\nabla}\rho\left(\vec{r}\right),\vec{\nabla}^2\rho\left(\vec{r}\right),\vec{r}\Big)d\vec{r}\,. \tag{2.62}$$

Several forms of such approximate functionals have been published [150–154]. Among them, the *Becke functional* for the exchange energy of a closed-shell system [152],

$$E_x^B[\rho] = -\frac{3}{4}\left(\frac{3}{\pi}\right)^{1/3} \int \rho(\vec{r})^{4/3}\, d\vec{r}$$

$$- 2^{-1/3}\beta \int \rho(\vec{r})^{4/3}\, \xi(\vec{r})^2 \left[1 + 6\beta\xi(\vec{r})\sinh^{-1}\xi(\vec{r})\right]^{-1} d\vec{r}, \quad (2.63)$$

where $\beta = 0.0042$ and

$$\xi(\vec{r}) = 2^{1/3}\rho(\vec{r})^{-4/3}\,|\vec{\nabla}\rho(\vec{r})|\,, \quad (2.64)$$

and the *Lee–Yang–Parr (LYP) functional* $E_{corr}^{LYP}[\rho]$ for the correlation energy [153, 154] have attracted considerable attention. When a combination of the Becke exchange-energy and Vosko–Wilk–Nusair correlation-energy functionals is used in conjunction with a fully numerical Kohn–Sham approach, the average error in atomization energies amounts to only 3.7 kcal/mol for a set of 55 test molecules — a significant improvement over the analogous LDA calculations that are found to overestimate the atomization energies by 36.2 kcal/mol on average [155].

It can be proven that both the total electronic energy and the electron correlation energy are functionals of the Hartree–Fock electron density $\rho_{HF}(\vec{r})$ [156–158]. Although the exact electron correlation functional of the Kohn–Sham theory is different from that giving $E_{corr}$ in terms of $\rho_{HF}(\vec{r})$, the same approximate functionals are often employed in both types of calculations. In the *hybrid approach* [159], the electron density is calculated with an ordinary SCF procedure (Section 2.5), and the computed HF energy is combined with the approximate electron correlation energy. The LYP functional yields only moderately accurate results when used in the context of the hybrid approach. However, significantly more accurate atomization energies, ionization potentials, and electron affinities are obtained with the *BLYP functional*, defined as [159]

$$E_{corr}^{BLYP}[\Gamma] = E_{corr}^{LYP}[\rho] + E_x^B[\rho] - E_x^{HF}[\rho]\,. \quad (2.65)$$

Since the expression (2.65) involves the exact Hartree–Fock exchange energy $E_x^{HF}$, the BLYP functional depends on the 1-matrix $\Gamma(\vec{x}_1', \vec{x}_1)$ (Section 2.2) rather than on the electron density itself. At present, it is unclear why augmenting the LYP functional with the difference between the exact exchange energy $E_x^{HF}[\Gamma]$ and its approximate counterpart $E_x^B[\rho]$ results in the observed improvement in accuracy.

Electronic structure methods based on the density functional theory include the electron correlation effects at a computational cost that is only a fraction of that involved in the conventional molecular-orbital approaches (Section 2.6). However, both the KS and the hybrid formalisms, which have been implemented in a variety of commercially available quantum-chemical software, require numerical integrations that limit the accuracy of the computed properties. Moreover, the approximate functionals contain adjustable parameters that confer a semiempirical character upon the DFT-based methods — a fact that should not be ignored.

## 2.18 Properties of Atoms in Molecules and Rigorous Interpretation of Electronic Wavefunctions

Prediction and understanding of electronic properties constitute the major goals of modern quantum-chemical calculations. Unfortunately, the common chemical concepts, such as atomic charges, bond orders, electronegativities, and molecular similarities, cannot be derived from the postulates of quantum mechanics alone. Instead, one has to resort to additional definitions that make it possible to compute atomic and bond properties from electronic wavefunctions. In the past, most of these definitions have been based on the *Hilbert space partitioning*, that is, assigning basis functions (often incorrectly called "atomic orbitals") to individual nuclei. An example of such an approach is provided by the *Mulliken population analysis* [160–163]. Since the choice of basis functions is arbitrary, atomic charges calculated with this method do not converge to a well-defined limit with the increasing basis set size. Therefore, the Mulliken charges do not reflect the actual distribution of electrons in a molecule.

An alternative to the Hilbert space partitioning is offered by the *theory of atoms in molecules (AIM)* [164]. In this approach, one postulates that atoms in molecules are bordered by *zero-flux surfaces* in electron density, that is, surfaces that are parallel everywhere to the electron density gradient and do not pass through any of the nuclei. This single additional quantum-mechanical postulate gives rise to a self-consistent formalism amenable to clear physical interpretation. In particular, the dominant interactions (such as chemical bonds and strong steric repulsions) are delineated within molecules by attractor interaction lines that pass through *bond critical points* (or simply *bond points*). At these bond points, $\rho(\vec{r})$ is a local maximum with respect to two orthogonal directions and a minimum with respect to the third one. Other types of critical points, called *ring and cage points*, are also known. Thanks to the recent progress in computational techniques [165], these critical points can be routinely located even in large systems such as fullerenes.

Properties of atoms in molecules are easily computed from 1-matrices (Section 2.2). Specifically, the contribution of the $I$th atom to the first-order property described by an operator $\hat{O}$ is given by

$$\langle O \rangle_I = \int_{\Omega_I} \left[ \hat{O}\Gamma(\vec{x}_1', \vec{x}_1) \right] \Bigg|_{\vec{x}_1'=\vec{x}_1} d\vec{x}_1 , \qquad (2.66)$$

where the region $\Omega_I$ is called the *atomic basin*. The atomic basins are delineated by zero-flux surfaces and fill the entire Cartesian space. The atomic properties defined in Eq. (2.66) possess several important properties. In particular, they sum up to the respective electronic properties of the entire molecule and satisfy the Ehrenfest's and virial theorems.

The definition of atomic charges $\{Q_I\}$,

$$Q_I = Z_I - \int_{\Omega_I} \rho(\vec{r})\, d\vec{r}\,, \qquad (2.67)$$

is a special case of Eq. (2.66). Although computations of these charges require time-consuming numerical integrations, they have recently become feasible even for fullerenes and their derivatives [166, 167]. Many other atomic and bond properties can be calculated within the AIM formalism, making it possible to analyze electronic wavefunctions in a rigorous manner. This analysis has been the subject of a recent review [168].

# References


1. M. J. Frisch, G. W. Trucks, M. Head-Gordon, P. M. W. Gill, M. W. Wong, J. B. Foresman, B. G. Johnson, H. B. Schlegel, M. A. Robb, E. S. Replogle, R. Gomperts, J. L. Andres, K. Raghavachari, J. S. Binkley, C. Gonzalez, R. L. Martin, D. J. Fox, D. J. Defrees, J. Baker, J. J. P. Stewart, and J. A. Pople, *GAUSSIAN 92*, Gaussian, Inc., Pittsburgh, PA, 1992.

2. R. Ahlrichs, M. Bär, M. Häser, H. Horn, and C. Kölmel, *Electronic Structure Calculations on Workstation Computers: The Program System TURBOMOLE*, Chem. Phys. Lett. **162**, 165 (1989).

3. I. N. Levine, *Quantum Chemistry*, Prentice-Hall, Englewood Cliffs, NJ, 1991.

4. M. Born and R. Oppenheimer, *Zur Quantentheorie der Molekeln*, Ann. Physik **84**, 457 (1927).

5. H. A. Jahn and E. Teller, *Stability of Polyatomic Molecules in Degenerate Electronic States I—Orbital Degeneracy*, Proc. Roy. Soc. (London) A **161**, 220 (1937).

6. E. R. Davidson, *Reduced Density Matrices in Quantum Chemistry*, Academic Press, New York, 1976.

7. T. Kato, *On the Eigenfunctions of Many-Particle Systems in Quantum Mechanics*, Commun. Pure Appl. Math. **10**, 151 (1957).

8. T. Hoffmann-Ostenhof, M. Hoffmann-Ostenhoff, and R. Ahlrichs, *"Schrödinger Inequalities" and Asymptotic Behavior of Many-Electron Densities*, Phys. Rev. A **18**, 328 (1978).

9. W. J. Hehre, L. Radom, P. v. R. Schleyer, and J. A. Pople, *Ab Initio Molecular Orbital Theory*, John Wiley & Sons, New York, 1986.

10. E. R. Davidson and D. Feller, *Basis Set Selection for Molecular Calculations*, Chem. Rev. **86**, 681 (1986).

11. R. Ditchfield, *Self-Consistent Perturbation Theory of Diamagnetism I. A Gauge-Invariant LCAO Method for N.M.R. Chemical Shifts*, Mol. Phys. **27**, 789 (1974).

12. A. Szabo and N. S. Ostlund, *Modern Quantum Chemistry. Introduction to Advanced Electronic Structure Theory*, McGraw-Hill, New York, 1989.

13. T. Koopmans, *Über die Zuordnung von Wellenfunktionen und Eigenwerten zu den Einzelnen Elektronen Eines Atoms*, Physica **1**, 104 (1934).

14. C. C. J. Roothaan, *New Developments in Molecular Orbital Theory*, Rev. Mod. Phys. **23**, 69 (1951).

15. G. G. Hall, *The Molecular Orbital Theory of Chemical Valency VIII. A Method of Calculating Ionization Potentials*, Proc. Roy. Soc. (London) A **205**, 541 (1951).

16. J. A. Pople and R. K. Nesbet, *Self-Consistent Orbitals for Radicals*, J. Chem. Phys. **22**, 571 (1954).

17. D. J. Thouless, *The Quantum Mechanics of Many-Body Systems*, Academic Press, New York, 1972.

18. J. Gerratt and I. M. Mills, *Force Constants and Dipole-Moment Derivatives of Molecules from Perturbed Hartree–Fock Calculations. I*, J. Chem. Phys. **49**, 1719 (1968).

19. C. Edmiston and K. Ruedenberg, *Localized Atomic and Molecular Orbitals*, Rev. Mod. Phys. **35**, 457 (1963).

20. M. R. Ibrahim and P. v. R. Schleyer, *Atom Equivalents for Relating ab Initio Energies to Enthalpies of Formation*, J. Comp. Chem. **6**, 157 (1985).

21. K. B. Wiberg, *Structures and Energies of the Tricyclo[4.1.0.0$^{1.3}$]heptanes and the Tetracyclo[4.2.1.0$^{2.9}$0$^{5.9}$]nonanes. Extended Group Equivalents for Converting ab Initio Energies to Heats of Formation*, J. Org. Chem. **50**, 5285 (1985).

22. J. M. Schulman, R. C. Peck, and R. L. Disch, *Ab Initio Heats of Formation of Medium-Sized Hydrocarbons. 11. The Benzenoid Aromatics*, J. Am. Chem. Soc. **111**, 5675 (1989).

23. P. Pulay, *Convergence Acceleration of Iterative Sequences. The Case of SCF Iteration*, Chem. Phys. Lett. **73**, 393 (1980).

24. J. Almlöf, K. Faegri, Jr., and K. Korsell, *Principles for a Direct SCF Approach to LCAO-MO ab-Initio Calculations*, J. Comp. Chem. **3**, 385 (1982).

25. V. Dyczmons, *No $N^4$-Dependence in the Calculation of Large Molecules*, Theor. Chim. Acta **28**, 307 (1973).

26. R. Ahlrichs, *Methods for Efficient Evaluation of Integrals for Gaussian Type Basis Sets*, Theor. Chim. Acta **33**, 157 (1974).

27. D. Cremer and J. Gauss, *An Unconventional SCF Method for Calculations on Large Molecules*, J. Comp. Chem. **7**, 274 (1986).

28. S. Obara and A. Saika, *Efficient Recursive Computation of Molecular*

*Integrals over Cartesian Gaussian Functions*, J. Chem. Phys. **84**, 3963 (1986).

29. M. Häser and R. Ahlrichs, *Improvements on the Direct SCF Method*, J. Comp. Chem. **10**, 104 (1989).

30. P. D. Dacre, *On the Use of Symmetry in SCF Calculations*, Chem. Phys. Lett. **7**, 47 (1970).

31. M. Elder, *Use of Molecular Symmetry in SCF Calculations*, Int. J. Quant. Chem. **7**, 75 (1973).

32. M. Dupuis and H.F. King, *Molecular Symmetry and Closed-Shell SCF Calculations. I*, Int. J. Quant. Chem. **11**, 613 (1977).

33. J. Cioslowski, *Ab Initio Calculations on Large Molecules: Methodology and Applications*, in *Reviews in Computational Chemistry*, Vol. 4 (K. B. Lipkowitz and D. B. Boyd, eds.), VCH Publishers, New York, 1993, Ch. 1, p. 1.

34. M. Häser, R. Ahlrichs, H. P. Baron, P. Weis, and H. Horn, *Direct Computation of Second-Order SCF Properties of Large Molecules on Workstation Computers with an Application to Large Carbon Clusters*, Theor. Chim. Acta **83**, 455 (1992).

35. H. Weiss, R. Ahlrichs, and M. Häser, *A Direct Algorithm For Self-Consistent-Field Linear Response Theory and Application to $C_{60}$: Excitation Energies, Oscillator Strengths, and Frequency-Dependent Polarizabilities*, J. Chem. Phys. **99**, 1262 (1993).

36. C. Møller and M. S. Plesset, *Note on an Approximation Treatment for Many-Electron Systems*, Phys. Rev. **46**, 618 (1934).

37. J. A. Pople, J. S. Binkley, and R. Seeger, *Theoretical Models Incorporating Electron Correlation*, Int. J. Quant. Chem. Quant. Chem. Symp. **10**, 1 (1976).

38. M. Head-Gordon, J. A. Pople, and M. J. Frisch, *MP2 Energy Evaluation by Direct Methods*, Chem. Phys. Lett. **153**, 503 (1988).

39. S. Sæbø and J. Almlöf, *Avoiding the Integral Storage Bottleneck in LCAO Calculations of Electron Correlation*, Chem. Phys. Lett. **154**, 83 (1989).

40. W. J. Hehre, R. Ditchfield, L. Radom, and J. A. Pople, *Molecular Orbital Theory of the Electronic Structure of Organic Compounds. V. Molecular Theory of Bond Separation*, J. Am. Chem. Soc. **92**, 4796 (1970).

41. R. J. Bartlett, *Many-Body Perturbation Theory and Coupled Cluster Theory for Electron Correlation in Molecules*, Ann. Rev. Phys. Chem. **32**, 359 (1981).

42. E. Hückel, *Quantentheoretische Beiträge zum Benzolproblem. I. Die Elektronenkonfiguration des Benzols und verwandter Verbindungen*, Z. Physik **70**, 204 (1931).

43. E. Hückel, *Quantentheoretische Beiträge zum Benzolproblem. II. Quantentheorie der induzierten Polaritäten*, Z. Physik **72**, 310 (1931).

44. E. Hückel, *Quantentheoretische Beiträge zum Problem der aromatischen und ungesättigten Verbindungen. III.*, Z. Physik **76**, 628 (1932).

45. A. Graovac, I. Gutman, and N. Trinajstić, *Topological Approach to the Chemistry of Conjugated Molecules*, in *Lecture Notes in Chemistry*, Vol. 4, Springer-Verlag, Berlin, 1977.

46. L. J. Schaad and B. A. Hess, Jr., *Hückel Molecular Orbital $\pi$ Resonance Energies. The Question of the $\sigma$ Structure*, J. Am. Chem. Soc. **94**, 3068 (1972).

47. M. Randić, *Conjugated Circuits and Resonance Energies of Benzenoid Hydrocarbons*, Chem. Phys. Lett. **38**, 68 (1976).

48. M. Randić, *A Graph Theoretical Approach to Conjugation and Resonance Energies of Hydrocarbons*, Tetrahedron **33**, 1905 (1977).

49. M. J. S. Dewar and C. de Llano, *Ground States of Conjugated Molecules. XI. Improved Treatment of Hydrocarbons*, J. Am. Chem. Soc. **91**, 789 (1969).

50. B. A. Hess, Jr., and L. J. Schaad, *Hückel Molecular Orbital $\pi$ Resonance Energies. A New Approach*, J. Am. Chem. Soc. **93**, 305 (1971).

51. B. A. Hess, Jr., and L. J. Schaad, *Hückel Molecular Orbital $\pi$ Resonance Energies. The Benzenoid Hydrocarbons*, J. Am. Chem. Soc. **93**, 2413 (1971).

52. I. Gutman, M. Milun, and N. Trinajstić, *Graph Theory and Molecular Orbitals. 19. Nonparametric Resonance Energies of Arbitrary Conjugated Systems*, J. Am. Chem. Soc. **99**, 1692 (1977).

53. C. A. Coulson and A. Gołębiewski, *On Perturbation Calculations for the $\pi$-Electrons and their Application to Bond Length Reconsiderations in Aromatic Hydrocarbons*, Proc. Phys. Soc. (London) **78**, 1310 (1961).

54. R. C. Haddon, *Hybridization and the Orientation and Alignment of $\pi$-Orbitals in Nonplanar Conjugated Organic Molecules: $\pi$-Orbital Axis Vector Analysis*, J. Am. Chem. Soc. **108**, 2837 (1986).

55. R. C. Haddon, *GVB and POAV Analysis of Rehybridization and $\pi$-Orbital Misalignment in Non-planar Conjugated Systems*, Chem. Phys. Lett. **125**, 231 (1986).

56. R. C. Haddon, L. E. Brus, and K. Raghavachari, *Electronic Structure and Bonding in Icosahedral $C_{60}$*, Chem. Phys. Lett. **125**, 459 (1986).

57. R. C. Haddon, L. E. Brus, and K. Raghavachari, *Rehybridization and $\pi$-Orbital Alignment: the Key to the Existence of Spheroidal Carbon Clusters*, Chem. Phys. Lett. **131**, 165 (1986).

58. D. Bakowies and W. Thiel, *MNDO Study of Large Carbon Clusters*, J. Am. Chem. Soc. **113**, 3704 (1991).

59. R. Pariser and R. G. Parr, *A Semi-Empirical Theory of the Electronic Spectra and Electronic Structure of Complex Unsaturated Molecules. I.*, J. Chem. Phys. **21**, 466 (1953).

60. R. Pariser and R. G. Parr, *A Semi-Empirical Theory of the Electronic*

*Spectra and Electronic Structure of Complex Unsaturated Molecules. II*, J. Chem. Phys. **21**, 767 (1953).

61. J. A. Pople, *Electron Interaction in Unsaturated Hydrocarbons*, Trans. Faraday Soc. **49**, 1375 (1953).

62. R. G. Parr, *A Method for Estimating Electronic Repulsion Integrals over LCAO MO's in Complex Unsaturated Molecules*, J. Chem. Phys. **20**, 1499 (1952).

63. G. Klopman, *A Semiempirical Treatment of Molecular Structures. II. Molecular Terms and Application to Diatomic Molecules*, J. Am. Chem. Soc. **86**, 4550 (1964).

64. K. Ohno, *Some Remarks on the Pariser-Parr-Pople Method*, Theor. Chim. Acta **2**, 219 (1964).

65. M. J. S. Dewar and N. L. Sabelli, *The Split p-Orbital (S.P.O.) Method. III. Relationship to Other M.O. Treatments and Application to Benzene, Butadiene, and Naphthalene*, J. Phys. Chem. **66**, 2310 (1962).

66. M. J. S. Dewar and N. L. Hojvat, *The SPO (Split p-Orbital) Method and Its Application to Ethylene*, J. Chem. Phys. **34**, 1232 (1961).

67. K. Nishimoto and N. Mataga, *Electronic Structure and Spectra of Some Nitrogen Heterocycles*, Z. Phys. Chemie **12**, 335 (1957).

68. J. Hubbard, *Electron Correlations in Narrow Energy Bonds*, Proc. Roy. Soc. (London) A **276**, 238 (1963).

69. M. C. Gutzwiller, *Effect of Correlation on the Ferromagnetism of Transition Metals*, Phys. Rev. A **134**, 923 (1964).

70. P. Joyes and R. J. Tarento, *Application of the Gutzwiller Method to Neutral and Ionic $C_{60}$ Aggregates*, Phys. Rev. B **45**, 12077 (1992).

71. A. Warshel and M. Karplus, *Calculation of Ground and Excited State Potential Surfaces of Conjugated Molecules. I. Formulation and Parametrization*, J. Am. Chem. Soc. **94**, 5612 (1972).

72. J. Kao and N. L. Allinger, *Conformational Analysis. 122. Heats of Formation of Conjugated Hydrocarbons by the Force Field Method*, J. Am. Chem. Soc. **99**, 975 (1977).

73. N. L. Allinger, Y. H. Yuh, and J.-H. Lii, *Molecular Mechanics. The MM3 Force Field for Hydrocarbons. 1*, J. Am. Chem. Soc. **111**, 8553 (1989).

74. J.-H. Lii and N. L. Allinger, *Molecular Mechanics. The MM3 Force Field for Hydrocarbons. 2. Vibrational Frequencies and Thermodynamics*, J. Am. Chem. Soc. **111**, 8566 (1989).

75. J.-H. Lii and N. L. Allinger, *Molecular Mechanics. The MM3 Force Field for Hydrocarbons. 3. The van der Waals' Potentials and Crystal Data for Aliphatic and Aromatic Hydrocarbons*, J. Am. Chem. Soc. **111**, 8576 (1989).

76. N. L. Allinger, F. Li, and L. Yan, *Molecular Mechanics. The MM3 Force Field for Alkenes*, J. Comp. Chem. **11**, 848 (1990).

77. N. L. Allinger, F. Li, L. Yan, and J. C. Tai, *Molecular Mechanics (MM3)*

*Calculations on Conjugated Hydrocarbons*, J. Comp. Chem. **11**, 868 (1990).

78. J. Tersoff, *Empirical Interatomic Potential for Carbon, with Applications to Amorphous Carbon*, Phys. Rev. Lett. **61**, 2879 (1988).

79. R. Hoffmann, *An Extended Hückel Theory. I. Hydrocarbons*, J. Chem. Phys. **39**, 1397 (1963).

80. M. Wolfsberg and L. Helmholz, *Spectra and Electronic Structure of the Tetrahedral Ions $MnO_4^-$, $CrO_4^-$, and $ClO_4^-$*, J. Chem. Phys. **20**, 837 (1952).

81. K. Ruedenberg, *An Approximate Relation between Orbital SCF Energies and Total SCF Energy in Molecules*, J. Chem. Phys. **66**, 375 (1977).

82. T. Anno and Y. Sakai, *A Remark on the Relation between Hartree–Fock Orbital Energies and the Hartree–Fock Total Energy in Molecules*, J. Chem. Phys. **67**, 4771 (1977).

83. C. H. Xu, C. Z. Wang, C. T. Chan, and K. M. Ho, *A Transferable Tight-Binding Potential for Carbon*, J. Phys. Condens. Matter **4**, 6047 (1992).

84. O. F. Sankey and D. J. Niklewski, *Ab Initio Multicenter Tight-Binding Model for Molecular-Dynamics Simulations and Other Applications in Covalent Systems*, Phys. Rev. B **40**, 3979 (1989).

85. J. A. Pople and D. L. Beveridge, *Approximate Molecular Orbital Theory*, McGraw-Hill, New York, 1970.

86. S. de Bruijn, *Analysis of the Inadequacies of Some Semi-Empirical MO Methods as Theories of Structure and Reactivity*, Int. J. Quant. Chem. **25**, 367 (1984).

87. K. F. Freed, *Theoretical Foundations of Purely Semiempirical Quantum Chemistry*, J. Chem. Phys. **60**, 1765 (1974).

88. J. A. Pople, D. P. Santry, and G. A. Segal, *Approximate Self-Consistent Molecular Orbital Theory. I. Invariant Procedures*, J. Chem. Phys. **43**, S129 (1965).

89. J. A. Pople and G. A. Segal, *Approximate Self-Consistent Molecular Orbital Theory. II. Calculations with Complete Neglect of Differential Overlap*, J. Chem. Phys. **43**, S136 (1965).

90. J. A. Pople and G. A. Segal, *Approximate Self-Consistent Molecular Orbital Theory. III. CNDO Results for $AB_2$ and $AB_3$ Systems*, J. Chem. Phys. **44**, 3289 (1966).

91. J. Del Bene and H. H. Jaffé, *Use of the CNDO Method in Spectroscopy. I. Benzene, Pyridine, and the Diazines*, J. Chem. Phys. **48**, 1807 (1968).

92. R. L. Ellis, G. Kuehnlenz, and H. H. Jaffé, *The Use of the CNDO Method in Spectroscopy VI. Further $n$–$\pi^*$ Transitions*, Theor. Chim. Acta **26**, 131 (1972).

93. J. A. Pople, D. L. Beveridge, and P. A. Dobosh, *Approximate Self-Consistent Molecular-Orbital Theory. V. Intermediate Neglect of Differential Overlap*, J. Chem. Phys. **47**, 2026 (1967).

94. J. Ridley and M. Zerner, *An Intermediate Neglect of Differential Overlap Technique for Spectroscopy: Pyrrole and the Azines*, Theor. Chim. Acta **32**, 111 (1973).

95. N. C. Baird and M. J. S. Dewar, *Ground States of σ-Bonded Molecules. IV. The MINDO Method and Its Application to Hydrocarbons*, J. Chem. Phys. **50**, 1262 (1969).

96. M. J. S. Dewar and E. Haselbach, *Ground States of σ-Bonded Molecules. IX. The MINDO/2 Method*, J. Am. Chem. Soc. **92**, 590 (1970).

97. R. C. Bingham, M. J. S. Dewar, and D. H. Lo, *Ground States of Molecules. XXV. MINDO/3. An Improved Version of the MINDO Semiempirical SCF-MO Method*, J. Am. Chem. Soc. **97**, 1285 (1975).

98. R. C. Bingham, M. J. S. Dewar, and D. H. Lo, *Ground States of Molecules. XXVI. MINDO/3 Calculations for Hydrocarbons*, J. Am. Chem. Soc. **97**, 1294 (1975).

99. R. C. Bingham, M. J. S. Dewar, and D. H. Lo, *Ground States of Molecules. XXVII. MINDO/3 Calculations for CHON Species*, J. Am. Chem. Soc. **97**, 1302 (1975).

100. R. C. Bingham, M. J. S. Dewar, and D. H. Lo, *Ground States of Molecules. XXVIII. MINDO/3 Calculations for Compounds Containing Carbon, Hydrogen, Fluorine, and Chlorine*, J. Am. Chem. Soc. **97**, 1307 (1975).

101. M. J. S. Dewar, D. H. Lo, and C. A. Ramsden, *Ground States of Molecules. XXIX. MINDO/3 Calculations of Compounds Containing Third Row Elements*, J. Am. Chem. Soc. **97**, 1311 (1975).

102. M. J. S. Dewar, *Quantum Organic Chemistry*, Science **187**, 1037 (1975).

103. J. A. Pople, *Some Deficiencies of MINDO/3 Semiempirical Theory*, J. Am. Chem. Soc. **97**, 5306 (1975).

104. W. J. Hehre, *MINDO/3. An Evaluation of Its Usefulness as a Structural Theory*, J. Am. Chem. Soc. **97**, 5308 (1975).

105. M. J. S. Dewar, *Concerning Criticisms of MINDO/3 by Pople and Hehre*, J. Am. Chem. Soc. **97**, 6591 (1975).

106. J. Chandrasekhar, P. K. Mehrotra, S. Subramanian, and P. T. Manoharan, *NDDO MO Calculations III. Detailed Results for Molecules Containing First-Row Atoms*, Theor. Chim. Acta **52**, 303 (1979).

107. J. Chandrasekhar, P. K. Mehrotra, S. Subramanian, and P. T. Manoharan, *NDDO MO Calculations I. Analysis of the Method*, Theor. Chim. Acta **41**, 243 (1976).

108. M. J. S. Dewar and W. Thiel, *Ground States of Molecules. 38. The MNDO Method. Approximations and Parameters*, J. Am. Chem. Soc. **99**, 4899 (1977).

109. M. J. S. Dewar and W. Thiel, *A Semiempirical Model for the Two-Center Repulsion Integrals in the NDDO Approximation*, Theor. Chim. Acta **46**, 89 (1977).

110. M. Kolb and W. Thiel, *MNDO Parameters for Helium: Optimization, Tests, and Application to Endohedral Fullerene-Helium Complexes*, J. Comp. Chem. **14**, 37 (1993).

111. M. J. S. Dewar and H. S. Rzepa, *Ground States of Molecules. 45. MNDO Results for Molecules Containing Beryllium*, J. Am. Chem. Soc. **100**, 777 (1978).

112. M. J. S. Dewar and M. L. McKee, *Ground States of Molecules. 41. MNDO Results for Molecules Coniaining Boron*, J. Am. Chem. Soc. **99**, 5231 (1977).

113. M. J. S. Dewar and H. S. Rzepa, *Ground States of Molecules. 40. MNDO Results for Molecules Containing Fluorine*, J. Am. Chem. Soc. **100**, 58 (1978).

114. M. J. S. Dewar, M. L. McKee, and H. S. Rzepa, *MNDO Parameters for Third Period Elements*, J. Am. Chem. Soc. **100**, 3607 (1978).

115. M. J. S. Dewar, J. Friedheim, G. Grady, E. F. Healy, and J. J. P. Stewart, *Revised MNDO Parameters for Silicon*, Organometallics **5**, 375 (1986).

116. M. J. S. Dewar and H. S. Rzepa, *Ground States of Molecules. 53. MNDO Calculations for Molecules Containing Chlorine*, J. Comp. Chem. **4**, 158 (1983).

117. M. J. S. Dewar and E. Healy, *Ground States of Molecules. 64. MNDO Calculations for Compounds Containing Bromine*, J. Comp. Chem. **4**, 542 (1983).

118. M. J. S. Dewar, E. F. Healy, and J. J. P. Stewart, *Ground States of Molecules. 67. MNDO Calculations for Compounds Containing Iodine*, J. Comp. Chem. **5**, 358 (1984).

119. M. J. S. Dewar and W. Thiel, *Ground States of Molecules. 39. MNDO Results for Molecules Containing Hydrogen, Carbon, Nitrogen, and Oxygen*, J. Am. Chem. Soc. **99**, 4907 (1977).

120. M. J. S. Dewar and H. S. Rzepa, *Calculations of Electron Affinities Using the MNDO Semiempirical SCF-MO Method*, J. Am. Chem. Soc. **100**, 784 (1978).

121. M. J. S. Dewar and J. J. P. Stewart, *A New Procedure for Calculating Molecular Polarizabilities; Applications Using MNDO*, Chem. Phys. Lett. **111**, 416 (1984).

122. M. J. S. Dewar, E. G. Zoebisch, E. F. Healy, and J. J. P. Stewart, *AM1: A New General Purpose Quantum Mechanical Molecular Model*, J. Am. Chem. Soc. **107**, 3902 (1985).

123. J. J. P. Stewart, *Optimization of Parameters for Semiempirical Methods I. Method*, J. Comp. Chem. **10**, 209 (1989).

124. J. J. P. Stewart, *Optimization of Parameters for Semiempirical Methods II. Applications*, J. Comp. Chem. **10**, 221 (1989).

125. W. Thiel, *The MNDOC Method, a Correlated Version of the MNDO Model*, J. Am. Chem. Soc. **103**, 1413 (1981).

126. W. Thiel, *MNDOC Study of Reactive Intermediates and Transition States*, J. Am. Chem. Soc. **103**, 1420 (1981).

127. M. J. S. Dewar, M. A. Fox, K. A. Campbell, C.-C. Chen, J. E. Friedheim, M. K. Holloway, S. C. Kim, P. B. Liescheski, A. M. Pakiari, T.-P. Tien, and E. G. Zoebisch, *Calculation of Energies of Excited States Using MNDO*, J. Comp. Chem. **5**, 480 (1984).

128. A. Schweig and W. Thiel, *MNDOC Study of Excited States*, J. Am. Chem. Soc. **103**, 1425 (1981).

129. G. Lauer, K.-W. Schulte, and A. Schweig, *LNDO/S, a Semiempirical SCF-CI Method for the Calculation of Ionization Potentials and Electronic Transition Energies of Valence Electrons*, J. Am. Chem. Soc. **100**, 4925 (1978).

130. W. Thiel, *Semiempirical NDDO Calculations with STO-3G and 4-31G Basis Sets*, Theor. Chim. Acta **59**, 191 (1981).

131. T. A. Halgren and W. N. Lipscomb, *Self-Consistent-Field Wavefunctions for Complex Molecules. The Approximation of Partial Retention of Diatomic Differential Overlap*, J. Chem. Phys. **58**, 1569 (1973).

132. D. S. Marynick and W. N. Lipscomb, *Extension of the Method of Partial Retention of Diatomic Differential Overlap to Second Row Atoms and Transition Metals*, Proc. Natl. Acad. Sci. USA **79**, 1341 (1982).

133. P. Hohenberg and W. Kohn, *Inhomogeneous Electron Gas*, Phys. Rev. B **136**, 864 (1964).

134. E. S. Kryachko and E. V. Ludeña, *Energy Density Functional Theory of Many-Electron Systems*, Kluwer Academic Publishers, Dordrecht, 1990.

135. R. G. Parr and W. Yang, *Density-Functional Theory of Atoms and Molecules*, Oxford University Press, New York, 1989.

136. M. Levy, *Universal Variational Functionals of Electron Densities, First-Order Density Matrices, and Natural Spin-Orbitals and Solution of the v-Representability Problem*, Proc. Natl. Acad. Sci. USA **76**, 6062 (1979).

137. J. Cioslowski, *Density Functionals for the Energy of Electronic Systems: Explicit Variational Construction*, Phys. Rev. Lett. **60**, 2141 (1988).

138. L. H. Thomas, *The Calculation of Atomic Fields*, Proc. Cambridge Phil. Soc. **23**, 542 (1927).

139. E. Fermi, *Eine statistische Methode zur Bestimmung einiger Eigenschaften des Atoms und ihre Anwendung auf die Theorie des periodischen Systems der Elemente*, Z. Physik **48**, 73 (1928).

140. W. Kohn and L. J. Sham, *Self-Consistent Equations Including Exchange and Correlation Effects*, Phys. Rev. A **140**, 1133 (1965).

141. J. Harris, *Simplified Method for Calculating the Energy of Weakly Interacting Fragments*, Phys. Rev. B **31**, 1770 (1985).

142. A. D. Becke, *A Multicenter Numerical Integration Scheme for Polyatomic Molecules*, J. Chem. Phys. **88**, 2547 (1988).

143. R. O. Jones and O. Gunnarsson, *The Density Functional Formalism, Its Applications and Prospects*, Rev. Mod. Phys. **61**, 689 (1989).

144. J. C. Slater, *A Simplification of the Hartree–Fock Method*, Phys. Rev. **81**, 385 (1951).

145. P. A. M. Dirac, *Note on Exchange Phenomena in the Thomas Atom*, Proc. Cambridge Phil. Soc. **26**, 376 (1930).

146. S. H. Vosko, L. Wilk, and M. Nusair, *Accurate Spin-Dependent Electron Liquid Correlation Energies for Local Spin Density Calculations: A Critical Analysis*, Can. J. Phys. **58**, 1200 (1980).

147. U. von Barth and L. Hedin, *A Local Exchange-Correlation Potential for the Spin Polarized Case: I*, J. Phys. C **5**, 1629 (1972).

148. L. Hedin and B. I. Lundqvist, *Explicit Local Exchange-Correlation Potentials*, J. Phys. C **4**, 2064 (1971).

149. J. P. Perdew and A. Zunger, *Self-Interaction Correction to Density-Functional Approximations for Many-Electron Systems*, Phys. Rev. B **23**, 5408 (1981).

150. A. D. Becke, *Correlation Energy of an Inhomogeneous Electron Gas: A Coordinate-Space Model*, J. Chem. Phys. **88**, 1053 (1988).

151. J. P. Perdew, *Density-Functional Approximation for the Correlation Energy of the Inhomogeneous Electron Gas*, Phys. Rev. B **33**, 8822 (1986).

152. A. D. Becke, *Density-Functional Exchange-Energy Approximation with Correct Asymptotic Behavior*, Phys. Rev. A **38**, 3098 (1988).

153. C. Lee, W. Yang, and R. G. Parr, *Development of the Colle-Salvetti Correlation-Energy Formula into a Functional of the Electron Density*, Phys. Rev. B **37**, 785 (1988).

154. B. Miehlich, A. Savin, H. Stoll, and H. Preuss, *Results Obtained with the Correlation Energy Density Functionals of Becke and Lee, Yang and Parr*, Chem. Phys. Lett. **157**, 200 (1989).

155. A. D. Becke, *Density-Functional Thermochemistry. I. The Effect of the Exchange-Only Gradient Correction*, J. Chem. Phys. **96**, 2155 (1992).

156. P. W. Payne, *Density Functionals in Unrestricted Hartree–Fock Theory*, J. Chem. Phys. **71**, 490 (1979).

157. J. Cioslowski, *Density-Driven Self-Consistent-Field Method: Density Functionals for Electron Correlation Energy*, Phys. Rev. A **41**, 3458 (1990).

158. R. A. Harris and L. R. Pratt, *A Method for Systematic Inclusion of Electron Correlation in Density Functionals*, J. Chem. Phys. **83**, 4024 (1985).

159. P. M. W. Gill, B. G. Johnson, and J. A. Pople, *An Investigation of the Performance of a Hybrid of Hartree–Fock and Density Functional Theory*, Int. J. Quant. Chem. Quant. Chem. Symp. **26**, 319 (1992).

160. R. S. Mulliken, *Electronic Population Analysis on LCAO-MO Molecular Wave Functions. I*, J. Chem. Phys. **23**, 1833 (1955).

161. R. S. Mulliken, *Electronic Population Analysis on LCAO-MO Molecular Wave Functions. II. Overlap Populations, Bond Orders, and Covalent Bond Energies*, J. Chem. Phys. **23**, 1841 (1955).

162. R. S. Mulliken, *Electronic Population Analysis on LCAO-MO Molecular Wave Functions. III. Effects of Hybridization on Overlap and Gross AO Populations*, J. Chem. Phys. **23**, 2338 (1955).

163. R. S. Mulliken, *Electronic Population Analysis on LCAO-MO Molecular Wave Functions. IV. Bonding and Antibonding in LCAO and Valence-Bond Theories*, J. Chem. Phys. **23**, 2343 (1955).

164. R. F. W. Bader, *Atoms in Molecules. A Quantum Theory*, Clarendon Press, Oxford, 1990.

165. J. Cioslowski and A. Nanayakkara, *A New Robust Algorithm for Fully Automated Determination of Attractor Interaction Lines in Molecules*, Chem. Phys. Lett. **219**, 151 (1994).

166. J. Cioslowski, *An Efficient Evaluation of Atomic Properties Using a Vectorized Numerical Integration With Dynamic Thresholding*, Chem. Phys. Lett. **194**, 73 (1992).

167. J. Cioslowski, A. Nanayakkara, and M. Challacombe, *Rapid Evaluation of Atomic Properties with Mixed Analytical/Numerical Integration*, Chem. Phys. Lett. **203**, 137 (1993).

168. J. Cioslowski and P. R. Surján, *An Observable-Based Interpretation of Electronic Wavefunctions: Application to "Hypervalent" Molecules*, J. Mol. Struct. (Theochem) **255**, 9 (1992).

# Chapter 3

# The $C_{60}$ Fullerene

The $C_{60}$ fullerene, also known as footballene, buckminsterfullerene, or [60]-fullerene, is the most prominent solvent-extractable constituent of the soot prepared by resistive heating of graphite under an inert atmosphere of the helium gas [1]. Conditions of this process, such as the gas pressure and the current density, have been optimized to maximize the yield of $C_{60}$ [2–4]. The $C_{60}$ fullerene can also be prepared by a variety of other methods, including electric discharges in liquid toluene [5] and pyrolysis of naphthalene [6]. The $C_{40}O_{10}$ carbon oxide loses 10 CO molecules upon laser desorption, the resulting transient $C_{30}$ cluster furnishing $C_{60}$ through dimerization and rearrangement [7, 8]. Sooting flames, such as those produced by benzene–oxygen mixtures, yield large quantities of fullerenes [9–12], offering a potential industrial source of $C_{60}$. The $C_{60}$ fullerene is also present in naturally occurring minerals such as fulgurite (a glassy rock that is formed when lightning strikes the ground) [13] and shungite (a carbonaceous rock found in Karelia, Russia) [14].

The $C_{60}$ molecule (Fig. 3.1) has the shape of a truncated icosahedron with $I_h$ symmetry. It possesses 32 rings, 12 of which are five-membered and 20 are six-membered. This structure, in which all carbon atoms are equivalent, has been the subject of early speculation [15]. Once macroscopic quantities of $C_{60}$ had become available [1], the observed single-line $^{13}C$ NMR spectrum [16–18] and the IR spectrum, in which only four lines are present (Section 3.2) [19–23], have provided the long awaited confirmation. The final, direct proof of the icosahedral structure of $C_{60}$ has arrived with the X-ray diffraction experiment on the $1,2\text{-}C_{60}(OsO_4)(4\text{-}t\text{-}BuC_5H_4N)_2$ derivative [24].

The electronic structure of the $C_{60}$ fullerene has been studied with a variety of quantum-chemical techniques. Accurate predictions of diverse properties, including geometry, vibrational frequencies, heat of formation, ionization potentials, electron affinity, energies of the excited states, polarizability, and magnetic shielding, have been obtained.

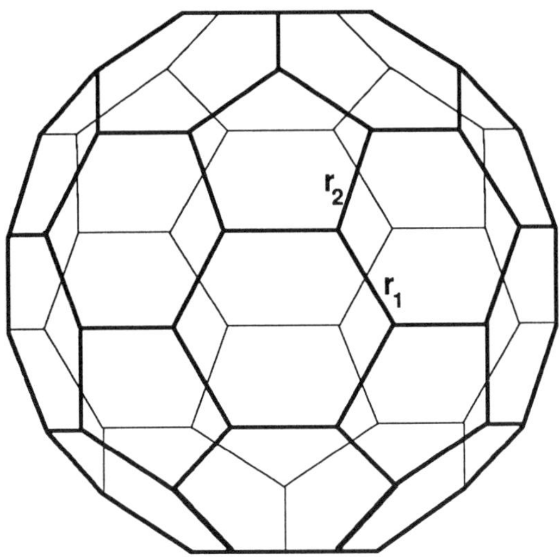


*Fig. 3.1. The IPR isomer of the $C_{60}$ fullerene.*

## 3.1  Geometry

There are two classes of bonds in the $C_{60}$ molecule. The first class consists of "double" bonds that connect the five-membered rings. The second class encompasses "single" bonds within each of the pentagons. Thanks to the high molecular symmetry, lengths of these bonds (denoted in Fig. 3.1 by $r_1$ and $r_2$, respectively) completely determine the geometry of the fullerene, including the cage radius $R$, which is given by [29]

$$R = \frac{1}{2} [\tau^2 (r_1 + 2r_2)^2 + r_1^2]^{1/2} .\tag{3.1}$$

In Eq. (3.1), $\tau$ is the golden ratio, equal to $(5^{1/2} + 1)/2 \approx 1.6180339$.

The early X-ray diffraction experiments that confirmed the icosahedral structure of $C_{60}$ and measured its bond lengths have been carried out not on the fullerene itself but on its osmium [24] and platinum [78] derivatives. The X-ray geometry determination for solid-state $C_{60}$, which is quite difficult to accomplish because of rotational disorder (Chapter 10), soon followed [76], providing values of bond lengths in rough agreement with those obtained from NMR experiments [80] (Table 3.1). Thus far, gas-phase electron

Table 3.1.  Bond lengths in the $C_{60}$ fullerene molecule

| Method | Bond length [Å][a] | |
| --- | --- | --- |
|  | $r_1$ | $r_2$ |
| MM2 [25] | 1.393 | 1.444 |
| MM3 [26] | 1.391 | 1.452 |
| Hückel [27] | 1.405 | 1.426 |
| Hückel [28] | 1.409 | 1.432 |
| Self-consistent $\beta$-HMO [29] | 1.403 | 1.434 |
| PPP [30, 31] | 1.398 | 1.439 |
| PPP [32] | 1.396 | 1.443 |
| QCFF/PI [33] | 1.411 | 1.471 |
| Tight-binding Hamiltonian [34] | 1.389 | 1.415 |
| Tight-binding Hamiltonian [35] | 1.396 | 1.458 |
| INDO [36] | 1.397 | 1.449 |
| INDO/1 [37, 38] | 1.398 | 1.451 |
| INDO/2 [39] | 1.398 | 1.451 |
| MINDO/3 [40] | 1.420 | 1.496 |
| PRDDO [41] | 1.360 | 1.436 |
| PRDDO [42] | 1.361 | 1.435 |
| MNDO [43–46] | 1.400 | 1.474 |
| AM1 [47–49] | 1.385 | 1.464 |
| PM3 [50] | 1.384 | 1.457 |
| HF/STO-3G [51] | 1.376 | 1.465 |
| HF/STO-3G [52] | 1.376 | 1.463 |
| HF/3-21G [53] | 1.365 | 1.453 |
| HF/3-21G [54, 55] | 1.367 | 1.453 |
| HF/4-31G [56, 57] | 1.370 | 1.450 |
| HF/dz [52] | 1.368 | 1.451 |
| HF/DZ [58] | 1.369 | 1.453 |
| HF/DZV [59] | 1.37 | 1.45 |
| HF/dzP [52] | 1.372 | 1.453 |
| HF/DZP [52] | 1.375 | 1.450 |
| HF/DZP [57] | 1.374 | 1.449 |

| Method | Bond length [Å][a] | |
| --- | --- | --- |
| | $r_1$ | $r_2$ |
| HF/TZV [60] | 1.373 | 1.455 |
| HF/6-311G* [61] | 1.372 | 1.448 |
| HF/TZP [52] | 1.370 | 1.448 |
| MP2/dz [62] | 1.407 | 1.470 |
| MP2/DZP [62] | 1.412 | 1.451 |
| MP2/TZP [62] | 1.406 | 1.446 |
| LDA [63] | 1.39 | 1.45 |
| LDA [64] | 1.398 | 1.442 |
| LDA [65] | 1.398 | 1.450 |
| LDA [42] | 1.40 | 1.48 |
| LDA [66] | 1.386 | 1.453 |
| LDA [67, 68] | 1.387 | 1.445 |
| LDA [69] | 1.385 | 1.444 |
| LDA [70] | 1.386 | 1.442 |
| LDA [71] | 1.41 | 1.45 |
| LDA [72] | 1.389 | 1.448 |
| LDA [73] | 1.39 | 1.45 |
| LDA/DZ [74] | 1.397 | 1.451 |
| LDA/DZP [74] | 1.395 | 1.445 |
| LDA/TZP [74] | 1.388 | 1.441 |
| BLYP/DZ [74] | 1.412 | 1.472 |
| BLYP/DZP [74] | 1.410 | 1.464 |
| BLYP/TZP [74] | 1.403 | 1.461 |
| Gas-phase el. diffr. [75] | 1.401±0.010 | 1.458±0.006 |
| X-ray [76] | 1.355±0.009 | 1.467±0.021 |
| X-ray [77] | 1.391±0.018 | 1.455±0.012 |
| X-ray (Os deriv.) [24] | 1.388±0.009 | 1.432±0.005 |
| X-ray (Pt deriv.) [78] | 1.388±0.030 | 1.445±0.030 |
| Neutron diffr. [79] | 1.39 | 1.46 |
| NMR [80] | 1.400±0.015 | 1.450±0.015 |

[a] See Fig. 3.1.

diffraction measurements [75] appear to yield the most accurate bond lengths, $r_1 = 1.401$ Å and and $r_2 = 1.458$ Å. Application of Eq. (3.1) in conjunction with these data gives 3.562 Å for the cage radius.

Extensive theoretical effort has been directed toward prediction of bond lengths in $C_{60}$. Results of these calculations, which are compiled in Table 3.1, demonstrate that reasonable estimates of bond lengths in $C_{60}$ are provided by a broad spectrum of electronic structure methods, although a closer inspection of the calculated bond lengths reveals that their accuracy varies widely. In particular, the experimentally determined bond alternation, which amounts to *ca.* 0.06 Å, is reasonably well reproduced by semiempirical approaches, including those of the MNDO family [43–50]. On the other hand, *ab initio* Hartree–Fock calculations, especially those carried out with smaller basis sets, tend to overestimate bond alternation (by predicting values of $r_1$ that are too small), whereas MP2 calculations underestimate it [62]. Bond lengths obtained with the BLYP density functional agree very well with their experimental counterparts, while those computed within the LDA formalism are usually slightly too short due to overbinding [74].

## 3.2 Vibrational Frequencies

The vibrational modes of the $C_{60}$ fullerene belong to the $2A_g + 3T_{1g} + 4T_{2g} + 6G_g + 8H_g + A_u + 4T_{1u} + 5T_{2u} + 6G_u + 7H_u$ manifold of irreducible representations [27]. Among these, only the four modes of $T_{1u}$ symmetry are IR-active. The Raman spectrum of $C_{60}$ consists of two polarized and eight depolarized lines arising from vibrations with $A_g$ and $H_g$ symmetries, respectively. The simplicity of the IR spectrum has been crucial in facilitating the initial search for $C_{60}$ in soot [81, 82]. The early theoretical predictions of vibrational frequencies have proved to be instrumental in ascribing the observed IR absorption lines to the $C_{60}$ fullerene.

The sparsity of the IR and Raman spectra makes the assignment of the calculated frequencies to the observed peaks a simple task. Among semiempirical methods, the QCFF approach [33, 83] fares the best in predicting vibrational frequencies of $C_{60}$ (Table 3.2). The LDA predictions [84] are also quite accurate (note that the data listed in [85] are incorrect), whereas calculations involving the tight-binding, MNDO-like, and *ab initio* Hamiltonians produce vibrational frequencies that are definitely too high. Interestingly, the inclusion of electron correlation effects at the MP2 level of theory results in only a minor improvement in the predicted frequencies of the $A_g$ "breathing" and "pentagonal-pitch" Raman-active modes, the remaining discrepancy being attributed to vibrational anharmonicities [62].

Several empirical force fields, involving from two to eight adjustable parameters, have been constructed in order to reproduce vibrational frequencies of the $C_{60}$ fullerene. The accuracy of the predicted frequencies ranges from

**Table 3.2. Theoretical and experimental vibrational frequencies of the $C_{60}$ fullerene**

| Method | $A_g$ (Raman) | | $H_g$ (Raman) | | | | $T_{1u}$ (IR) | |
|---|---|---|---|---|---|---|---|---|
| QCFF/PI [33, 83] | 513 | 1442 | 258 | 440 | 691 | 801 | 544 | 607 |
| | | | 1154 | 1265 | 1465 | 1644 | 1212 | 1437 |
| Tight-binding Hamilt. [34] | 682 | 1974 | | n/a | | | | n/a |
| Tight-binding Hamilt. [35] | 590 | 1610 | 237 | 392 | 690 | 786 | 485 | 602 |
| | | | 1135 | 1273 | 1538 | 1653 | 1201 | 1573 |
| MNDO [46, 86, 87] | 610 | 1667 | 263 | 447 | 771 | 924 | 578 | 718 |
| | | | 1261 | 1407 | 1596 | 1722 | 1352 | 1628 |
| AM1 [88] | 660 | 1798 | | n/a | | | 574 | 776 |
| | | | | | | | 1405 | 1753 |
| HF/STO-3G [44, 51] | 570 | 1772 | | n/a | | | | n/a |
| HF/DZP [62] | 487 | 1615 | | n/a | | | | n/a |
| HF/TZP [62] | 483 | 1614 | | n/a | | | | n/a |
| MP2/DZP [62] | 437 | 1614 | | n/a | | | | n/a |
| MP2/TZP [62] | 437 | 1586 | | n/a | | | | n/a |
| LDA [64] | 537 | 1680 | 249 | 413 | 681 | 845 | 494 | 643 |
| | | | 1209 | 1453 | 1624 | 1726 | 1358 | 1641 |
| LDA [84] | 482 | 1447 | 261 | 435 | 730 | 775 | 541 | 566 |
| | | | 1098 | 1208 | 1394 | 1573 | 1158 | 1399 |
| LDA [65] | 550 | 1500 | | n/a | | | | n/a |
| LDA [73] | 454 | 1368 | 246 | 405 | 688 | n/a | 530 | 555 |
| | | | 1110 | 1314 | 1500 | n/a | 1105 | 1345 |
| IR (gas phase) [19] | | n/a | | n/a | | | 527 | 570 |
| | | | | | | | 1169 | 1407 |
| IR/Raman (solution) [20] | 497 | 1469 | 267 | 433 | n/a | 774 | 528 | 578 |
| | | | 1101 | 1252 | n/a | 1575 | 1183 | 1429 |
| IR/Raman (film) [21, 22] | 496 | 1470 | 273 | 437 | 710 | 774 | 527 | 577 |
| | | | 1099 | 1250 | 1428 | 1575 | 1183 | 1428 |
| Raman (solid) [23] | 495 | 1468 | 272 | 432 | 710 | 772 | | n/a |
| | | | 1100 | 1249 | 1425 | 1574 | | |
| NIS (solid) [89] | 488 | 1448 | 271 | 432 | 715 | 765 | 526 | 563 |
| | | | 1089 | 1217 | 1448 | 1563 | 1217 | 1448 |

Vibrational frequencies [cm$^{-1}$]

very poor [90, 91] and poor [28, 92–94] to reasonable [95] and excellent [96–98]. The performance of a bond-charge model [98], which uses only four adjustable parameters and reproduces the experimental IR and Raman frequencies within 3.0%, is quite impressive.

## 3.3 Heat of Formation

The recent combustion experiments on the $C_{60}$ fullerene provide benchmark data for its standard enthalpy of formation, against which the accuracy of theoretical predictions can be judged. When combined with the experimentally determined enthalpy of sublimation [99, 100] (corrected for the temperature difference), three independent measurements [101–103] yield values of $\Delta H_f^0(\text{gas})$ approximately equal to 600 kcal/mol [104]. In light of this close agreement, the data obtained from two other combustion experiments [113, 114] appear to be less reliable.

Comparison of the experimental data with the calculated values of the standard enthalpy of formation (Table 3.3) reveals a mixed performance

### Table 3.3.  The standard enthalpy of formation of the $C_{60}$ fullerene

| Method | $\Delta H_f^0(\text{gas})$ [kcal/mol] | Method | $\Delta H_f^0(\text{gas})$ [kcal/mol] |
|---|---|---|---|
| MM2 [25] | 519.8 | MMP2 [47] | 285.9 |
| MM3 [26] | 573.1 | MM3 [105, 106] | 573.8 |
| Tight-binding Hamilt. [107, 108] | 555 | PRDDO [41] | 978 |
| MNDO [45, 46, 50] | 869.3 | MNDO [47] | 868.7 |
| MNDO [43] | 870 | AM1 [109] | 972.7 |
| AM1 [47] | 972.6 | AM1 [48, 50] | 973.3 |
| PM3 [50, 110] | 811.7 | HF/STO-3G [48] | 582 |
| HF/STO-3G [26] | 625.0 | HF/STO-3G [111] | 720.6 |
| HF/6-31G*//HF/STO-3G [112] | 672 | Exp.[a] [101] | 597.5±4.0 |
| Exp.[a] [102] | 599.6±2.1 | Exp.[a] [103] | 598.7±3.8 |
| Exp.[a] [113] | 633.1±3.8 | Exp. [114] | 609.6±3.6 |

[a] $\Delta H_f^0(\text{solid})$ data from original references combined with the standard enthalpy of sublimation from [103].

of the molecular mechanics methods. Semiempirical approaches, such as PRDDO, MNDO, AM1, and PM3, overestimate $\Delta H_f^0(C_{60},\text{gas})$ by a wide margin, most probably because of the exaggerated strain energy. On the other hand, when used in conjunction with the empirical carbon-atom equivalents, the HF/STO-3G and HF/6-31G* total energies produce reasonable estimates for the standard enthalpy of formation of the $C_{60}$ fullerene [26, 48, 111, 112].

## 3.4 Orbital Energy Levels, Ionization Potentials, and Electron Affinities

The molecular orbitals of the $C_{60}$ fullerene fall into two broad categories [61]. The $\pi$-like orbitals possess nodal surfaces that approximately follow the outline of the cluster cage, whereas for the $\sigma$-like orbitals the nodal surfaces are absent. Approximate energy levels of the $\pi$-like orbitals can be obtained by diagonalization of the Hückel Hamiltonian (Section 3.10), which yields a fivefold degenerate $h_u$ HOMO, a triply degenerate $t_{1u}$ LUMO, and a triply degenerate $t_{1g}$ LUMO+1 [27, 128–131] (Fig. 3.2). This picture of frontier orbitals is upheld at more sophisticated, all-electron levels of theory such as CNDO/S [132], INDO/S [39], HF/STO-3G, HF/STO-3G*, HF/6-31G, HF/6-31G* [118], and

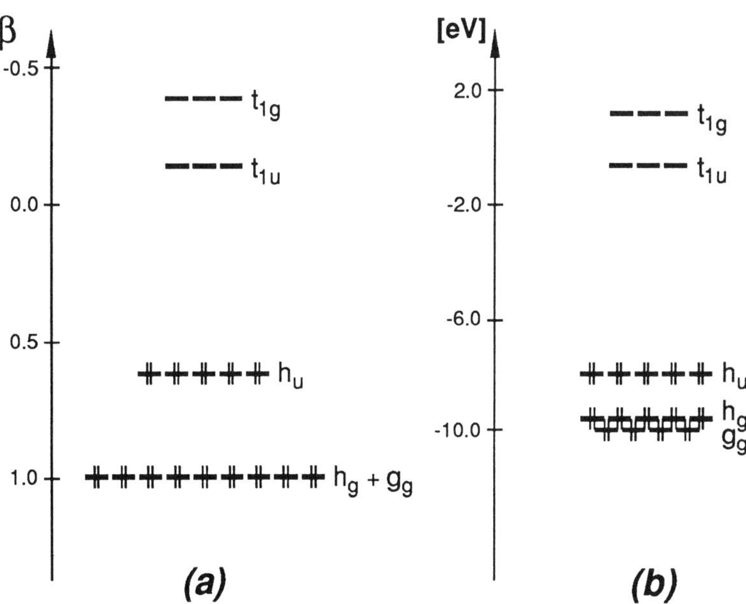



*Fig. 3.2. The frontier orbital energy levels of the $C_{60}$ fullerene: (a) the Hückel Hamiltonian, (b) the HF/6-311G* level of theory.*

**Table 3.4. Theoretical and experimental values
of the first vertical ionization potential
of the $C_{60}$ fullerene**

| Method | $IP_1$ [eV] | Method | $IP_1$ [eV] |
|---|---|---|---|
| 2D electron gas [115] | 8.86 | Gutzwiller approx. [116] | 7.2 |
| VEH[a] [49] | 7.52 | QCFF/PI[a] [33] | 8.12 |
| INDO/S[b] [37] | 6.36 | INDO/S[a] [117] | 6.57 |
| PRDDO[a] [41] | 4.5 | MNDO[a] [43, 44, 46, 47] | 9.13 |
| MNDO[b] [45, 46] | 8.95 | AM1[a] [47, 49] | 9.64 |
| PM3[a] [50, 110] | 9.48 | HF/SZ[a] [72] | 10.80 |
| HF/STO-3G[a] [48, 52] | 5.46 | HF/STO-3G[a] [118] | 5.62 |
| HF/STO-3G*[a] [118] | 5.13 | HF/3-21G[a] [55] | 8.3 |
| HF/4-31G[a] [56] | 7.97 | HF/6-31G//MNDO[a] [119] | 8.02 |
| HF/6-31G[a] [118] | 8.08 | HF/dz[a] [52] | 8.27 |
| HF/DZ[a] [58] | 8.24 | HF/DZ[a] [72] | 9.08 |
| HF/DZ[b] [58] | 7.92 | HF/DZ[b] [72] | 9.21 |
| HF/6-31G*[a] [118] | 7.72 | HF/dzP[a] [52] | 7.86 |
| HF/DZP[a] [57] | 8.03 | HF/DZP//MNDO[a] [119] | 8.04 |
| HF/DZP[b] [72] | 8.38 | HF/TZV[a] [60] | 8.27 |
| HF/TZV[a] [120] | 8.33 | HF/TZV[b] [60] | 8.07 |
| HF/TZV[b] [120] | 8.14 | HF/6-311G*[a] [61] | 7.94 |
| HF/TZP[a] [52] | 7.97 | LDA[c] [121] | 6.9 |
| LDA[c] [121] | 7.8 | LDA[b] [66] | 7.74 |
| LDA[b] [63, 68] | 7.60 | LDA[c] [63] | 7.49 |
| Photoion.[d] [122] | 7.57±0.01 | Synchr. rad.[d] [123] | $7.58^{+0.04}_{-0.02}$ |
| Synchr. rad.[d] [124] | 7.54±0.04 | PES[d] [125] | 7.61±0.02 |
| CT bracketing[d] [126] | 7.61±0.11 | CT bracketing[d] [127] | 7.6±0.1 |

[a] Koopmans' theorem.

[b] $\Delta$SCF calculation.

[c] Transition-state approximation.

[d] Experimental.

## Table 3.5. Theoretical and experimental values of the second vertical ionization potential of the $C_{60}$ fullerene

| Method | $IP_2$ [eV] | Method | $IP_2$ [eV] |
|---|---|---|---|
| Zero-range pot. [138] | 11.2 | Gutzwiller approx. [116] | 9.68 |
| MNDO[a] [45] | 12.2 | HF/3-21G[a] [55] | 12.0 |
| HF/DZ[a] [139] | 11.2 | LDA[b] [121] | 10.1 |
| LDA[b] [121, 134] | 10.8 | Photoion.[c] [140] | 11.46±0.05 |
| CT bracketing[c] [141] | <11.8 | CT bracketing[c] [127] | 9.7±0.2 |
| Charge strip.[c] [142] | 11.9 | Charge strip.[c] [143] | 12.25±0.5 |

[a] $\Delta$SCF calculation.

[b] Transition-state approximation.

[c] Experimental.

LDA [63, 121, 133–135]. The $h_g$ HOMO–1 and the $g_g$ HOMO–2 energy levels, which are degenerate within the Hückel approximation, remain closely spaced upon inclusion of bond alternation and electron–electron repulsion. The entire manifold of occupied orbitals corresponds to a local Hartree–Fock minimum with respect to orbital rotations [136]. However, as in the case of many $\pi$-electron systems, the Hartree–Fock wavefunction is found to be triplet unstable at the HF/SV level of theory. Similar instability occurs within the MNDO approximation, as reflected by the energy lowering of 57 kcal/mol observed upon going from the RHF wavefunction [46] to the UHF one [137].

The first vertical ionization potential of the $C_{60}$ fullerene ($IP_1$) has been the subject of many experimental measurements and theoretical predictions (Table 3.4). The experimental values of $IP_1$ cluster around 7.6 eV, whereas the estimates based on Koopmans' theorem converge to *ca.* 8.0 eV with increasing basis set size. The $\Delta$SCF values are *ca.* 0.2–0.3 eV lower, indicating marginal orbital relaxation effects [58, 60, 120]. Semiempirical methods of the MNDO family, the PRDDO approach, and *ab initio* calculations employing minimal basis sets perform poorly in predicting the first vertical ionization potential of $C_{60}$, whereas the LDA formalism yields excellent results.

The discrepancies between the early charge-transfer bracketing [127] and the other [140, 142, 143] measurements of the second vertical ionization potential of $C_{60}$ ($IP_2$, Table 3.5) have been resolved in favor of the latter [141]. Although charge transfer between the $C_{60}^{2+}$ cation and neutral molecules with ionization potentials greater than 9.7 eV does not occur, the Coulombic repul-

## Table 3.6. Theoretical and experimental values of the first vertical electron affinity of the $C_{60}$ fullerene

| Method | $EA_1$ [eV] | Method | $EA_1$ [eV] |
|---|---|---|---|
| Zero-range pot. [138] | 1.9 | Gutzwiller approx. [116] | 2.23 |
| QCFF/PI[a] [33] | 1.52 | QCFF/PI[b] [144] | 1.69 |
| MINDO/3[b] [40] | 0.76 | MNDO[a] [44, 46] | 2.56 |
| MNDO[b] [46, 154] | 2.68 | MNDO[b] [145] | 2.7 |
| PM3[a] [50, 110] | 2.89 | HF/STO-3G[a] [118] | $-3.51$ |
| HF/STO-3G[a] [52] | $-3.05$ | HF/STO-3G*[a] [118] | $-3.35$ |
| HF/3-21G[b] [54] | 0.87 | HF/4-31G[a] [56] | 0.34 |
| HF/6-31G//MNDO[a] [119] | 0.64 | HF/6-31G[a] [118] | 0.45 |
| HF/DZV[b] [151] | 1.13 | HF/dz[a] [52] | 0.63 |
| HF/DZ[a] [58] | 0.60 | HF/DZ[b] [58] | 0.80 |
| HF/6-31G*[a] [118] | 0.19 | HF/dzP[a] [52] | 0.38 |
| HF/DZP//MNDO[a] [119] | 0.82 | HF/DZP[a] [52] | 0.76 |
| HF/TZV[a] [60] | 0.76 | HF/TZV[a] [120] | 0.74 |
| HF/TZV[b] [60] | 0.96 | HF/TZV[b] [120] | 0.92 |
| HF/TZP[a] [52] | 0.65 | LDA[b] [63, 68] | 2.82 |
| LDA[b] [66] | 2.78 | LDA[b] [65] | 2.88 |
| LDA[c] [63] | 2.72 | LDA[c] [121, 134] | 2.7 |
| LDA[c] [121] | 2.0 | UPS[d] [146] | 2.7±0.1 |
| PES[d] [147] | 2.65±0.05 | CT bracketing[d] [148] | 2.6±0.1 |

[a] Koopmans' theorem.

[b] $\Delta$SCF calculation.

[c] Transition-state approximation.

[d] Experimental.

sion of *ca.* 2.1 eV between $C_{60}^+$ and the ionized species at the instant of charge transfer has to be taken into account in order to arrive at the correct estimate for $IP_2$ [149]. Several electronic structure calculations carried out to date also support the higher value of $IP_2$ [45, 55, 139].

Charge-stripping experiments locate the third vertical ionization potential of $C_{60}$ ($IP_3$) at 17.0±0.7 eV [143]. Applying the aforementioned Coulombic correction (in this case estimated at 4.5±0.4 eV) to the threshold ionization potential of 11.2±0.2 eV [127] yields the value of 15.6±0.5 eV for $IP_3$ [150].

In contrast to the results obtained for the first vertical ionization potential, very inaccurate values of the first vertical electron affinity of $C_{60}$ ($EA_1$) are furnished by Koopmans' theorem (Table 3.6). Hartree–Fock calculations that employ minimal basis sets are particularly inadequate for predicting $EA_1$, yielding affinities that are too negative by over 6 eV. Inclusion of the orbital relaxation effects through the $\Delta$SCF formalism results in only a marginal improvement, indicating the importance of electron correlation in the description of the $C_{60}^-$ ion. This observation is confirmed by the superior performance of the LDA method. In light of these considerations, it is rather surprising to find out that the MNDO, AM1, and PM3 approaches, when used in conjunction with Koopmans' theorem, reproduce $EA_1$ rather accurately.

Although the $C_{60}^{2-}$ anion can be detected in the gas phase [152, 153], the second vertical electron affinity of $C_{60}$ ($EA_2$) is consistently predicted to be negative at diverse levels of theory, including the Gutzwiller method [116], QCFF/PI ($\Delta$SCF) [144], MNDO ($\Delta$SCF) [145, 154], HF/DZV ($\Delta$SCF) [151], HF/TZV ($\Delta$SCF) [120], LDA ($\Delta$SCF) [65, 66], and LDA (transition-state approximation) [121, 134]. The possible explanations for this discrepancy involve a substantial geometry relaxation that stabilizes the $C_{60}^{2-}$ ion enough to be detectable and/or a high barrier to electron detachment.

## 3.5  Positive Ions

Removal of one electron from the $C_{60}$ fullerene results in the $C_{60}^+$ cation, which is subject to a Jahn–Teller distortion because of its $h_u^9$ electronic configuration [37, 155]. Symmetry considerations lead to the prediction of either a pentagonal ($D_{5d}$) or a trigonal ($D_{3d}$) distortion [155]. Which of these distortions actually results in an energy lowering can only be decided on the basis of electronic structure calculations. A detailed INDO/1 investigation of the $C_{60}^+$ cation has concluded that the six equivalent $D_{5d}$ structures are energy minima related through 15 $D_{2h}$ transition states [37]. On the other hand, each of the 10 $D_{3d}$ structures is a cusp on the potential energy hypersurface. The energy lowering due to symmetry breaking is substantial, amounting to 8.1 kcal/mol. The pseudorotation between the equivalent $D_{5d}$ minima (corresponding to $^2A_{1u}$ electronic states) is hindered by the energy barrier of only 2.2 kcal/mol, suggesting the apparent $I_h$ symmetry of $C_{60}^+$ at ambient temperatures. The energy of the first allowed transition to the electronic excited state of $C_{60}^+$, which is estimated by INDO/S calculations at 1.07 eV [37], agrees fairly well with the position of the absorption band that is observed at 1.27 eV [156, 157].

The $C_{60}^{2+}$ cation, which is efficiently produced by collisions between $He^+$ or $Ne^+$ and $C_{60}$ [158], reacts readily with ammonia [159, 160], amines [160], alcohols [161], and alkenes [162, 163]. As in the case of $C_{60}^+$, the $I_h$ geometry of $C_{60}^{2+}$ is subject to a Jahn–Teller distortion. Indeed, HF/DZ calculations find the $D_{5d}$ distorted geometry to be an energy minimum [139]. At the HF/3-21G level of theory, the energy lowering due to symmetry breaking is calculated at 16 kcal/mol [55]. The fivefold degenerate $h_u$ HOMO is split into a pair of doubly occupied $e_{1u}$ and $e_{2u}$ orbitals, and an empty $a_{1u}$ LUMO. The optimized geometry exhibits marked deviations from $I_h$ symmetry, with lengths of the C–C bonds located near the cage poles barely different from those in $C_{60}$, and lengths of the bonds spanning the cage equator almost completely equalized. Overall, the Jahn–Teller distortion is found to elongate the $C_{60}^+$ cage by *ca.* 0.15 Å along the $C_5$ axis. Interestingly, the symmetry breaking is less severe in the first triplet state of $C_{60}^{2+}$, which is predicted to lie 0.46 eV above the ground state [55].

Stability of highly charged cations is limited by the phenomenon of Coulomb explosion. As a molecule is stripped of electrons, the bonding orbitals are depopulated and the positive charges on atoms result in a net atom–atom repulsion. Beyond a certain critical value of the molecular electric charge, bonding is no longer sustainable, and either ejection of small fragments or total dissociation occurs. A simple electrostatic model predicts the $C_{60}^{n+}$ cations to be stable up to $n = 6$ [149], in agreement with the experimental observations of $C_{60}^{3+}$ [160, 164], $C_{60}^{4+}$ [164], and $C_{60}^{5+}$ [165].

According to HF/6-311G* calculations, the $C_{60}^{10+}$, $C_{60}^{18+}$, and $C_{60}^{28+}$ (but not $C_{60}^{36+}$) cations are stable against Coulomb explosion, provided $I_h$ symmetry is maintained in the process [61]. Because of their closed-shell electronic configurations (compare Fig. 3.2 and note that the ordering of the $h_g$ and $g_g$ orbital energy levels is reversed in $C_{60}^{10+}$), none of the $C_{60}^{10+}$, $C_{60}^{18+}$, and $C_{60}^{28+}$ cations undergoes a Jahn–Teller distortion. However, the question of whether geometries of these species are subject to symmetry lowering, which could lead to a lower-symmetry channel for Coulomb explosion, remains open. Indeed, MNDO, AM1, and PM3 calculations predict $C_{60}^{10+}$ to be a local minimum, whereas $C_{60}^{18+}$ and $C_{60}^{28+}$ are found to be higher-order saddle points, as indicated by the presence of several imaginary frequencies in the computed vibrational spectra [61]. The progressive removal of electrons from the $C_{60}$ cage affects the "double" bonds (Section 3.1) more than the "single" ones. This phenomenon is easily understood by noticing that the frontier orbitals of $C_{60}$ are $\pi$-like. Occupying these orbitals gives rise to higher electron densities around the "double" bonds, as reflected in the directions of atomic dipole moments [166]. For this reason, removal of electrons from the frontier orbitals of $C_{60}$ weakens the "double" bonds more than the "single" ones. This effect is nicely illustrated in the $C_{60}^{28+}$ cation, in which the calculated lengths of the "double" and "single" bonds equal 1.651 and 1.609 Å, respectively [61].

## 3.6  Negative Ions

A plethora of experimental and theoretical studies on the negative ions of the $C_{60}$ fullerene has been published. Attachment of one electron to $C_{60}$ results in the $t_{1u}^1$ electronic configuration, implying a Jahn–Teller distortion of the $C_{60}^-$ anion. MINDO/3 calculations predict the $D_{3d}$ distorted geometry of $C_{60}^-$ to be an energy minimum [40]. The symmetry breaking splits the triply degenerate $t_{1u}$ LUMO of $C_{60}$ into a singly occupied $a_{2u}$ orbital and an $e_u$ LUMO. A similar picture of bonding in $C_{60}^-$ is obtained at the UHF/3-21G level of theory [54]. The barriers to pseudorotations that relate 10 equivalent $D_{3d}$ structures are found to be less than 0.1 kcal/mol high. The $D_{3d}$ energy minima, each corresponding to the $^2A_{2u}$ electronic state, lie 2.0 kcal/mol below the fully symmetrical $I_h$ structure. Energies of the $D_{5d}$ structures are only 0.02 kcal/mol higher than those of the minima. At the ROHF/3-21G level of theory, however, the order of stabilities is reversed and the energy minima are attained at six equivalent $D_{5d}$ geometries, the energy difference between the $D_{5d}$ and $D_{3d}$ structures amounting to a mere 0.15 kcal/mol [54]. Both the UHF/3-21G and ROHF/3-21G calculations predict changes in bond lengths due to the distortion not to exceed $\pm 0.02$ Å. The $D_{5d}$ minima are also favored at the LDA [66] and ROHF/DZV [151] levels of theory, which estimate the energy lowering from $I_h$ symmetry at 0.6 and 3.2 kcal/mol, respectively. The overall conclusion to be reached from these calculations is that the potential energy surface of $C_{60}^-$ is extremely flat along the distortion modes, indicating a profound susceptibility of the distortion pattern to temperature, solvents, and counterions. It is most probable that this sensitivity is responsible for the unusual, temperature-dependent features observed in the EPR spectra of the $C_{60}^-$ anion [167–171].

Although, among the multiply charged negative ions of $C_{60}$ only $C_{60}^{2-}$ is (meta)stable in the gas phase (Section 3.4), the $C_{60}^{3-}$ [172–175], $C_{60}^{4-}$ [176], $C_{60}^{5-}$ [177], and $C_{60}^{6-}$ anions [178] have been observed in solution. Except for the $C_{60}^{6-}$ anion, all of these species are subject to a Jahn–Teller distortion. The presence of this distortion has been confirmed by X-ray diffraction measurements on the $(PPN^+)_2 C_{60}^{2-}$ ($PPN^+$ = bis(triphenylphosphine)iminium cation) salt [179]. The $D_{5d}$ geometry is found to be an energy minimum for $C_{60}^{2-}$ at the MNDO [180] and ROHF/DZV [151] levels of theory. The barrier to pseudorotation computed with the former method amounts to only 0.5 kcal/mol. HF/TZV calculations carried out within $I_h$ symmetry for the $C_{60}^{n-}$ $(1 \leq n \leq 6)$ species predict the $r_1$ and $r_2$ bond lengths (Section 3.1) to depend almost linearly on $n$, with the bond alternation diminishing progressively from the monoanion to the hexaanion [60]. As expected from the gradual occupation of the $\pi$-like antibonding $t_{1u}$ LUMO of $C_{60}$, the equalization of bond lengths proceeds mostly through elongation of the "double" bonds. The MNDO-optimized bond lengths exhibit similar behavior, although their dependence on the total charge is somewhat less smooth [145]. According to QCFF/PI calculations,

## Table 3.7. Theoretical and experimental energies of the allowed $^1A_g \rightarrow {}^1T_{1u}$ electronic transitions in the $C_{60}$ fullerene

| Method | Excitation energy [eV] | | | | | | | | |
|---|---|---|---|---|---|---|---|---|---|
| PPP [30] | n/a | n/a | 4.23 | n/a | n/a | 4.75 | n/a | 6.52 | n/a |
| PPP [31] | n/a | n/a | 4.00 | n/a | n/a | 4.68 | n/a | 6.69 | n/a |
| PPP [32] | n/a | n/a | 4.01 | n/a | n/a | 4.64 | n/a | n/a | n/a |
| QCFF/PI [33] | n/a | n/a | 4.08 | n/a | n/a | 4.53 | n/a | 6.19 | n/a |
| CNDO/S [132] | n/a | 3.48 | 4.39 | n/a | n/a | 5.24 | n/a | 5.78 | n/a |
| CNDO/S [184] | n/a | 3.36 | 4.26 | 4.00 | 4.59 | 4.97 | 5.49 | 5.57 | 6.27 |
| CNDO/S [185] | 3.4 | n/a | 4.40 | n/a | n/a | 5.23 | n/a | 5.79 | n/a |
| INDO/S-CIS [117] | n/a | 3.38 | 4.08 | n/a | n/a | 4.95 | 5.25 | 5.81 | n/a |
| INDO/S-CISD [117] | n/a | 3.49 | 4.19 | n/a | n/a | 5.05 | 5.38 | 5.92 | n/a |
| INDO/S-CIS [38] | n/a | 3.20 | 3.92 | 4.00 | 4.32 | 4.70 | 5.38 | 5.91 | 6.22 |
| INDO/S-RPA [38] | n/a | 3.20 | 3.88 | 4.08 | 4.33 | 4.69 | 5.39 | 6.06 | 6.27 |
| RPA/6-31G+s [136] | n/a | n/a | 5.22 | n/a | 6.19 | 6.69 | 8.04 | 8.46 | n/a |
| UV/MCD spectra[a] [183] | n/a | n/a | 3.81 | n/a | n/a | 4.90 | n/a | 5.96 | n/a |
| UV spectrum[b] [182] | 3.04 | 3.30 | 3.78 | 4.06 | 4.35 | 4.84 | 5.46 | 5.88 | 6.36 |

[a] Argon matrix at 5 K.

[b] Hexane solution at room temperature.

the bond alternation is marginal in the $C_{60}^{6-}$ anion, which has computed bond lengths equal to 1.453 and 1.464 Å [181]. These changes in the cage geometry are indicative of bond weakening that is also manifested in softening of the vibrational modes.

Experimentally, the $C_{60}^-$ anion is known to absorb at 1.16, 1.25, and 1.35 eV [157, 172]. The UHF/3-21G calculations place the first excited state of $C_{60}^-$ 0.27 eV above the ground state [54], whereas the corresponding excitation energy computed at the HF/TZV level of theory equals 1.67 eV [120]. The excitation energy of 0.64 eV obtained with the QCFF/PI method is also in poor agreement with the experimental data [144]. The number of low-lying excited states is predicted by the same approach to increase dramatically from the monoanion to the hexaanion of the $C_{60}$ fullerene, making the assignment of individual absorption bands in the electronic spectra a difficult, if not impossible, task.

## 3.7  Excited States

High-resolution UV spectra of the $C_{60}$ cluster exhibit three strong absorption bands at *ca.* 3.8, 4.8, and 5.9 eV [182, 183]. These very intense absorptions can be unequivocally ascribed to the allowed $^1A_g \rightarrow {}^1T_{1u}$ transitions (Table 3.7) that have substantial contributions stemming from the HOMO$\rightarrow$LUMO+1 single-electron excitations.   Calculations involving all-electron, spectroscopically-parameterized semiempirical methods, such as CNDO/S [132, 183, 184] or INDO/S [38, 117], carried out in conjunction with a limited configuration interaction, reproduce the positions of the intense absorptions quite accurately.

## Table 3.8.  Theoretical and experimental energies of the lowest singlet and triplet excited states of the $C_{60}$ fullerene

| Method | Excitation energy [eV][a] | |
| --- | --- | --- |
| | $S_1$ | $T_1$ |
| Gutzwiller approximation [116] | 1.74 | 1.67 |
| Tight-binding Hamiltonian [186] | n/a | 1.6  $(^3T_{2g})$ |
| PPP [30] | 2.87  $(^1T_{1g})$ | 2.46  $(^3T_{2g})$ |
| PPP [31] | 2.61  $(^1T_{1g})$ | 2.23  $(^3T_{2g})$ |
| PPP [32] | 2.69  $(^1T_{1g})$ | n/a |
| QCFF/PI [33] | 2.58  $(^1T_{2g})$ | 2.06  $(^3T_{2g})$ |
| CNDO/S [184] | 2.29  $(^1T_{2g})$ | n/a |
| INDO/S-CIS [38] | 2.10  $(^1T_{1g})$ | n/a |
| INDO/S-CIS [117] | 2.22  $(^1T_{1g})$ | n/a |
| INDO/S-CISD [117] | 2.28  $(^1T_{1g})$ | n/a |
| MNDO-CI [145] | 3.17  $(^1T_{2g})$ | 3.03  $(^3T_{2g})$ |
| CIS/6-31G+s [136] | 2.97  $(^1T_{2g})$ | 1.69  $(^3T_{2g})$ |
| Triplet quenching [187] | 2.00 | 1.63$\pm$0.20 |
| Phosphorescence [188] | n/a | 1.57 |
| Photoelectron spectrum [189] | n/a | 1.7 |
| UV/MCD spectra [183] | 1.92 | n/a |

[a] Symmetries of the excited states given in parentheses.

On the other hand, excitation energies computed at the *ab initio* RPA/6-31+s level of theory are substantially overestimated, most probably because of the limited basis set employed [136].

Assignment of the less intense absorption bands is more problematic. The transitions at *ca.* 4.1 and 4.4 eV, observed in the UV spectra of $C_{60}$ dissolved in hexane [182], are most probably due to excited states of $^1T_{1u}$ symmetry. On the other hand, the weak absorption around 3.0 eV has been attributed to either the allowed $^1A_g \to {}^1T_{1u}$ transition with a low oscillator strength [182] or the forbidden $^1A_g \to {}^1T_{2u}$ transition [183]. The former assignment is supported by CNDO/S calculations [184], whereas the latter one is in agreement with the *ab initio* prediction of the first $^1A_g \to {}^1T_{1u}$ transition possessing substantial oscillator strength [136].

### Table 3.9. Polarizability and the second-order hyperpolarizability of the $C_{60}$ fullerene

| Method | $\alpha\ [\text{Å}^3]$ | $10^{36} \cdot \gamma\ [\text{esu}]$ |
|---|---|---|
| PPP (CPHF) [202] | n/a | 95.6 |
| VEH//AM1 (sum-over-states) [49] | 154.0 | 202. |
| INDO (CPHF) [203] | 81.7 | n/a |
| INDO/S (sum-over-states) [117] | n/a | 710. |
| PM3 (finite field) [50, 110] | 63.9 | 21.1 |
| HF/STO-3G (CPHF) [118] | 45.3 | n/a |
| STO-3G* (CPHF) [118] | 49.5 | n/a |
| HF/6-31G (CPHF) [118] | 63.7 | n/a |
| HF/6-31G* (CPHF) [118] | 65.5 | n/a |
| HF/6-31+G* (CPHF) [136, 204] | 78.8 | n/a |
| LDA (sum-over-states) [65] | 311.2 | n/a |
| LDA (finite field) [65, 198] | 82.7 | 7.0 |
| LDA (finite field) [70] | 77.9 | 15.9 |
| Dielectric const. (exp. est.) [198] | 84.2±0.7 | n/a |
| Third-harmonic gener. (exp.) [205] | n/a | 430. |
| Second-harmonic gener. (exp.) [206] | n/a | 750±200 |
| Four-wave mixing (exp.)$^a$ [207] | n/a | 300. |

$^a$ As reported in [70].

Group-theoretical considerations lead to the conclusion that the HOMO$\rightarrow$LUMO single-electron excitations in $C_{60}$ participate only in the symmetry-forbidden transitions to the ${}^1T_{1g}$, ${}^1T_{2g}$, ${}^1G_g$, ${}^1H_g$, ${}^3T_{1g}$, ${}^3T_{2g}$, ${}^3G_g$, and ${}^3H_g$ excited states [27, 30, 39]. The lowest-energy triplet excited state of $C_{60}$ is predicted to possess ${}^3T_{2g}$ symmetry by a broad spectrum of electronic structure methods (Table 3.8). The computed $S_0 \rightarrow T_1$ excitation energies are in surprisingly good agreement with the experimental data, the MNDO result being the only exception. On the other hand, no consensus on the symmetry of the lowest-energy singlet excited state of $C_{60}$ is reached among theoretical predictions as the ${}^1T_{1g}$ and ${}^1T_{2g}$ excited states are calculated to be very close in energy.

Several photochemical properties of the $S_1$ and $T_1$ excited states, such as their absorption spectra, have been measured [190–194]. The experimental data provide ample evidence for a Jahn–Teller distortion of the $T_1$ state [195, 196]. A simple model based on the Hubbard Hamiltonian predicts the $T_1$ state to possess $D_{5d}$ symmetry [197].

## 3.8 Electric Polarizabilities

With its 60 $\pi$-like electrons, the $C_{60}$ fullerene is highly polarizable. Measurements of the dielectric constant of solid $C_{60}$ (Chapter 10) yield the estimate of *ca.* 84 Å$^3$ for the electric polarizability ($\alpha$) of buckminsterfullerene [198]. In general, electronic structure calculations on $C_{60}$ yield values of $\alpha$ that are too low (Table 3.9). However, in the case of *ab initio* approaches the agreement between the computed polarizabilities and their experimental counterpart improves steadily with increasing quality of basis sets, the discrepancy amounting to only 5% at the HF/6-31+G$^*$ level of theory. The data presented in Table 3.9 also demonstrate that the finite-field and CPHF calculations are clearly preferable to the sum-over-states ones.

The first-order hyperpolarizability ($\beta$) of $C_{60}$ is zero by symmetry. The second-order polarizability ($\gamma$) has been measured with several techniques. It has been shown [199, 200] that some of the experimental values of $\gamma$ [201] are grossly in error. Reliable experimental estimates and accurate theoretical computations of $\gamma$, such as those carried out within the HF or LDA formalism, are scarce. They are also in poor agreement among themselves (Table 3.9), the source of this discrepancy being presently unknown.

The electric polarizabilities of selected anions of $C_{60}$ have been calculated and found to rise with the increasing negative charge [65]. The frequency-dependent polarizabilities of the parent fullerene have also been computed [136].

## 3.9 The $^{13}$C NMR Spectrum and Other Magnetic Properties

The $^{13}$C NMR spectrum of the $C_{60}$ fullerene consists of only one line with the chemical shift ($\delta$) measured at 142.68 [16], 142.8 (in $C_6H_6$) [208], 143.2 (in $C_6D_6$) [17], 143.2 (in $CDCl_2CDCl_2$) [106], 142.5 (in $CCl_4$) [18], and 143.0 ppm (in $CCl_4$ or toluene) [175]. The chemical shift tensor is very anisotropic, as reflected in its experimentally determined principal components of 220, 186, and 40 ppm [209]. The known values of the $^{13}$C chemical shift and the absolute shielding in benzene can be used to convert the $^{13}$C chemical shift in $C_{60}$ to the absolute shielding ($\sigma$) of 43 ppm and the components of the chemical shift tensor to those of the shielding tensor, equaling $-34$, 0, and 146 ppm [210].

Several *ab initio* electronic structure calculations have been carried out with the goal of predicting the NMR spectrum of $C_{60}$. Early CPHF studies, employing an approach that is not gauge invariant, have produced the extrapolated average shielding of 98 [118] and 46 ppm [211], and the principal components of the shielding tensor equal to $-51.2$, 10.4, and 179.3 ppm [210]. Gauge-invariant GIAO HF/DZP and HF/TZP calculations, carried out at the MP2/DZP optimized geometry, fare much better, yielding the absolute shielding of 44.5 and 39.4 ppm, respectively [204]. However, the observed agreement with experimental data may be fortuitous as the shielding in $C_{60}$ appears to be very sensitive to the cage geometry. This sensitivity is well illustrated by the results of GIAO HF/6-31G calculations that yield values of $\sigma$ equal to 55.2 and 47.1 ppm at the HF/6-31G and MNDO geometries, respectively [212].

Buckminsterfullerene is a diamagnetic molecule. The experimentally determined molar magnetic susceptibilities of $-252$ [213] and $-260$ ppm cgs/mol [214] are equivalent to the susceptibilities of $-2.8 \cdot 10^{-3}$ and $-2.9 \cdot 10^{-3}$ au per molecule of $C_{60}$. These experimental values should be compared with the susceptibility of $-4.0 \cdot 10^{-3}$ au obtained from extrapolated results of *ab initio* calculations involving the CPHF formalism [118]. On a per-atom basis, the magnetic susceptibility of $C_{60}$ amounts to only *ca.* 5% of that of a graphite sheet [213].

The vanishingly small susceptibility of the $C_{60}$ fullerene is the result of an almost perfect cancellation between the diamagnetic and the paramagnetic contributions [215–218]. A qualitative description of this rather unusual phenomenon is furnished by the London theory. Within the Hückel approximation, the ratio of the diamagnetic and the paramagnetic terms is found to depend strongly on the relative magnitudes of the resonance integrals assigned to the "single" and "double" bonds, and the near-cancellation is attained only for weak bond alternation [215, 216]. Another way of accounting for the vanishingly small magnetic susceptibility of $C_{60}$ is to analyze ring currents [217, 218]. Strong paramagnetic currents are found to flow around the five-membered rings, whereas ring currents associated with the bonds connecting pentagons are diamagnetic and much weaker. Again, the overall ring-current response to the external magnetic field is almost nonexistent, meaning that the observed magnetic susceptibility is dominated by the local diamagnetic contributions [214].

The lack of strong ring currents in $C_{60}$ is a manifestation of its weakly aromatic character (Section 3.10) [219] (note, however, a dissenting opinion [220]) and has far-reaching consequences for magnetic shielding of guests in endohedral complexes (Chapter 8). Interestingly, the London theory predicts the $C_{60}^{6-}$ anion to be strongly diamagnetic due to diamagnetic ring currents that flow around both the hexagons and the pentagons of the fullerene cage [215, 217, 218].

## 3.10 Simple Theoretical Descriptions of Electronic Structure

Although far from being rigorous, simple theoretical descriptions of the $C_{60}$ fullerene play an important role in qualitative understanding of its electronic structure. One such description is offered by the topological Hückel Hamiltonian, expressed in the basis of the 60 $2p_z$-like carbon orbitals of $C_{60}$ that span the $a_g + t_{1g} + 2t_{1u} + t_{2g} + 2g_g + 3h_g + 2t_{2u} + 2g_u + 2h_u$ manifold of irreducible representations [27, 130, 131]. Thanks to the high symmetry of the $C_{60}$ cage, the Hückel matrix can be easily factorized and all of its 15 distinct eigenvalues can be expressed in the form of exact expressions (Table 3.10) [129, 130] — an accomplishment that was reported as early as in 1981 [128]. The total $\pi$-electron energy of $C_{60}$, which is obtained by summing the energies of all occupied orbitals, amounts to $[29 + 125^{1/2} + 272^{1/2} + 325^{1/2} + (3/2)(38 - 20^{1/2})^{1/2} + (3/2)(38 + 20^{1/2})^{1/2}]\beta \approx 93.161602\beta$ [27, 130, 131, 221]. The bond orders of the "single" and "double" C–C bonds (Section 3.1) equal 0.476 and 0.601, respectively [27, 222, 223].

The question of whether the $C_{60}$ fullerene is an aromatic molecule has been raised several times in the chemical literature. Contrary to several published claims [221, 224, 225], the fact that the delocalization energy per carbon atom of $C_{60}$ (equal to $0.5527\beta$) is greater than that of benzene does not imply aromaticity of buckminsterfullerene. Actually, computation of the topological resonance energy per electron (TREPE), which is a proper measure of aromaticity, reveals that $C_{60}$ is no more aromatic than heptacene [226]. Similar conclusions are reached from calculations of the Hess–Schaad resonance energy per electron [227-229]. One should also be reminded that, in order to account for the diminished magnitude of the $\beta$ integral due to the curvature of the $C_{60}$ cage, the above stability estimates should be reduced even further [27, 230]. The marginally aromatic nature of buckminsterfullerene is reflected in its chemical reactions, in which it exhibits the characteristics of an electron-poor alkene (Chapter 9). The weak diamagnetism of $C_{60}$ (Section 3.9) is also in agreement with this contention [215].

## Table 3.10. Eigenvalues of the Hückel Hamiltonian of the $C_{60}$ fullerene

|  |  | Orbital energy [$\beta$ units] | |
| --- | --- | --- | --- |
| Orbital[a] | Degeneracy | Exact expression | Approx. value |
| $a_g$ | 1 | 3 | 3.000000 |
| $t_{1u}$ | 3 | $(1/4)\,[3 + 5^{1/2} + (38 - 20^{1/2})^{1/2}]$ | 2.756598 |
| $h_g$ | 5 | $(1/2)\,(1 + 13^{1/2})$ | 2.302776 |
| $t_{2u}$ | 3 | $(1/4)\,[3 - 5^{1/2} + (38 + 20^{1/2})^{1/2}]$ | 1.820249 |
| $g_u$ | 4 | $(1/2)\,(-1 + 17^{1/2})$ | 1.561553 |
| $g_g + h_g$ | 9 | 1 | 1.000000 |
| $h_u$ | 5 | $(1/2)\,(-1 + 5^{1/2})$ | 0.618034 |
| $t_{1u}$ | 3 | $(1/4)\,[3 + 5^{1/2} - (38 - 20^{1/2})^{1/2}]$ | $-0.138564$ |
| $t_{1g}$ | 3 | $(1/2)\,(-3 + 5^{1/2})$ | $-0.381966$ |
| $h_g$ | 5 | $(1/2)\,(1 - 13^{1/2})$ | $-1.302776$ |
| $t_{2u}$ | 3 | $(1/4)\,[3 - 5^{1/2} - (38 + 20^{1/2})^{1/2}]$ | $-1.438283$ |
| $h_u$ | 5 | $(1/2)\,(-1 - 5^{1/2})$ | $-1.618034$ |
| $g_g$ | 4 | $-2$ | $-2.000000$ |
| $g_u$ | 4 | $(1/2)\,(-1 - 17^{1/2})$ | $-2.561553$ |
| $t_{2g}$ | 3 | $(1/2)\,(-3 - 5^{1/2})$ | $-2.618034$ |

[a] The horizontal gap denotes the Fermi level.

The topological Hückel approach and the valence-bond resonance theory are closely related, often yielding similar results when used in qualitative description of conjugated $\pi$-electron systems. This is also the case for the $C_{60}$ fullerene. Although the $C_{60}$ cage admits as many as 12 500 Kekulé structures [228, 231, 232], arguments based on the conjugated circuits theory find the aromaticity of buckminsterfullerene to be less pronounced than that of benzene [228, 233, 234].

The spherical shape of the $C_{60}$ fullerene has prompted several attempts to describe its electronic structure in terms of a two-dimensional electron gas. In its simplest version, such an approach invokes the picture of $\pi$ electrons confined to a sphere with a radius equal to that of the $C_{60}$ cage [27, 29, 131]. The eigenvectors of the resulting rigid-rotor Hamiltonian are given by spherical harmonics and the corresponding eigenvalues read [29]

$$E_L = 3\beta - \frac{\beta}{10} L(L+1), \qquad L = 0, 1, \ldots, \tag{3.2}$$

where $L$ is the angular momentum number. The singly degenerate $L = 0$ energy level corresponds to the $a_g$ orbital in the Hückel theory, the triply degenerate $L = 1$ level to the $t_{1u}$ orbital, and the fivefold degenerate $L = 2$ level to the $h_g$ orbital. One should note that none of these levels are split by the icosahedral perturbation introduced by the presence of 60 carbon cores (nuclei and $\sigma$ electrons). However, the $L = 3$ level is split, giving rise to the $t_{2u}$ and $g_u$ orbitals. The $L = 4$ level is ninefold degenerate at the Hückel level of theory, but the degeneracy is lifted once bond alternation is taken into account (Section 3.4). The $h_u$ HOMO orbital of the $C_{60}$ fullerene constitutes one of the $L = 5$ sublevels.

The above approximation has been refined by including electron–electron repulsion [115] and replacing the infinite potential of the rigid-rotor Hamiltonian with more realistic Dirac bubble [235] and zero-range potentials [138]. In a related development, the $C_{60}$ cage has been modeled with the help of the Thomas–Fermi theory [236]. With only $\pi$ electrons considered, the total energy has been found to possess a minimum corresponding to the cage radius of 3.89 Å, which is *ca.* 10% greater than the experimental value (Section 3.1). The Thomas–Fermi theory has also been invoked to explain the positive valuedness of the endohedral potential (Chapter 8) [57, 237]. Finally, the localized molecular orbital (LMO) description of $C_{60}$ should be mentioned. Such a description furnishes a particularly simple picture of bonding in buckminsterfullerene. A pattern of 60 two-center LMOs describing the $\sigma$ "single" bonds, together with 30 two-center LMOs accounting for the $\sigma$ "double" bonds and 30 two-center LMOs corresponding to the $\pi$ "double" bonds, is obtained with both the Boys localization of PRDDO orbitals [41] and the Edmiston–Ruedenberg localization of CNDO/2 orbitals [238]. This pattern is reminiscent of the dominant Kekulé structure of the $C_{60}$ fullerene, in which each of the five-membered rings is spanned exclusively by single bonds and each six-membered ring has alternating single and double bonds. The lack of three-center orbitals in the localized description once again provides evidence for the weakly aromatic character of $C_{60}$.

## 3.11 Mechanism of Formation

Despite numerous empirical and theoretical studies, the mechanism of fullerene formation is not yet fully understood [239, 240]. Isotope scrambling experiments, in which mixtures of isotopically pure $^{12}$C and $^{13}$C graphite are ablated and mass distributions of the resulting $C_{60}$ are measured, establish unequivocally that fullerenes are formed in the gas phase and that the initial precursors are carbon atoms rather than small carbon clusters or fragments of the graphite lattice [80, 241, 242]. With the results of a few molecular dynamics simulations [243, 244] being the only available clues, several speculative theories of fullerene formation were proposed in the early days of $C_{60}$ research. These hypotheses fall into two broad categories encompassing theories that invoke the growth of curved graphite sheets and those that employ the concept of building blocks.

The icospiral particle nucleation theory, which belongs to the first of these categories, assumes that the formation of $C_{60}$ and other fullerenes begins with small carbon clusters [245–247]. These clusters attempt to minimize the number of dangling bonds, which inevitably results in curved surfaces incorporating some five-membered rings. The precursor surfaces, such as that of the $C_{20}$ corannulene-like cluster (Chapter 7), grow through accretion of carbon atoms and small carbon clusters at their edges, producing nautilus-like curved carbon shells. The growth can either terminate, furnishing fullerenes, or continue, resulting in giant soot particles. In other words, the icospiral particle nucleation theory views fullerenes as by-products of soot formation. This contention is a weak point of the theory, as there is experimental evidence that different mechanisms are responsible for the production of soot particles and fullerenes [248].

A somewhat different mechanism of fullerene formation, called the "pentagonal road", has been proposed [249]. It assumes that the most stable carbon sheets are composed solely of five- and six-membered rings with a maximum number of pentagons that do not abut. As these sheets grow by accretion, they have ample time to anneal, thus never departing from the "pentagonal road". Initially, this growth is accompanied by an increase in the number of dangling bonds. However, beginning at $C_{15}$, the number of dangling bonds stabilizes at 10 and remains constant up to $C_{50}$, where it begins to fall, dropping to zero for $C_{60}$. Although the "pentagon road" mechanism is appealing in its simplicity, it fails to account for the experimentally observed preference for fullerenes among clusters with more than 36 carbon atoms (Chapter 7) [250, 251].

Theories that belong to the second category invoke the concept of building blocks in order to explain the formation of fullerenes. One of these theories employs the $C_{10}$ naphthalene-like cluster as such a block [252–254]. This precursor presumably polymerizes, producing fullerenes as the result. Unfortunately, this hypothesis is in variance with the experimentally observed structural preferences among carbon clusters (Chapter 7) [250, 251] and the fact that electronic structure calculations, carried out at the MP4/6-31G* level of theory, find the bicyclic $C_{10}$ cluster to be less stable than its monocyclic counterpart by 62 kcal/mol [255].

A variant of this theory also assumes the bicyclic $C_{10}$ cluster to be a fullerene precursor [256, 257]. Stacking of this precursor with even-membered carbon rings is supposed to produce fullerenes. In addition to suffering from all of the aforementioned shortcomings, this theory predicts the formation of a wrong $D_2$ isomer of the $C_{84}$ fullerene (Chapter 5) [257].

The powerful technique of "ion chromatography" (Chapter 7) [251] has shed some light on the fullerene formation process. Experimental results obtained with this technique, together with the observation that the transient monocyclic $C_{30}$ cluster forms predominantly the $C_{60}$ fullerene, whereas the analogous $C_{18}$ and $C_{24}$ species furnish mainly $C_{70}$ (apparently via a $C_{72}$ intermediate) [7], have led to the current belief that fullerenes are formed through coalescence of smaller carbon rings. Possessing excess energies, the coalescence products spontaneously rearrange to fullerene cages, if necessary ejecting small fragments (such as $C_2$) in the process [258–261].

## 3.12  Other $C_{60}$ Carbon Clusters

Among the 1812 fullerene cages composed of 60 carbon atoms, buckminsterfullerene is the only IPR structure [262, 263]. Within the Hückel approximation, the absence of abutting pentagons confers an exquisite stability upon the $C_{60}$ fullerene, which has the highest total $\pi$-electron energy and the largest HOMO–LUMO gap among its isomers [264–266].

Formation of unspecified metastable $C_{60}$ isomers in sooting flames has been reported [10]. It is conceivable that these isomers are related to the $C_{60}$ fullerene through the Stone–Wales transformation (or a series of them), in which a $C_2$ fragment of the carbon cage is rotated by 90° [267]. LDA calculations predict a barrier of 161 kcal/mol for this concerted pericyclic process [71], in approximate agreement with the estimates of 189, 221, 214, 205, 219, 201, and 194 kcal/mol obtained at the MNDO, HF/DZ, LDA/DZ//HF/DZ, BLYP/DZ//HF/DZ, HF/DZP, LDA/DZP//HF/DZP, and BLYP/DZP//HF/DZP levels of theory, respectively [268]. An alternative mechanism, involving an $sp^3$-hybridized carbon rather than a quasi-planar transition state, has a lower activation energy, amounting to 136, 145, 171, 164, 141, 164, and 159 kcal/mol at the same levels of theory.

The $C_{2v}$ isomer of the $C_{60}$ fullerene that results from a single Stone–Wales transformation has two pairs of abutting pentagons, each pair forming a pentalene fragment (Fig. 3.3). It lies 44 (MNDO) [268–270], 49 (HF/3-21G) [269], 74 (BLYP/DZ//HF/DZ) [268], and 37 kcal/mol (LDA) [71] above the $I_h$ IPR isomer (buckminsterfullerene) and possesses significant bond alternation. Application of two successive Stone–Wales transformations leads to several non-IPR structures. Among those, the $D_{2h}$ isomer with a total of four pentalene motifs located pairwise at the antipodes of the cage (Fig. 3.4) is less stable than the $I_h$ isomer by 86 (MNDO) [269, 270] and 98 kcal/mol (HF/3-21G) [269].

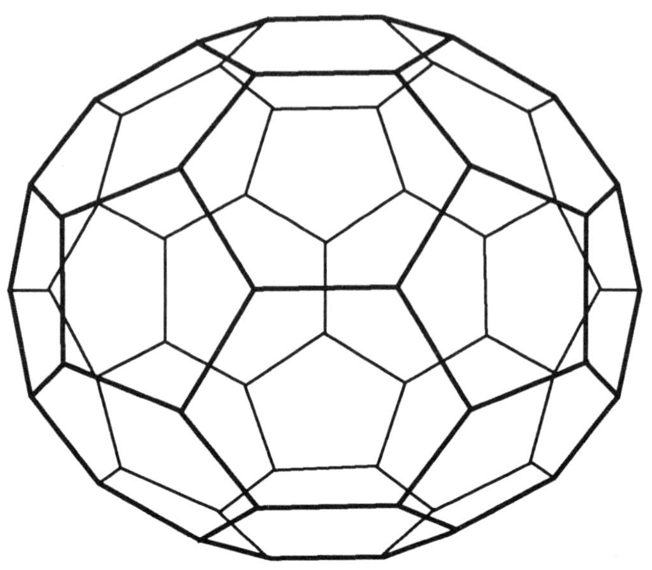

*Fig. 3.3. The $C_{2v}$ non-IPR isomer of the $C_{60}$ fullerene.*

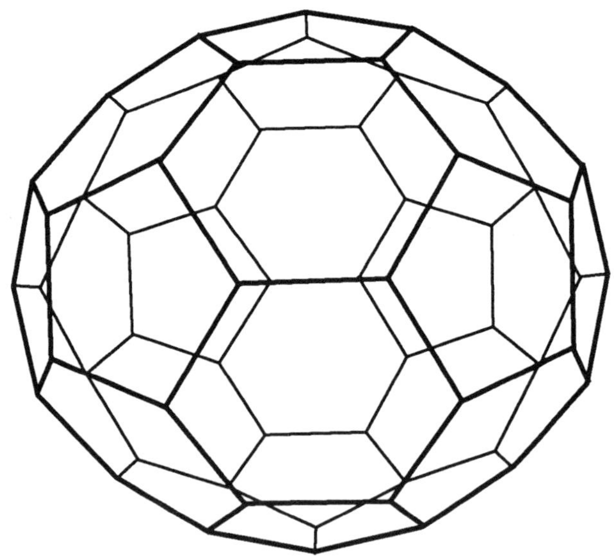

*Fig. 3.4. The $D_{2h}$ non-IPR isomer of the $C_{60}$ fullerene.*

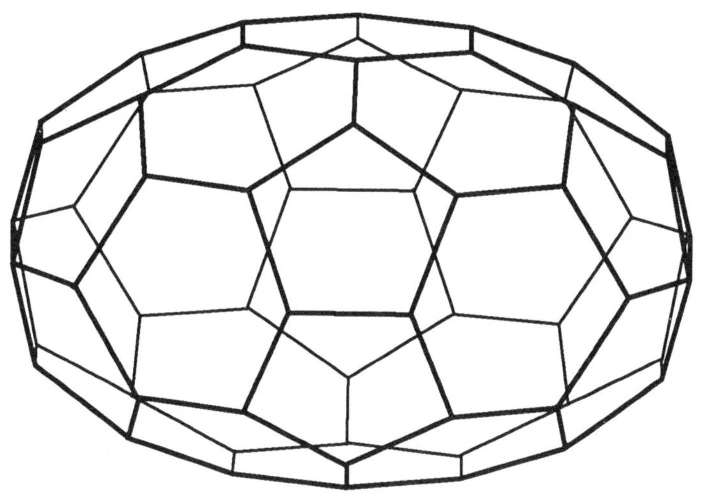

*Fig. 3.5. The "graphitene" non-IPR isomer of the $C_{60}$ fullerene.*

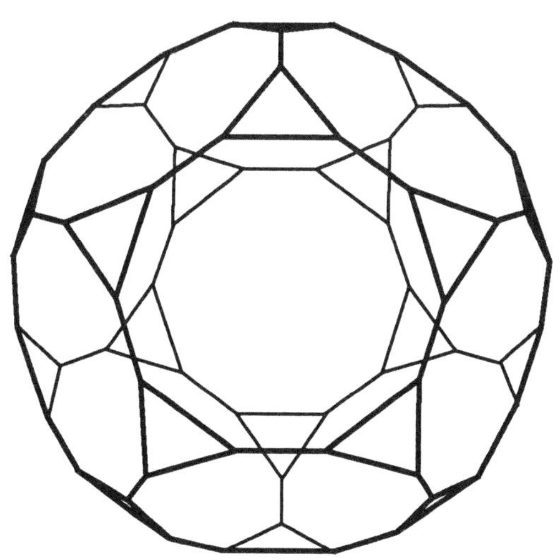

*Fig. 3.6. The $I_h$ non-fullerene $C_{60}$ cluster.*

Three subsequent Stone–Wales transformations result in, among others, the $D_3$ and $D_{2d}$ non-IPR isomers [269]. With three pentagon–pentagon pairs, the first structure is 69 (MNDO) and 76 kcal/mol (HF/3-21G) higher in energy than the $I_h$ isomer of the $C_{60}$ fullerene. For the second structure, which possesses four such pairs, the analogous values are 89 and 98 kcal/mol. Analysis of the aforementioned relative energies leads to the conclusion that each pair of abutting pentagons destabilizes the fullerene cage by 23–25 kcal/mol [269, 270].

A $D_{6h}$ structure composed of two $C_{24}$ coronene sheets joined through a belt of six $C_2$ fragments (Fig. 3.5) has been proposed as an alternative to the $C_{60}$ fullerene [221, 224]. This "graphitene" cluster, which possesses six pentagon–pentagon pairs, is less stable than buckminsterfullerene by 231 kcal/mol at the MNDO level of theory, which is more than the value predicted by the above additive scheme [43, 44, 270]. The discrepancy can be attributed to the additional strain present in graphitene due to the bending of the flat coronene sheets. Other fullerene isomers, such as the barrel cage [271] and the $C$-type structure [46], are predicted to be even higher in energy.

Other spherical non-fullerene $C_{60}$ carbon clusters have been investigated. An elegant $I_h$ structure of the truncated dodecahedron (Fig. 3.6) with three- and ten-membered rings gives rise to a closed-shell system within the Hückel approximation [131]. Orbital energy levels of this unusual system have been calculated at the INDO level of theory and compared with those of the $C_{60}$ fullerene [36]. A $C_{2v}$ $C_{60}$ cage with two pairs of pentagons replaced by two hexagons and two tetragons has also been considered [137]. The MNDO method finds this isomer to lie 94 kcal/mol above buckminsterfullerene, whereas the energy difference of 78 kcal/mol is obtained with molecular mechanics calculations.

## References

1. W. Krätschmer, L. D. Lamb, K. Fostiropoulos, and D. R. Huffman, *Solid $C_{60}$: A New Form of Carbon*, Nature **347**, 354 (1990).

2. A. Mittelbach, W. Hönle, H. G. von Schnering, J. Carlsen, R. Janiak, and H. Quast, *Optimization of the Production and Separation of Fullerenes*, Angew. Chem. Int. Ed. Engl. **31**, 1640 (1992).

3. R. E. Haufler, Y. Chai, L. P. F. Chibante, J. Conceicao, C. Jin, L.-S. Wang, S. Maruyama, and R. E. Smalley, *Carbon Arc Generation of $C_{60}$*, Mat. Res. Soc. Symp. Proc. **206**, 627 (1991).

4. R. E. Haufler, J. Conceicao, L. P. F. Chibante, Y. Chai, N. E. Byrne, S. Flanagan, M. M. Haley, S. C. O'Brien, C. Pan, Z. Xiao, W. E. Billups, M. A. Ciufolini, R. H. Hauge, J. L. Margrave, L. J. Wilson, R. F. Curl, and R. E. Smalley, *Efficient Production of $C_{60}$ (Buckminsterfullerene), $C_{60}H_{36}$, and the Solvated Buckide Ion*, J. Phys. Chem. **94**, 8634 (1990).

5. M. T. Beck, Z. Dinya, S. Kéki, and L. Papp, *Formation of $C_{60}$ and Polycyclic Aromatic Hydrocarbons upon Electric Discharges in Liquid Toluene*, Tetrahedron **49**, 285 (1993).

6. R. Taylor, G. J. Langley, H. W. Kroto, and D. R. M. Walton, *Formation of $C_{60}$ by Pyrolysis of Naphthalene*, Nature **366**, 728 (1993).

7. S. W. McElvany, M. M. Ross, N. S. Goroff, and F. Diederich, *Cyclocarbon Coalescence: Mechanisms for Tailor-Made Fullerene Formation*, Science **259**, 1594 (1993).

8. Y. Rubin, M. Kahr, C. B. Knobler, F. Diederich, and C. L. Wilkins, *The Higher Oxides of Carbon $C_{8n}O_{2n}$ ($n = 3 - 5$): Synthesis, Characterization, and X-ray Crystal Structure. Formation of Cyclo[n]carbon Ions $C_n^+$ ($n = 18, 24$), $C_n^-$ ($n = 18, 24, 30$), and Higher Carbon Ions Including $C_{60}^+$ in Laser Desorption Fourier Transform Mass Spectrometric Experiments*, J. Am. Chem. Soc. **113**, 495 (1991).

9. J. B. Howard, J. T. McKinnon, M. E. Johnson, Y. Makarovsky, and A. L. Lafleur, *Production of $C_{60}$ and $C_{70}$ Fullerenes in Benzene–Oxygen Flames*, J. Phys. Chem. **96**, 6657 (1992).

10. J. F. Anacleto, H. Perreault, R. K. Boyd, S. Pleasance, M. A. Quilliam, P. G. Sim, J. B. Howard, Y. Makarovsky, and A. L. Lafleur, *$C_{60}$ and $C_{70}$ Fullerene Isomers Generated in Flames — Detection and Verification by Liquid Chromatography Mass Spectrometry Analyses*, Rapid Commun. Mass Spectrom. **6**, 214 (1992).

11. C. J. Pope, J. A. Marr, and J. B. Howard, *Chemistry of Fullerenes $C_{60}$ and $C_{70}$ Formation in Flames*, J. Phys. Chem. **97**, 11001 (1993).

12. J. B. Howard, J. T. McKinnon, Y. Makarovsky, A. L. Lafleur, and M. E. Johnson, *Fullerenes $C_{60}$ and $C_{70}$ in Flames*, Nature **352**, 139 (1991).

13. T. K. Daly, P. R. Buseck, P. Williams, and C. F. Lewis, *Fullerenes from a Fulgurite*, Science **259**, 1599 (1993).

14. P. R. Buseck, S. J. Tsipursky, and R. Hettich, *Fullerenes from the Geological Environment*, Science **257**, 215 (1992).

15. H. W. Kroto, J. R. Heath, S. C. O'Brien, R. F. Curl, and R. E. Smalley, *$C_{60}$: Buckminsterfullerene*, Nature **318**, 162 (1985).

16. R. Taylor, J. P. Hare, A. K. Abdul-Sada, and H. W. Kroto, *Isolation, Separation and Characterization of the Fullerenes $C_{60}$ and $C_{70}$: The Third Form of Carbon*, J. Chem. Soc. Chem. Comm. 1423 (1990).

17. H. Ajie, M. M. Alvarez, S. J. Anz, R. D. Beck, F. Diederich, K. Fostiropoulos, D. R. Huffman, W. Krätschmer, Y. Rubin, K. E. Schriver, D. Sensharma, and R. L. Whetten, *Characterization of the Soluble All-Carbon Molecules $C_{60}$ and $C_{70}$*, J. Phys. Chem. **94**, 8630 (1990).

18. R. D. Johnson, G. Meijer, and D. S. Bethune, *$C_{60}$ Has Icosahedral Symmetry*, J. Am. Chem. Soc. **112**, 8983 (1990).

19. C. I. Frum, R. Engleman, Jr., H. G. Hedderich, P. F. Bernath, L. D. Lamb, and D. R. Huffman, *The Infrared Emission Spectrum of Gas-Phase $C_{60}$ (Buckminsterfullerene)*, Chem. Phys. Lett. **176**, 504 (1991).

20. B. Chase, N. Herron, and E. Holler, *Vibrational Spectroscopy of $C_{60}$ and $C_{70}$ Temperature-Dependent Studies*, J. Phys. Chem. **96**, 4262 (1992).

21. J. P. Hare, T. J. Dennis, H. W. Kroto, R. Taylor, A. W. Allaf, S. Balm, and D. R. M. Walton, *The IR Spectra of Fullerene-60 and -70*, J. Chem. Soc. Chem. Commun. 412 (1991).

22. D. S. Bethune, G. Meijer, W. C. Tang, H. J. Rosen, W. G. Golden, H. Seki, C. A. Brown, and M. S. de Vries, *Vibrational Raman and Infrared Spectra of Chromatographically Separated $C_{60}$ and $C_{70}$ Fullerene Clusters*, Chem. Phys. Lett. **179**, 181 (1991).

23. B. Chase and P. J. Fagan, *Substituted $C_{60}$ Molecules: A Study in Symmetry Reduction*, J. Am. Chem. Soc. **114**, 2252 (1992).

24. J. M. Hawkins, A. Meyer, T. A. Lewis, S. Loren, and F. J. Hollander, *Crystal Structure of Osmylated $C_{60}$: Confirmation of The Soccer Ball Framework*, Science **252**, 312 (1991).

25. M. Froimowitz, *Molecular Geometries and Heats of Formation of $C_{60}$ and $C_{70}$ as Computed by MM2-87*, J. Comp. Chem. **12**, 1129 (1991).

26. R. L. Murry, J. R. Colt, and G. E. Scuseria, *How Accurate Are Molecular Mechanics Predictions for Fullerenes? A Benchmark Comparison with Hartree–Fock Self-Consistent Field Results*, J. Phys. Chem. **97**, 4954 (1993).

27. R. C. Haddon, L. E. Brus, and K. Raghavachari, *Electronic Structure and Bonding in Icosahedral $C_{60}$*, Chem. Phys. Lett. **125**, 459 (1986).

28. S. J. Cyvin, E. Brendsdal, B. N. Cyvin, and J. Brunvoll, *Molecular Vibrations of Footballene*, Chem. Phys. Lett. **143**, 377 (1988).

29. M. Ozaki and A. Takahashi, *On Electronic States and Bond Lengths of the Truncated Icosahedral $C_{60}$ Molecule*, Chem. Phys. Lett. **127**, 242 (1986).

30. I. László and L. Udvardi, *On the Geometrical Structure and UV Spectrum of the Truncated Icosahedral $C_{60}$ Molecule*, Chem. Phys. Lett. **136**, 418 (1987).

31. I. László and L. Udvardi, *A Study of the UV Spectrum of the Truncated Icosahedral $C_{60}$ Molecule*, J. Mol. Struct. (Theochem) **183**, 271 (1989).

32. M. Kataoka and T. Nakajima, *Geometrical Structures and Spectra of Corannulene and Icosahedral $C_{60}$*, Tetrahedron **42**, 6437 (1986).

33. F. Negri, G. Orlandi, and G. Zerbetto, *Quantum-Chemical Investigation of Franck-Condon and Jahn–Teller Activity in the Electronic Spectra of Buckminsterfullerene*, Chem. Phys. Lett. **144**, 31 (1988).

34. M. Menon and K. R. Subbaswamy, *Universal Parameter Tight-Binding Molecular Dynamics: Application to $C_{60}$*, Phys. Rev. Lett. **67**, 3487 (1991).

35. C. Z. Wang, C. T. Chan, and K. M. Ho, *Structure and Dynamics of $C_{60}$ and $C_{70}$ from Tight-Binding Molecular Dynamics*, Phys. Rev. B **46**, 9761 (1992).

36. T. I. Shibuya and M. Yoshitani, *Two Icosahedral Structures for the $C_{60}$ Cluster*, Chem. Phys. Lett. **137**, 13 (1987).

37. R. D. Bendale, J. F. Stanton, and M. C. Zerner, *Investigation of the Electronic Structure and Spectroscopy of Jahn–Teller Distorted $C_{60}^+$*, Chem. Phys. Lett. **194**, 467 (1992).

38. R. D. Bendale, J. D. Baker, and M. C. Zerner, *Calculations on the Electronic Structure and Spectroscopy of $C_{60}$ and $C_{70}$ Cage Structures*, Int. J. Quant. Chem. Quant. Chem. Symp. **25**, 557 (1991).

39. J. Feng, J. Li, Z. Wang, and M. C. Zerner, *Quantum-Chemical Investigation of Buckminsterfullerene and Related Carbon Clusters (I): The Electronic Structure and UV Spectra of Buckminsterfullerene, and Other $C_{60}$ Cages*, Int. J. Quant. Chem. **37**, 599 (1991).

40. K. Tanaka, M. Okada, K. Okahara, and T. Yamabe, *Structure and Electronic State of $C_{60}^-$*, Chem. Phys. Lett. **193**, 101 (1992).

41. D. S. Marynick and S. K. Estreicher, *Localized Molecular Orbitals and Electronic Structure of Buckminsterfullerene*, Chem. Phys. Lett. **132**, 383 (1986).

42. S. K. Estreicher, C. D. Latham, M. I. Heggie, R. Jones, and S. Öberg, *Stable and Metastable States of $C_{60}H$: Buckminsterfullerene Monohydride*, Chem. Phys. Lett. **196**, 311 (1992).

43. M. D. Newton and R. E. Stanton, *Stability of Buckminsterfullerene and Related Carbon Clusters*, J. Am. Chem. Soc. **108**, 2469 (1986).

44. M. L. McKee and W. C. Herndon, *Calculated Properties of $C_{60}$ Isomers and Fragments*, J. Mol. Struct. (Theochem) **153**, 75 (1987).

45. M. Kolb and W. Thiel, *MNDO Parameters for Helium: Optimization, Tests, and Application to Endohedral Fullerene-Helium Complexes*, J. Comp. Chem. **14**, 37 (1993).

46. D. Bakowies and W. Thiel, *MNDO Study of Large Carbon Clusters*, J. Am. Chem. Soc. **113**, 3704 (1991).

47. J. M. Rudziński, Z. Slanina, M. Togasi, E. Ōsawa, and T. Iizuka, *Computational Study of Relative Stabilities of $C_{60}(I_h)$ and $C_{70}(D_{5h})$ Gas-Phase Clusters*, Thermochim. Acta **125**, 155 (1988).

48. J. M. Schulman, R. L. Disch, M. A. Miller, and R. C. Peck, *Symmetrical Clusters of Carbon Atoms: The $C_{24}$ and $C_{60}$ Molecules*, Chem. Phys. Lett. **141**, 45 (1987).

49. Z. Shuai and J. L. Brédas, *Electronic Structure and Nonlinear Optical Properties of the Fullerenes $C_{60}$ and $C_{70}$: A Valence-Effective-Hamiltonian Study*, Phys. Rev. B **46**, 16135 (1992).

50. N. Matsuzawa and D. A. Dixon, *Semiempirical Calculations of the Polarizability and Second-Order Hyperpolarizability of $C_{60}$, $C_{70}$, and Model Aromatic Compounds*, J. Phys. Chem. **96**, 6241 (1992).

51. R. L. Disch and J. M. Schulman, *On Symmetrical Clusters of Carbon Atoms: $C_{60}$*, Chem. Phys. Lett. **125**, 465 (1986).

52. G. E. Scuseria, *Ab Initio Theoretical Predictions of the Equilibrium Geometries of $C_{60}$, $C_{60}H_{60}$ and $C_{60}F_{60}$*, Chem. Phys. Lett. **176**, 423 (1991).

53. K. Raghavachari and C. M. Rohlfing, *Structures and Vibrational Frequencies of $C_{60}$, $C_{70}$, and $C_{84}$*, J. Phys. Chem. **95**, 5768 (1991).

54. N. Koga and K. Morokuma, *Ab Initio MO Study of the $C_{60}$ Anion Radical: The Jahn–Teller Distortion and Electronic Structure*, Chem. Phys. Lett. **196**, 191 (1992).

55. J. Hrušák and H. Schwarz, *Ab Initio MO Calculations on the Electronic and Geometric Structure of Fullerene Dications, $C_{60}^{2+}$, and the Double-Ionization Energy of $C_{60}$*, Chem. Phys. Lett. **205**, 187 (1993).

56. J. Cioslowski and E. D. Fleischmann, *Endohedral Complexes: Atoms and Ions Inside the $C_{60}$ Cage*, J. Chem. Phys. **94**, 3730 (1991).

57. J. Cioslowski, *Ab Initio Electronic Structure Calculations on Endohedral Complexes of the $C_{60}$ Cluster*, in *Spectroscopic and Computational Studies of Supramolecular Systems* (J. E. D. Davies, ed.), Kluwer Academic Publishers, Dordrecht, 1992, Ch. 10, p. 269.

58. H. P. Lüthi and J. Almlöf, *Ab Initio Studies on the Thermodynamic Stability of the Icosahedral $C_{60}$ Molecule "Buckminsterfullerene"*, Chem. Phys. Lett. **135**, 357 (1987).

59. M. Feyereisen, M. Gutowski, J. Simons, and J. Almlöf, *Relative Stabilities of Fullerene, Cumulene, and Polyacetylene Structures for $C_n$: $n = 18 - 60$*, J. Chem. Phys. **96**, 2926 (1992).

60. J. Hutter and H. P. Lüthi, *The Structure of n-Fold Negatively Charged $C_{60}$ (n = 1, 2, ..., 6)*, Int. J. Quant. Chem. **46**, 81 (1993).

61. J. Cioslowski, W. Thiel, and S. Pachkovski, unpublished.

62. M. Häser, J. Almlöf, and G. E. Scuseria, *The Equilibrium Geometry of $C_{60}$ as Predicted by Second-Order (MP2) Perturbation Theory*, Chem. Phys. Lett. **181**, 497 (1991).

63. B. I. Dunlap, D. W. Brenner, J. W. Mintmire, R. C. Mowrey, and C. T. White, *Local Density Functional Electronic Structures of Three Stable Icosahedral Fullerenes*, J. Phys. Chem. **95**, 8737 (1991).

64. G. B. Adams, J. B. Page, O. F. Sankey, K. Sinha, J. Menendez, and D. R. Huffman, *First-Principles Quantum-Molecular-Dynamics Study of the Vibrations of Icosahedral $C_{60}$*, Phys. Rev. B **44**, 4052 (1991).

65. M. R. Pederson and A. A. Quong, *Polarizabilities, Charge States, and Vibrational Modes of Isolated Fullerene Molecules*, Phys. Rev. B **46**, 13584 (1992).

66. V. de Coulon, J. L. Martins, and F. Reuse, *Electronic Structure of Neutral and Charged $C_{60}$ Clusters*, Phys. Rev. B **45**, 13671 (1992).

67. L. Ye, A. J. Freeman, and B. Delley, *Local Density Functional Study of the Structural and Electronic Properties of $C_{60}$ and $XC_{60}$ (X = K, Rb, Cs)*, Chem. Phys. **160**, 415 (1992).

68. B. I. Dunlap, *Isomerization and Icosahedral Fullerenes*, in *Physics and Chemistry of Finite Systems: From Clusters to Crystals*, Vol. II (P. Jena *et al.*, eds.), Kluwer, Dordrecht, 1992, p. 1295.

69. X.-Q. Wang, C. Z. Wang, B. L. Zhang, and K. M. Ho, *Structural and Electronic Properties of $C_{84}$: A First-Principles Study*, Phys. Rev. Lett. **69**, 69 (1992).

70. N. Matsuzawa and D. A. Dixon, *Local Density Functional Calculations of the Polarizability and Second-Order Hyperpolarizability of $C_{60}$*, J. Phys. Chem. **96**, 6872 (1992).

71. J.-Y. Yi and J. Bernholc, *Isomerization of $C_{60}$ Fullerenes*, J. Chem. Phys. **96**, 8634 (1992).

72. G. Corongiu and E. Clementi, *Comments on the Carbon Cluster $C_{60}$ and on Its Complexes with Alkaline Elements*, Int. J. Quant. Chem. **42**, 1185 (1992).

73. B. P. Feuston, W. Andreoni, M. Parrinello, and E. Clementi, *Electronic and Vibrational Properties of $C_{60}$ at Finite Temperature from ab Initio Molecular Dynamics*, Phys. Rev. B **44**, 4056 (1991).

74. C. van Wüllen, *An Implementation of a Kohn-Sham Density Functional Program Using a Gaussian-Type Basis Set. Application to the Equilibrium Geometry of $C_{60}$ and $C_{70}$*, Chem. Phys. Lett. **219**, 8 (1994).

75. K. Hedberg, L. Hedberg, D. S. Bethune, C. A. Brown, H. C. Dorn, R. D. Johnson, and M. de Vries, *Bond Lengths in Free Molecules of Buckminsterfullerene, $C_{60}$, from Gas-Phase Electron Diffraction*, Science **254**, 410 (1991).

76. S. Liu, Y.-J. Lu, M. M. Kappes, and J. A. Ibers, *The Structure of the $C_{60}$ Molecule: X-ray Crystal Structure Determination of a Twin at 110 K*, Science **254**, 408 (1991).

77. W. I. F. David, R. M. Ibberson, J. C. Matthewman, K. Prassides, T. J. S. Dennis, J. P. Hare, H. W. Kroto, R. Taylor, and D. R. M. Walton, *Crystal Structure and Bonding of Ordered $C_{60}$*, Nature **353**, 147 (1991).

78. P. J. Fagan, J. C. Calabrese, and B. Malone, *The Chemical Nature of Buckminsterfullerene ($C_{60}$) and the Characterization of a Platinum Derivative*, Science **252**, 1160 (1991).

79. F. Li, D. Ramage, J. S. Lannin, and J. Conceicao, *Radial Distribution Function of $C_{60}$: Structure of Fullerene*, Phys. Rev. B **44**, 13167 (1991).

80. C. S. Yannoni, P. P. Bernier, D. S. Bethune, G. Meijer, and J. R. Salem, *NMR Determination of the Bond Lengths in $C_{60}$*, J. Am. Chem. Soc. **113**, 3190 (1991).

81. W. Krätschmer, K. Fostiropoulos, and D. R. Huffman, *The Infrared and Ultraviolet Absorption Spectra of Laboratory-Produced Carbon Dust: Evidence for the Presence of the $C_{60}$ Molecule*, Chem. Phys. Lett. **170**, 167 (1990).

82. D. R. Huffman and W. Krätschmer, *Solid $C_{60}$ — How We Found It*, Mat. Res. Soc. Symp. Proc. **206**, 601 (1991).

83. F. Negri, G. Orlandi, and F. Zerbetto, *QCFF/PI Vibrational Frequencies of Some Spherical Carbon Clusters*, J. Am. Chem. Soc. **113**, 6037 (1991).

84. G. Onida, W. Andreoni, J. Kohanoff, and M. Parrinello, *Ab Initio Molecular Dynamics of $C_{70}$. Intramolecular Vibrations and Zero-Point Motion Effects*, Chem. Phys. Lett. **219**, 1 (1994).

85. J. Kohanoff, W. Andreoni, and M. Parrinello, *Zero-Point-Motion Effects on the Structure of $C_{60}$*, Phys. Rev. B **46**, 4371 (1992).

86. R. E. Stanton and M. D. Newton, *Normal Vibrational Modes of Buckminsterfullerene*, J. Phys. Chem. **92**, 2141 (1988).

87. D. Bakowies and W. Thiel, *Theoretical Infrared Spectra of Large Carbon Clusters*, Chem. Phys. **151**, 309 (1991).

88. Z. Slanina, J. M. Rudziński, M. Togasi, and E. Ōsawa, *Quantum-Chemically Supported Vibrational Analysis of Giant Molecules: the $C_{60}$ and $C_{70}$ Clusters*, J. Mol. Struct. (Theochem) **202**, 169 (1989).

89. C. Coulombeau, H. Jobic, P. Bernier, C. Fabre, D. Schütz, and A. Rassat, *Neutron Inelastic Scattering Spectrum of Footballene $C_{60}$*, J. Phys. Chem. **96**, 22 (1992).

90. D. E. Weeks and W. G. Harter, *Vibrational Frequencies and Normal Modes of Buckminsterfullerene*, Chem. Phys. Lett. **144**, 366 (1988).

91. D. E. Weeks and W. G. Harter, *Rotation-Vibration Spectra of Icosahedral Molecules. II. Icosahedral Symmetry, Vibrational Eigenfrequencies, and Normal Modes of Buckminsterfullerene*, J. Chem. Phys. **90**, 4744 (1989).

92. E. Brendsdal, B. N. Cyvin, J. Brunvoll, and S. J. Cyvin, *Normal Coordinate Analysis of "Footballene" $C_{60}$*, Spectrosc. Lett. **21**, 313 (1988).

93. E. Brendsdal, *Symmetry Coordinates of Molecular Vibrations of "Footballene" $C_{60}$*, Spectrosc. Lett. **21**, 319 (1988).

94. Q. Jiang, H. Xia, Z. Zhang, and D. Tian, *Vibrational Spectrum of $C_{60}$*, Chem. Phys. Lett. **192**, 93 (1992).

95. Z. C. Wu, D. A. Jelski, and T. F. George, *Vibrational Motions of Buckminsterfullerene*, Chem. Phys. Lett. **137**, 291 (1987).

96. R. A. Jishi, R. M. Mirie, and M. S. Dresselhaus, *Force-Constant Model for the Vibrational Modes in $C_{60}$*, Phys. Rev. B **45**, 13685 (1992).

97. J. L. Feldman, J. Q. Broughton, L. L. Boyer, D. E. Reich, and M. D. Kluge, *Intramolecular-Force-Constant Model for $C_{60}$*, Phys. Rev. B **46**, 12731 (1992).

98. G. Onida and G. Benedek, *Vibrational Spectrum of $C_{60}$: A Bond-Charge Model Calculation*, Europhys. Lett. **18**, 403 (1992).

99. C. K. Mathews, M. Sai Baba, T. S. Lakshmi Narasimhan, R. Balasubramanian, N. Sivaraman, T. G. Srinivasan, and P. R. Vasudeva Rao, *Vaporization Studies on Buckminsterfullerene*, J. Phys. Chem. **96**, 3566 (1992).

100. C. Pan, M. P. Sampson, Y. Chai, R. H. Hauge, and J. L. Margrave, *Heats of Sublimation from a Polycrystalline Mixture of $C_{60}$ and $C_{70}$*, J. Phys. Chem. **95**, 2944 (1991).

101. T. Kiyobayashi and M. Sakiyama, *Combustion Calorimetric Studies on $C_{60}$ and $C_{70}$*, Full. Sci. Tech. **1**, 269 (1993).

102. H.-D. Beckhaus, C. Rüchardt, M. Kao, F. Diederich, and C. S. Foote, *The Stability of Buckminsterfullerene ($C_{60}$): Experimental Determination of the Heat of Formation*, Angew. Chem. Int. Ed. Engl. **31**, 63 (1992).

103. H. P. Diogo, M. E. Minas da Piedade, T. J. S. Dennis, J. P. Hare, H. W. Kroto, R. Taylor, and D. R. M. Walton, *Enthalpies of Formation of Buckminsterfullerene ($C_{60}$) and of the Parent Ions $C_{60}^{+}$, $C_{60}^{2+}$, $C_{60}^{3+}$, and $C_{60}^{-}$*, J. Chem. Soc. Faraday Trans. **89**, 3541 (1993).

104. J. Cioslowski, *Heats of Formation of Higher Fullerenes from ab Initio Hartree–Fock and Correlation Energy Functional Calculations*, Chem. Phys. Lett. **216**, 389 (1993).

105. F. Diederich, R. L. Whetten, C. Thilgen, R. Ettl, I. Chao, and M. M. Alvarez, *Fullerene Isomerism: Isolation of $C_{2v}$-$C_{78}$ and $D_3$-$C_{78}$*, Science **254**, 1768 (1991).

106. F. Diederich and R. L. Whetten, *Beyond $C_{60}$: The Higher Fullerenes*, Acc. Chem. Res. **25**, 119 (1992).

107. B. L. Zhang, C. Z. Wang, and K. M. Ho, *Structures of Large Fullerenes: $C_{60}$ to $C_{94}$*, Chem. Phys. Lett. **193**, 225 (1992).

108. B. L. Zhang, C. Z. Wang, K. M. Ho, C. H. Xu, and C. T. Chan, *The Geometry of Small Fullerene Cages: $C_{20}$ to $C_{70}$*, J. Chem. Phys. **97**, 5007 (1992).

109. Z. Slanina, J. M. Rudziński, M. Togasi, and E. Ōsawa, *On Relative Stability Reasoning for Clusters of Different Dimensions: An Illustration with the $C_{60}$–$C_{70}$ System*, Thermochim. Acta **140**, 87 (1989).

110. N. Matsuzawa, D. A. Dixon, and T. Fukunaga, *Semiempirical Calculations of Dihydrogenated Buckminsterfullerenes, $C_{60}H_2$*, J. Phys. Chem. **96**, 7594 (1992).

111. J. M. Schulman, R. C. Peck, and R. L. Disch, *Ab Initio Heats of Formation of Medium-Sized Hydrocarbons. 11. The Benzenoid Aromatics*, J. Am. Chem. Soc. **111**, 5675 (1989).

112. J. M. Schulman and R. L. Disch, *The Heat of Formation of Buckminsterfullerene, $C_{60}$*, J. Chem. Soc. Chem. Commun. 411 (1991).

113. W. V. Steele, R. D. Chirico, N. K. Smith, W. E. Billups, P. R. Elmore, and A. E. Wheeler, *Standard Enthalpy of Formation of Buckminsterfullerene*, J. Phys. Chem. **96**, 4731 (1992).

114. H.-D. Beckhaus, S. Verevkin, C. Rüchardt, F. Diederich, C. Thilgen, H.-U. ter Meer, H. Mohn, and W. Müller, *$C_{70}$ Is More Stable than $C_{60}$: Experimental Determination of the Heat of Formation of $C_{70}$*, Angew. Chem. Int. Ed. Engl. **33**, 996 (1994).

115. L. Wang, P. S. Davids, A. Saxena, and A. R. Bishop, *Quasiparticle Energy Spectra and Magnetic Response of Certain Curved Graphitic Geometries*, Phys. Rev. B **46**, 7175 (1992).

116. P. Joyes and R. J. Tarento, *Application of the Gutzwiller Method to Neutral and Ionic $C_{60}$ Aggregates*, Phys. Rev. B **45**, 12077 (1992).

117. J. Li, J. Feng, and J. Sun, *Quantum Chemical Calculations on the Spectra and Nonlinear Third-Order Optical Susceptibility of $C_{60}$*, Chem. Phys. Lett. **203**, 560 (1993).

118. P. W. Fowler, P. Lazzeretti, and R. Zanasi, *Electric and Magnetic Properties of the Aromatic Sixty-Carbon Cage*, Chem. Phys. Lett. **165**, 79 (1990).

119. J. Cioslowski and K. Raghavachari, *Electrostatic Potential, Polarization, Shielding, and Charge Transfer in Endohedral Complexes of the $C_{60}$, $C_{70}$, $C_{76}$, $C_{78}$, $C_{82}$, and $C_{84}$ Clusters*, J. Chem. Phys. **98**, 8734 (1993).

120. A. H. H. Chang, W. C. Ermler, and R. M. Pitzer, *$C_{60}$ and Its Ions: Electronic Structure, Ionization Potentials, and Excitation Energies*, J. Phys. Chem. **95**, 9288 (1991).

121. A. Rosén and B. Wästberg, *Calculations of the Ionization Thresholds and Electron Affinities of the Neutral, Positively and Negatively Charged $C_{60}$ – "Follene-60"*, J. Chem. Phys. **90**, 2525 (1989).

122. R. K. Yoo, B. Ruscic, and J. Berkowitz, *Vacuum Ultraviolet Photoionization Mass Spectrometric Study of $C_{60}$*, J. Chem. Phys. **96**, 911 (1992).

123. J. de Vries, H. Steger, B. Kamke, C. Menzel, B. Weisser, W. Kamke, and I. V. Hertel, *Single-Photon Ionization of $C_{60}$- and $C_{70}$- Fullerene with Synchrotron Radiation: Determination of the Ionization Potential of $C_{60}$*, Chem. Phys. Lett. **188**, 159 (1992).

124. I. V. Hertel, H. Steger, J. de Vries, B. Weisser, C. Menzel, B. Kamke, and W. Kamke, *Giant Plasmon Excitations in Free $C_{60}$ and $C_{70}$ Molecules Studied by Photoionization*, Phys. Rev. Lett. **68**, 784 (1992).

125. D. L. Lichtenberger, M. E. Jatcko, K. W. Nebesny, C. D. Ray, D. R. Huffman, and L. D. Lamb, *The Ionizations of $C_{60}$ in the Gas Phase and in Thin Solid Films*, Mat. Res. Soc. Symp. Proc. **206**, 673 (1991).

126. J. A. Zimmerman, J. R. Eyler, S. B. H. Bach, and S. W. McElvany, *"Magic Number" Carbon Clusters: Ionization Potentials and Selective Reactivity*, J. Chem. Phys. **94**, 3556 (1991).

127. S. W. McElvany and S. B. H. Bach, *First and Second Ionization Potentials and Reactions of $C_{60}$ and Related Clusters*, in *Proc. 39th ASMS Conf. on Mass Spectrometry and Allied Topics*, 422 (1991).

128. R. A. Davidson, *Spectral Analysis of Graphs by Cyclic Automorphism Subgroups*, Theor. Chim. Acta **58**, 193 (1981).

129. J. R. Dias, *A Facile Hückel Molecular Orbital Solution of Buckminsterfullerene Using Chemical Graph Theory*, J. Chem. Ed. **66**, 1012 (1989).

130. W. Byers Brown, *High Symmetries in Quantum Chemistry*, Chem. Phys. Lett. **136**, 128 (1987).

131. P. W. Fowler and J. Woolrich, $\pi$-*Systems in Three Dimensions*, Chem. Phys. Lett. **127**, 78 (1986).

132. M. Braga, A. Rosén, and S. Larsson, *Electronic Transitions in $C_{60}$ and Its Ions*, Z. Phys. D **19**, 435 (1991).

133. A. Rosén and B. Wästberg, *First-Principle Calculations of the Ionization Potentials and Electron Affinities of the Spheroidal Molecules $C_{60}$ and $LaC_{60}$*, J. Am. Chem. Soc. **110**, 8701 (1988).

134. A. Rosén and B. Wästberg, *Electronic Structure of Spheroidal Metal Containing Carbon Shells: Study of the $LaC_{60}$ and $C_{60}$ Clusters and Their Ions within the Local Density Approximation*, Z. Phys. D **12**, 387 (1989).

135. A. Rosén and B. Wästberg, *Buckminsterfullerene $C_{60}$ — A Surface with Curvature and Interesting Properties*, Surf. Sci. **269**, 1121 (1992).

136. H. Weiss, R. Ahlrichs, and M. Häser, *A Direct Algorithm for Self-Consistent-Field Linear Response Theory and Application to $C_{60}$: Excitation Energies, Oscillator Strengths, and Frequency-Dependent Polarizabilities*, J. Chem. Phys. **99**, 1262 (1993).

137. Y.-D. Gao and W. C. Herndon, *Fullerenes with Four-Membered Rings*, J. Am. Chem. Soc. **115**, 8459 (1993).

138. G. A. Gallup, *The Application of Zero-Range Potentials to the Electronic Properties of Footballene, $C_{60}$*, Chem. Phys. Lett. **187**, 187 (1991).

139. J. Cioslowski, unpublished results of HF/DZ calculations (full geometry optimizations) quoted in [55].

140. H. Steger, J. de Vries, B. Kamke, and W. Kamke, *Direct Double Ionization of $C_{60}$ and $C_{70}$ Fullerenes Using Synchrotron Radiation*, Chem. Phys. Lett. **194**, 452 (1992).

141. S. Petrie, G. Javahery, J. Wang, and D. K. Bohme, *Selected-Ion Flow Tube Study of Charge Transfer from Fullerene Dications: "Bracketing" the Second Ionization Energies of $C_{60}$ and $C_{70}$*, J. Phys. Chem. **96**, 6121 (1992).

142. K. A. Caldwell, D. E. Giblin, and M. L. Gross, *High-Energy Collisions of Fullerene Radical Cations with Target Gases: Capture of the Target Gas and Charge Stripping of $C_{60}^{\bullet+}$, $C_{70}^{\bullet+}$, and $C_{84}^{\bullet+}$*, J. Am. Chem. Soc. **114**, 3743 (1992).

143. C. Lifshitz, M. Iraqi, T. Peres, and J. E. Fischer, *Charge Stripping $C_{60}^{+}$*, Rapid Commun. Mass Spectrom. **5**, 238 (1991).

144. F. Negri, G. Orlandi, and F. Zerbetto, *Low-Lying Electronic Excited States of Buckminsterfullerene Anions*, J. Am. Chem. Soc. **114**, 2909 (1992).

145. R. Saito, G. Dresselhaus, and M. S. Dresselhaus, *Multiplet Structures of $C_{60}$ Ions*, Chem. Phys. Lett. **210**, 159 (1993).

146. S. H. Yang, C. L. Pettiette, J. Conceicao, O. Cheshnovsky, and R. E. Smalley, *UPS of Buckminsterfullerene and Other Large Clusters of Carbon*, Chem. Phys. Lett. **139**, 233 (1987).

147. L. S. Wang, J. Conceicao, C. Jin, and R. E. Smalley, *Threshold Photode-tachment of Cold $C_{60}^-$*, Chem. Phys. Lett. **182**, 5 (1991).

148. S. B. H. Bach, J. E. Bruce, R. Ramanathan, C. H. Watson, J. A. Zimmerman, and J. R. Eyler, *Ionization Potentials and Electron Affinities of Semiconductor Clusters from Charge Transfer Reactions*, in *On Clusters and Clustering, from Atoms to Fractals* (P. J. Reynolds, ed.), Elsevier Science Publishers, New York, 1993, p. 59.

149. S. Petrie, J. Wang, and D. K. Bohme, *Charge Transfer from Polycharged Ions: $C_{60}^{n+}$ as a Model System*, Chem. Phys. Lett. **204**, 473 (1993).

150. G. Javahery, H. Wincel, S. Petrie, and D. K. Bohme, *Charge-Transfer Reactions of $C_{60}^{3+\bullet}$: "Bracketing" the Third Ionization Energy of $C_{60}$*, Chem. Phys. Lett. **204**, 467 (1993).

151. P. Weis, R. D. Beck, G. Bräuchle, and M. M. Kappes, *Properties of Size and Composition Selected Gas Phase Alkali Fulleride Clusters*, J. Chem. Phys. **100**, 5684 (1994).

152. P. A. Limbach, L. Schweikhard, K. A. Cowen, M. T. McDermott, A. G. Marshall, and J. V. Coe, *Observation of the Doubly Charged, Gas-Phase Fullerene Anions $C_{60}^{2-}$ and $C_{70}^{2-}$*, J. Am. Chem. Soc. **113**, 6795 (1991).

153. R. L. Hettich, R. N. Compton, and R. H. Ritchie, *Doubly Charged Negative Ions of Carbon-60*, Phys. Rev. Lett. **67**, 1242 (1991).

154. R. L. Martin and J. P. Ritchie, *Coulomb and Exchange Interactions in $C_{60}^{n-}$*, Phys. Rev. B **48**, 4845 (1993).

155. A. Ceulemans and P. W. Fowler, *The Jahn–Teller Instability of Five-fold Degenerate States in Icosahedral Molecules*, J. Chem. Phys. **93**, 1221 (1990).

156. T. Kato, T. Kodama, T. Shida, T. Nakagawa, Y. Matsui, S. Suzuki, H. Shiromaru, K. Yamauchi, and Y. Achiba, *Electronic Absorption Spectra of the Radical Anions and Cations of Fullerenes: $C_{60}$ and $C_{70}$*, Chem. Phys. Lett. **180**, 446 (1991).

157. Z. Gasyna, L. Andrews, and P. N. Schatz, *Near-Infrared Absorption Spectra of $C_{60}$ Radical Cations and Anions Prepared Simultaneously in Solid Argon*, J. Phys. Chem. **96**, 1525 (1992).

158. G. Javahery, S. Petrie, J. Wang, and D. K. Bohme, *Fullerene Cation and Dication Production by Novel Thermal Energy Reactions of $He^+$, $Ne^+$, and $Ar^+$ with $C_{60}$*, Chem. Phys. Lett. **195**, 7 (1992).

159. J. J. Stry, M. T. Coolbaugh, E. Turos, and J. F. Garvey, *Novel Ion-Molecule Reactions of $C_{60}^{2+}$ with $NH_3$*, J. Am. Chem. Soc. **114**, 7914 (1992).

160. G. Javahery, S. Petrie, H. Wincel, J. Wang, and D. K. Bohme, *Experimental Study of Reactions of the Buckminsterfullerene Cations $C_{60}^{\bullet+}$, $C_{60}^{2+}$, and $C_{60}^{\bullet 3+}$ with Ammonia and Amines in the Gas Phase*, J. Am. Chem. Soc. **115**, 5716 (1993).

161. S. Petrie, G. Javahery, and D. K. Bohme, *Attaching Handles to $C_{60}^{2+}$: The*

*Double-Derivatization of* $C_{60}^{2+}$, J. Am. Chem. Soc. **115**, 1445 (1993).

162. S. Petrie, G. Javahery, J. Wang, and D. K. Bohme, *Derivatization of the Fullerene Dications* $C_{60}^{2+}$ *and* $C_{70}^{2+}$ *by Ion-Molecule Reactions in the Gas Phase*, J. Am. Chem. Soc. **114**, 9177 (1992).

163. J. Wang, G. Javahery, S. Petrie, and D. K. Bohme, *Fullerene Dications as Initiators of Polymerization with 1,3-Butadiene in the Gas Phase: Chemistry Directed by Electrostatics?*, J. Am. Chem. Soc. **114**, 9665 (1992).

164. C. W. Walter, Y. K. Bae, D. C. Lorents, and J. R. Peterson, *Production and Stability of Multiply Charged* $C_{60}$ *and* $C_{70}$ *Fullerene Ions*, Chem. Phys. Lett. **195**, 543 (1992).

165. P. Scheier, R. Robl, B. Schiestl, and T. D. Märk, *Electron-Impact-Induced Production and Isotope-Resolved Identification of* $C_{60}^{5+}$, Chem. Phys. Lett. **220**, 141 (1994).

166. J. Cioslowski, *An Efficient Evaluation of Atomic Properties Using a Vectorized Numerical Integration with Dynamic Thresholding*, Chem. Phys. Lett. **194**, 73 (1992).

167. J. Stinchcombe, A. Pénicaud, P. Bhyrappa, P. D. W. Boyd, and C. A. Reed, *Buckminsterfulleride(1−) Salts: Synthesis, EPR, and the Jahn–Teller Distortion of* $C_{60}^-$, J. Am. Chem. Soc. **115**, 5212 (1993).

168. S. G. Kukolich and D. R. Huffman, *EPR Spectra of* $C_{60}$ *Anion and Cation Radicals*, Chem. Phys. Lett. **182**, 263 (1991).

169. P.-M. Allemand, G. Srdanov, A. Koch, K. Khemani, F. Wudl, Y. Rubin, F. Diederich, M. M. Alvarez, S. J. Anz, and R. L. Whetten, *The Unusual Electron Spin Resonance of Fullerene* $C_{60}^{\bullet-}$, J. Am. Chem. Soc. **113**, 2780 (1991).

170. P. N. Keizer, J. R. Morton, K. F. Preston, and A. K. Sugden, *EPR Spectrum of* $C_{60}^-$ *Trapped in Molecular Sieve 13X*, J. Phys. Chem. **95**, 7117 (1991).

171. A. J. Schell-Sorokin, F. Mehran, G. R. Eaton, S. S. Eaton, A. Viehbeck, T. R. O'Toole, and C. A. Brown, *Electron Spin Relaxation Times of* $C_{60}^-$ *in Solution*, Chem. Phys. Lett. **195**, 225 (1992).

172. M. A. Greaney and S. M. Gorun, *Production, Spectroscopy, and Electronic Structure of Soluble Fullerene Ions*, J. Phys. Chem. **95**, 7142 (1991).

173. P. Bhyrappa, P. Paul, J. Stinchcombe, P. D. W. Boyd, and C. A. Reed, *Synthesis and Electronic Characterization of Discrete Buckminsterfulleride Salts:* $C_{60}^{2-}$ *and* $C_{60}^{3-}$, J. Am. Chem. Soc. **115**, 11004 (1993).

174. D. Dubois, M. T. Jones, and K. M. Kadish, *Electroreduction of Buckminsterfullerene,* $C_{60}$, *in Aprotic Solvents: Electron Spin Resonance Characterization of Singly, Doubly and Triply Reduced* $C_{60}$ *in Frozen Solutions*, J. Am. Chem. Soc. **114**, 6446 (1992).

175. D. M. Cox, S. Behal, M. Disko, S. M. Gorun, M. Greaney, C. S. Hsu, E.

B. Kollin, J. Millar, J. Robbins, W. Robbins, R. D. Sherwood, and P. Tindall, *Characterization of $C_{60}$ and $C_{70}$ Clusters*, J. Am. Chem. Soc. **113**, 2940 (1991).

176. D. Dubois, K. M. Kadish, S. Flanagan, R. E. Haufler, L. P. F. Chibante, and L. J. Wilson, *Spectroelectrochemical Study of the $C_{60}$ and $C_{70}$ Fullerenes and Their Mono-, Di-, Tri- and Tetraanions*, J. Am. Chem. Soc. **113**, 4364 (1991).

177. D. Dubois, K. M. Kadish, S. Flanagan, and L. J. Wilson, *Electrochemical Detection of Fulleronium and Highly Reduced Fulleride ($C_{60}^{5-}$) Ions in Solution*, J. Am. Chem. Soc. **113**, 7773 (1991).

178. Q. Xie, E. Pérez-Cordero, and L. Echegoyen, *Electrochemical Detection of $C_{60}^{6-}$ and $C_{70}^{6-}$: Enhanced Stability of Fullerides in Solution*, J. Am. Chem. Soc. **114**, 3978 (1992).

179. P. Paul, Z. Xie, R. Bau, P. D. W. Boyd, and C. A. Reed, *Ordered Structure of a Distorted $C_{60}^{2-}$ Fulleride Ion*, J. Am. Chem. Soc. **116**, 4145 (1994).

180. A. Tachibana, S. Ishikawa, and T. Yamabe, *Vibronic Attractive Interaction for Superconductivity in a Local Model of $C_{60}$*, Chem. Phys. Lett. **201**, 315 (1993).

181. F. Negri, G. Orlandi, and F. Zerbetto, *The Infrared and Raman Active Vibrational Frequencies of $C_{60}$ Hexaanion*, Chem. Phys. Lett. **196**, 303 (1992).

182. S. Leach, M. Vervloet, A. Desprès, E. Bréheret, J. P. Hare, T. J. Dennis, H. W. Kroto, R. Taylor, and D. R. M. Walton, *Electronic Spectra and Transitions of the Fullerene $C_{60}$*, Chem. Phys. **160**, 451 (1992).

183. Z. Gasyna, P. N. Schatz, J. P. Hare, T. J. Dennis, H. W. Kroto, R. Taylor, and D. R. M. Walton, *The Magnetic Circular Dichroism and Absorption Spectra of $C_{60}$ Isolated in Ar Matrices*, Chem. Phys. Lett. **183**, 283 (1991).

184. F. Negri, G. Orlandi, and F. Zerbetto, *Interpretation of the Vibrational Structure of the Emission and Absorption Spectra of $C_{60}$*, J. Chem. Phys. **97**, 6496 (1992).

185. M. Braga, S. Larsson, A. Rosén, and A. Volosov, *Electronic Transitions in $C_{60}$. On the Origin of the Strong Interstellar Absorption at 217 nm*, Astron. Astrophys. **245**, 232 (1991).

186. K. Yabana and G. F. Bertsch, *Forbidden Transitions in the Absorption Spectra of $C_{60}$*, Chem. Phys. Lett. **197**, 32 (1992).

187. J. W. Arbogast, A. P. Darmanyan, C. S. Foote, Y. Rubin, F. N. Diederich, M. M. Alvarez, S. J. Anz, and R. L. Whetten, *Photophysical Properties of $C_{60}$*, J. Phys. Chem. **95**, 11 (1991).

188. Y. Zeng, L. Biczok, and H. Linschitz, *External Heavy Atom Induced Phosphorescence Emission of Fullerenes: The Energy of Triplet $C_{60}$*, J. Phys. Chem. **96**, 5237 (1992).

189. R. E. Haufler, L.-S. Wang, L. P. F. Chibante, C. Jin, J. Conceicao, Y. Chai, and R. E. Smalley, *Fullerene Triplet State Production and Decay: R2PI Probes of $C_{60}$ and $C_{70}$ in a Supersonic Beam*, Chem. Phys. Lett. **179**, 449 (1991).

190. T. W. Ebbesen, K. Tanigaki, and S. Kuroshima, *Excited-State Properties of $C_{60}$*, Chem. Phys. Lett. **181**, 501 (1991).

191. D. K. Palit, A. V. Sapre, J. P. Mittal, and C. N. R. Rao, *Photophysical Properties of the Fullerenes, $C_{60}$ and $C_{70}$*, Chem. Phys. Lett. **195**, 1 (1992).

192. M. R. Wasielewski, M. P. O'Neil, K. R. Lykke, M. J. Pellin, and D. M. Gruen, *Triplet States of Fullerenes $C_{60}$ and $C_{70}$: Electron Paramagnetic Resonance Spectra, Photophysics, and Electronic Structures*, J. Am. Chem. Soc. **113**, 2774 (1991).

193. R. V. Bensasson, T. Hill, C. Lambert, E. J. Land, S. Leach, and T. G. Truscott, *Pulse Radiolysis Study of Buckminsterfullerene in Benzene Solution. Assignment of the $C_{60}$ Triplet-Triplet Absorption Spectrum*, Chem. Phys. Lett. **201**, 326 (1993).

194. C. A. Steren, P. R. Levstein, H. van Willigen, H. Linschitz, and L. Biczok, *FT-EPR Study of Triplet State $C_{60}$. Spin Dynamics and Electron Transfer Quenching*, Chem. Phys. Lett. **204**, 23 (1993).

195. G. L. Closs, P. Gautam, D. Zhang, P. J. Krusic, S. A. Hill, and E. Wasserman, *Steady-State and Time-Resolved Direct Detection EPR Spectra of Fullerene Triplets in Liquid Solution and Glassy Matrices. Evidence for a Dynamic Jahn–Teller Effect in Triplet $C_{60}$*, J. Phys. Chem. **96**, 5228 (1992).

196. M. Terazima, N. Hirota, H. Shinohara, and Y. Saito, *Time-Resolved EPR Investigation of the Triplet States of $C_{60}$ and $C_{70}$*, Chem. Phys. Lett. **195**, 333 (1992).

197. P. R. Surján, K. Németh, and L. Udvardi, *Jahn–Teller Distorted Excited States of $C_{60}$*, in *Electronic Properties of Fullerenes* (H. Kuzmany, J. Fink, M. Mehring, and S. Roth, eds.), Springer-Verlag, Berlin, 1993, p. 126.

198. A. A. Quong and M. R. Pederson, *Density-Functional-Based Linear and Nonlinear Polarizabilities of Fullerene and Benzene Molecules*, Phys. Rev. B **46**, 12906 (1992).

199. R. J. Knize and J. P. Partanen, *Comment on "Large Infrared Nonlinear Optical Response of $C_{60}$"*, Phys. Rev. Lett. **68**, 2704 (1992).

200. Z. H. Kafafi, F. J. Bartoli, J. R. Lindle, and R. G. S. Pong, *Comment on "Large Infrared Nonlinear Optical Response of $C_{60}$"*, Phys. Rev. Lett. **68**, 2705 (1992).

201. W. J. Blau, H. J. Byrne, D. J. Cardin, T. J. Dennis, J. P. Hare, H. W. Kroto, R. Taylor, and D. R. M. Walton, *Large Infrared Nonlinear Optical Response of $C_{60}$*, Phys. Rev. Lett. **67**, 1423 (1992).

202. A. Takahashi, H. X. Wang, and S. Mukamel, *Collective Charge Density Fluctuations and Nonlinear Optical Response of $C_{60}$*, Chem. Phys. Lett. **216**, 394 (1993).

203. G. B. Talapatra, N. Manickam, M. Samoc, M. E. Orczyk, S. P. Karna, and P. N. Prasad, *Nonlinear Optical Properties of the $C_{60}$ Molecule: Theoretical and Experimental Studies*, J. Phys. Chem. **96**, 5206 (1992).

204. M. Häser, R. Ahlrichs, H. P. Baron, P. Weis, and H. Horn, *Direct Computation of Second-Order SCF Properties of Large Molecules on Workstation Computers with an Application to Large Carbon Clusters*, Theor. Chim. Acta **83**, 455 (1992).

205. J. S. Meth, H. Vanherzeele, and Y. Wang, *Dispersion of the Third-Order Optical Nonlinearity of $C_{60}$. A Third-Harmonic Generation Study*, Chem. Phys. Lett. **197**, 26 (1992).

206. Y. Wang and L.-T. Cheng, *Nonlinear Optical Properties of Fullerenes and Charge-Transfer Complexes of Fullerenes*, J. Phys. Chem. **96**, 1530 (1992).

207. Z. H. Kafafi, J. R. Lindle, R. G. S. Pong, F. J. Bartoli, L. J. Lingg, and J. Milliken, *Off-Resonant Nonlinear Optical Properties of $C_{60}$ Studied by Degenerate Four-Wave Mixing*, Chem. Phys. Lett. **188**, 492 (1992).

208. R. D. Johnson, C. S. Yannoni, J. Salem, G. Meijer, and D. S. Bethune, *Solution and Solid State NMR Studies of the Structure and Dynamics of $C_{60}$ and $C_{70}$*, Mat. Res. Soc. Symp. Proc. **206**, 715 (1991).

209. C. S. Yannoni, R. D. Johnson, G. Meijer, D. S. Bethune, and J. R. Salem, *$^{13}C$ NMR Study of the $C_{60}$ Cluster in the Solid State: Molecular Motion and Carbon Chemical Shift Anisotropy*, J. Phys. Chem. **95**, 9 (1991).

210. P. W. Fowler, P. Lazzeretti, M. Malagoli, and R. Zanasi, *Anisotropic Nuclear Magnetic Shielding in $C_{60}$*, J. Phys. Chem. **95**, 6404 (1991).

211. P. W. Fowler, P. Lazzeretti, M. Malagoli, and R. Zanasi, *Magnetic Properties of $C_{60}$ and $C_{70}$*, Chem. Phys. Lett. **179**, 174 (1991).

212. J. Cioslowski, unpublished.

213. R. S. Ruoff, D. Beach, J. Cuomo, T. McGuire, R. L. Whetten, and F. Diederich, *Confirmation of a Vanishingly Small Ring-Current Magnetic Susceptibility of Icosahedral $C_{60}$*, J. Phys. Chem. **95**, 3457 (1991).

214. R. C. Haddon, L. F. Schneemeyer, J. V. Waszczak, S. H. Glarum, R. Tycko, G. Dabbagh, A. R. Kortan, A. J. Muller, A. M. Mujsce, M. J. Rosseinsky, S. M. Zahurak, A. V. Makhija, F. A. Thiel, K. Raghavachari, E. Cockayne, and V. Elser, *Experimental and Theoretical Determination of the Magnetic Susceptibility of $C_{60}$ and $C_{70}$*, Nature **350**, 46 (1991).

215. V. Elser and R. C. Haddon, *Icosahedral $C_{60}$: An Aromatic Molecule with a Vanishingly Small Ring Current Magnetic Susceptibility*, Nature **325**, 792 (1987).

216. V. Elser and R. C. Haddon, *Magnetic Behavior of Icosahedral $C_{60}$*, Phys. Rev. A **36**, 4579 (1987).

217. A. Pasquarello, M. A. Schlüter, and R. C. Haddon, *Ring Currents in Icosahedral* $C_{60}$, Science **257**, 1660 (1992).

218. A. Pasquarello, M. A. Schlüter, and R. C. Haddon, *Ring Currents in Topologically Complex Molecules: Application to* $C_{60}$, $C_{70}$, *and Their Hexa-Anions*, Phys. Rev. A **47**, 1783 (1993).

219. R. C. Haddon and V. Elser, *Icosahedral* $C_{60}$ *Revisited: An Aromatic Molecule with a Vanishingly Small Ring Current Magnetic Susceptibility*, Chem. Phys. Lett. **169**, 362 (1990).

220. T. G. Schmalz, *The Magnetic Susceptibility of Buckminsterfullerene*, Chem. Phys. Lett. **175**, 3 (1990).

221. A. D. J. Haymet, *Footballene: A Theoretical Prediction for the Stable, Truncated Icosahedral Molecule* $C_{60}$, J. Am. Chem. Soc. **108**, 319 (1986).

222. D. Amić and N. Trinajstić, *On the Lack of Reactivity of Buckminsterfullerene. A Theoretical Study*, J. Chem. Soc. Perkin Trans. 2 1595 (1990).

223. R. Taylor, $C_{60}$, $C_{70}$, $C_{76}$, $C_{78}$ *and* $C_{84}$: *Numbering,* $\pi$-*Bond Order Calculations and Addition Pattern Considerations*, J. Chem. Soc. Perkin Trans. 2 813 (1993).

224. A. D. J. Haymet, $C_{120}$ *and* $C_{60}$: *Archimedean Solids Constructed from* $sp^2$ *Hybridized Carbon Atoms*, Chem. Phys. Lett. **122**, 421 (1985).

225. K. Balasubramanian and X. Liu, *Computer Generation of Spectra of Graphs: Applications to* $C_{60}$ *Clusters and Other Systems*, J. Comp. Chem. **9**, 406 (1988).

226. J. Aihara and H. Hosoya, *Spherical Aromaticitiy of Buckminsterfullerene*, Bull. Chem. Soc. Jpn. **61**, 2657 (1988).

227. B. A. Hess, Jr., and L. J. Schaad, *The Stability of Footballene*, J. Org. Chem. **51**, 3902 (1986).

228. D. J. Klein, W. A. Seitz, and T. G. Schmalz, *Icosahedral Symmetry Carbon Cage Molecules*, Nature **323**, 703 (1986).

229. A. Moyano and J.-C. Paniagua, *A Simple Approach for the Evaluation of Local Aromaticities*, J. Org. Chem. **56**, 1858 (1991).

230. R. C. Haddon, L. E. Brus, and K. Raghavachari, *Rehybridization and* $\pi$-*Orbital Alignment: The Key to the Existence of Spheroidal Carbon Clusters*, Chem. Phys. Lett. **131**, 165 (1986).

231. H. Hosoya, *Matching and Symmetry of Graphs*, Comp. & Maths. Appl. **12B**, 271 (1986).

232. E. Brendsdal and S. J. Cyvin, *Kekulé Structures of Footballene*, J. Mol. Struct. (Theochem) **188**, 55 (1989).

233. D. J. Klein, T. G. Schmalz, G. E. Hite, and W. A. Seitz, *Resonance in* $C_{60}$, *Buckminsterfullerene*, J. Am. Chem. Soc. **108**, 1301 (1986).

234. M. Randić, S. Nikolić, and N. Trinajstić, *On the Aromatic Stability of a Conjugated* $C_{60}$ *Cluster*, Croat. Chem. Acta **60**, 595 (1987).

235. L. L. Lohr and S. M. Blinder, *Electron Photodetachment from a Dirac Bubble Potential. A Model for the Fullerene Negative Ion* $C_{60}^-$, Chem. Phys. Lett. **198**, 100 (1992).
236. F. Siringo, G. Piccitto, and R. Pucci, *Thomas-Fermi Model for the* $C_{60}$ *Molecule*, Phys. Rev. A **46**, 4048 (1992).
237. J. Cioslowski and A. Nanayakkara, *Endohedral Effect in Inclusion Complexes of the* $C_{60}$ *Cluster*, J. Chem. Phys. **96**, 8354 (1992).
238. J. Li, C.-W. Liu, and J.-X. Lu, *Localized Molecular Orbital Studies of Fullerenes:* $C_{60}$ *and* $C_{70}$, Full. Sci. Tech. **2**, 35 (1994).
239. H. Schwarz, *The Mechanism of Fullerene Formation*, Angew. Chem. Int. Ed. Engl. **32**, 1412 (1993).
240. R. F. Curl, *On the Formation of the Fullerenes*, Phil. Trans. R. Soc. Lond. A **343**, 19 (1993).
241. T. W. Ebbesen, J. Tabuchi, and K. Tanigaki, *The Mechanistics of Fullerene Formation*, Chem. Phys. Lett. **191**, 336 (1992).
242. J. M. Hawkins, A. Meyer, S. Loren, and R. Nunlist, *Statistical Incorporation of* $^{13}C_2$ *Units into* $C_{60}$, J. Am. Chem. Soc. **113**, 9394 (1991).
243. D. H. Robertson, D. W. Brenner, and C. T. White, *On the Way to Fullerenes: Molecular Dynamics Study of the Curling and Closure of Graphitic Ribbons*, J. Phys. Chem. **96**, 6133 (1992).
244. J. R. Chelikowsky, *Nucleation of* $C_{60}$ *Clusters*, Phys. Rev. Lett. **67**, 2970 (1991).
245. H. Kroto, *Space, Stars,* $C_{60}$, *and Soot*, Science **242**, 1139 (1988).
246. H. W. Kroto and K. McKay, *The Formation of Quasi-icosahedral Spiral Shell Carbon Particles*, Nature **331**, 328 (1988).
247. Q. L. Zhang, S. C. O'Brien, J. R. Heath, Y. Liu, R. F. Curl, H. W. Kroto, and R. E. Smalley, *Reactivity of Large Carbon Clusters: Spheroidal Carbon Shells and Their Possible Relevance to the Formation and Morphology of Soot*, J. Phys. Chem. **90**, 525 (1986).
248. R. Pallasser, L. S. K. Pang, L. Prochazka, D. Rigby, and M. A. Wilson, *Isotopic Fractionation During Fullerene, Nanotube, and Nanopolyhedra Formation*, J. Am. Chem. Soc. **115**, 11634 (1993).
249. R. E. Smalley, *Self-Assembly of the Fullerenes*, Acc. Chem. Res. **25**, 98 (1992).
250. G. von Helden, M.-T. Hsu, N. Gotts, and M. T. Bowers, *Carbon Cluster Cations with up to 84 Atoms: Structures, Formation Mechanism, and Reactivity*, J. Phys. Chem. **97**, 8182 (1993).
251. G. von Helden, M.-T. Hsu, P. R. Kemper, and M. T. Bowers, *Structures of Carbon Cluster Ions from 3 to 60 Atoms: Linears to Rings to Fullerenes*, J. Chem. Phys. **95**, 3835 (1991).
252. A. Goeres and E. Sedlmayr, *On the Nucleation Mechanism of Effective Fullerite Condensation*, Chem. Phys. Lett. **184**, 310 (1991).
253. A. Goeres and E. Sedlmayr, *Hydrogen-Blocking in* $C_{60}$-*Formation Theories*, Full. Sci. Tech. **1**, 563 (1993).

254. M. Broyer, A. Goeres, M. Pellarin, E. Sedlmayr, J. L. Vialle, and L. Wöste, *Experimental Studies on the Formation Process of $C_{60}$*, Chem. Phys. Lett. **198**, 128 (1992).

255. K. Raghavachari and J. S. Binkley, *Structure, Stability, and Fragmentation of Small Carbon Clusters*, J. Chem. Phys. **87**, 2191 (1987).

256. T. Wakabayashi and Y. Achiba, *A Model for the $C_{60}$ and $C_{70}$ Growth Mechanism*, Chem. Phys. Lett. **190**, 465 (1992).

257. T. Wakabayashi, H. Shiromaru, K. Kikuchi, and Y. Achiba, *A Selective Isomer Growth of Fullerenes*, Chem. Phys. Lett. **201**, 470 (1993).

258. G. von Helden, N. G. Gotts, and M. T. Bowers, *Experimental Evidence for the Formation of Fullerenes by Collisional Heating of Carbon Rings in the Gas Phase*, Nature **363**, 60 (1993).

259. G. von Helden, N. G. Gotts, and M. T. Bowers, *Annealing of Carbon Cluster Cations: Rings to Rings and Rings to Fullerenes*, J. Am. Chem. Soc. **115**, 4363 (1993).

260. M. T. Bowers, P. R. Kemper, G. von Helden, and P. A. M. van Koppen, *Gas-Phase Ion Chromatography: Transition Metal State Selection and Carbon Cluster Formation*, Science **260**, 1446 (1993).

261. J. Hunter, J. Fye and M. F. Jarrold, *Annealing $C_{60}^{+}$: Synthesis of Fullerenes and Large Carbon Rings*, Science **260**, 784 (1993).

262. D. E. Manolopoulos, *Comment on "Favourable Structures for Higher Fullerenes"*, Chem. Phys. Lett. **192**, 330 (1992).

263. X. Liu, T. G. Schmalz, and D. J. Klein, *Reply to Comment on "Favourable Structures for Higher Fullerenes"*, Chem. Phys. Lett. **192**, 331 (1992).

264. D. E. Manolopoulos, J. C. May, and S. E. Down, *Theoretical Studies of the Fullerenes: $C_{34}$ to $C_{70}$*, Chem. Phys. Lett. **181**, 105 (1991).

265. X. Liu, D. J. Klein, W. A. Seitz, and T. G. Schmalz, *Sixty-Atom Carbon Cages*, J. Comp. Chem. **12**, 1265 (1991).

266. T. G. Schmalz, W. A. Seitz, D. J. Klein, and G. E. Hite, *$C_{60}$ Carbon Cages*, Chem. Phys. Lett. **130**, 203 (1986).

267. A. J. Stone and D. J. Wales, *Theoretical Studies of Icosahedral $C_{60}$ and Some Related Species*, Chem. Phys. Lett. **128**, 501 (1986).

268. R. L. Murry, D. L. Strout, G. K. Odom, and G. E. Scuseria, *Role of $sp^{3}$ Carbon and 7-Membered Rings in Fullerene Annealing and Fragmentation*, Nature **366**, 665 (1993).

269. K. Raghavachari and C. M. Rohlfing, *Imperfect Fullerene Structures: Isomers of $C_{60}$*, J. Phys. Chem. **96**, 2463 (1992).

270. R. E. Stanton, *Fullerene Structures and Reactions: MNDO Calculations*, J. Phys. Chem. **96**, 111 (1992).

271. L. A. Chernosatonskii, *A Non-fullerene Form of $C_{60}$ and Layered Metal-Doped $C_{60}$ Solid*, Phys. Lett. A **160**, 392 (1991).

# Chapter 4

# The $C_{70}$ Fullerene

The $C_{70}$ fullerene is the second most abundant extractable carbon cluster in the soot produced by the vaporization of graphite [1]. It usually constitutes about 15% of the fullerene extract, although conditions have been described in which fullerene mixtures containing up to 50% of $C_{70}$ are produced [2]. Like $C_{60}$, $C_{70}$ is formed in benzene–oxygen flames [3, 4] and can be detected in the minerals shungite [5] and fulgurite [6].

At present, the reasons for the preferential formation of the $C_{60}$ and $C_{70}$ fullerenes in the process of graphite vaporization are unknown. In the absence of a definitive explanation, a working hypothesis assuming a ring-stacking mechanism of fullerene formation has been put forward [7].

## 4.1  General Structural Features

Topological considerations (Chapter 7) lead to the conclusion that the $C_{70}$ fullerene cage must possess 12 pentagonal and 25 hexagonal faces sharing a total of 105 edges. Several structures are possible within this constraint but, as in the case of $C_{60}$, only one of them satisfies the IPR requirement [8]. The $D_{5h}$ symmetry of this structure (Fig. 4.1) gives rise to five classes of symmetry-unequivalent carbon atoms ($A - E$). Each of the atoms $A$, $B$, $C$, and $D$ is shared by one five- and two six-membered rings, whereas each of the atoms belonging to the class $E$ is located at a junction of three six-membered rings. There are eight classes of symmetry-unequivalent bonds ($a - h$, Fig. 4.1). One pentagon and one hexagon are fused along each of the bonds $a$, $c$, $e$, and $f$, whereas each of the bonds $b$, $d$, $g$, and $h$ is shared by two hexagons. Such an arrangement of bonds is compatible with 52 168 distinct Kekulé structures [9].

There is no doubt that the experimentally observed $C_{70}$ fullerene has the IPR structure discussed above. An unequivocal proof of such a structure is pro-

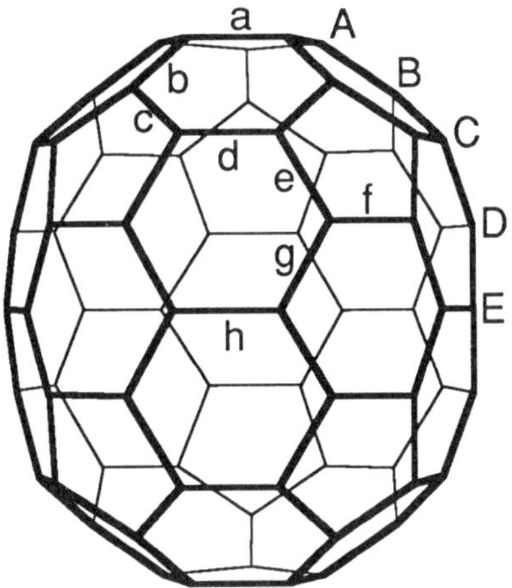



*Fig. 4.1.   The IPR isomer of the $C_{70}$ fullerene.*

vided by the 2D NMR INADEQUATE experiment that delineates all expected bond connectivities in $C_{70}$ [10]. Direct confirmation of the $D_{5h}$ structure is also afforded by a gas-phase electron diffraction study of the pristine fullerene [11] and X-ray diffraction measurements of the $C_{70} \cdot (S_8)_6$ solid-state adduct [12].

## 4.2  Energetics and Stability

In consonance with the general trend among fullerenes, $C_{70}$ is more stable on a per-atom basis than $C_{60}$. This fact is reflected in the exothermicity of the fullerene interconversion reaction

$$\frac{70}{60}\, C_{60} \rightarrow C_{70} \tag{4.1}$$

The standard enthalpy of the isodesmic reaction (4.1) has been estimated at $-73$ kcal/mol [26] from the measured standard enthalpies of formation of solid $C_{60}$ and $C_{70}$ [28] and the corresponding enthalpies of sublimation [29]. Taking into account that the experimental value of $\Delta H_r^0$ [Eq. (4.1)] has an uncertainty of about 10 kcal/mol, one finds several electronic structure methods capable of predicting the energetics of reaction (4.1) in a quantitative manner. Inspection of Table 4.1 reveals that the approaches of molecular mechanics badly underestimate $\Delta H_r^0$ [Eq. (4.1)]. Values that are most probably too low are also afforded by the tight-binding Hamiltonian [16–18], the Harris functional

## Table 4.1. The standard enthalpy of formation of the $C_{70}$ fullerene and the standard enthalpy of the fullerene interconversion reaction, Eq. (4.1)

| Method | $\Delta H_f^0$(gas) [Eq. (4.1)] [kcal/mol] | $\Delta H_r^0$ [Eq. (4.1)] [kcal/mol] |
|---|---|---|
| MM2 [13] | 582.6 | −23.8 |
| MMP2 [14] | 344.9 | −11.1 |
| MM3 [15] | 639.6 | −29.0 |
| Tight-binding Hamilt. [16–18] | 589.2 | −58.1 |
| MNDO [19–21] | 939.3 | −74.9 |
| AM1 [14] | 1071.9 | −63.5 |
| PM3 [22, 23] | 884.2 | −62.8 |
| HF/STO-3G [15] | 666.8 | −62.4 |
| HF/3-21G [21] | n/a | −50.8 |
| HF/6-31G//MNDO [24] | n/a | −62.4 |
| HF/DZ [25] | n/a | −62.7 |
| HF/DZP//MNDO [24, 26] | 639.9 | −60.1 |
| LYP/DZP//MNDO [26] | 636.1 | −63.9 |
| BLYP/DZ [25] | n/a | −64.0 |
| LDA [18] | n/a | −70.4 |
| LDA/DZ [25] | n/a | −68.6 |
| Harris functional [27] | n/a | −48.4 |
| Estimated exp. [26] | 627.0 | −73.0 |
| Exp. [30] | 675.5 | −35.7 |

[27], and HF/3-21G [21] calculations. Results of other calculations are in much better consensus, both among themselves and with respect to the experimental data, although the agreement with the HF/STO-3G value [15] is most probably fortuitous. In light of the consistency in the predicted values of $\Delta H_r^0$ [Eq. (4.1)], the recently reported results of combustion experiments on $C_{70}$ [30] appear to be in error.

The estimated gas-phase standard enthalpy of formation $\Delta H_f^0$ equals $ca.$ 627 kcal/mol for $C_{70}$ [26]. As in the case of $C_{60}$, semiempirical methods of the MNDO family grossly overestimate $\Delta H_f^0$ [14, 19–23]. On the other hand, the energy estimates based on *ab initio* approaches fare much better, with the LYP/DZP//MNDO value [26] falling within error brackets of the experimental data.

The $C_{70}$ species constitutes an "island of stability" among small fullerenes, as its interconversion to either $C_{68}$ or $C_{72}$ is endothermic (Section 4.10). Although this local stability may account for the preferential formation of $C_{70}$, almost certainly, kinetic factors also play an important role there.

## 4.3 Geometry

The pattern of bond lengths in the $C_{70}$ cage can be easily understood by regarding $C_{70}$ as two halves of the $C_{60}$ cluster joined by a belt of five benzenoid rings [31]. In accordance with this description, bonds $a$, $c$, and $e$ (Fig. 4.1) are analogous to the "single" C–C bonds of $C_{60}$ (Chapter 3), whereas bonds $b$ and $d$ correspond to the "double" bonds. Bonds $f$ and $g$ have lengths close to those in benzene and the length of bond $h$ mirrors that of the interring C–C linkage in biphenyl.

The calculated bond lengths (Table 4.2) are in agreement with the above model and follow the order $d < b < g < f < c < a < e < h$ in almost all cases, the only exceptions being bond lengths obtained from the tight-binding Hamiltonian [32], INDO/1 [33], AM1 [14, 34], BLYP [25], and LDA [25, 37] calculations. There is a close (although not perfect) correlation between increasing bond lengths and decreasing Hückel $\pi$-electron bond orders $(d > b > g > f > h > c > a > e)$ [38]. The theoretical predictions are confirmed by the geometry of $C_{70}$ obtained from X-ray diffraction experiments on the $C_{70} \cdot (S_8)_6$ solid-state adduct [12]. Statistical analysis of the available data shows [23] that the experimental bond lengths in the $C_{70}$ fullerene are reproduced most faithfully by the PM3 method, although the HF/STO-3G and HF/dzP values are also quite accurate.

The geometry of $C_{70}$ has also been studied with gas-phase electron diffraction [11]. Although most of the experimental bond lengths that have been measured within an error of $\pm 0.01$ Å follow closely the theoretical predictions listed in Table 4.2, substantial discrepancies for bonds $f$ and $h$ are observed. Taking into account that reasonable agreement is present among bond lengths calculated with a broad spectrum of theoretical methods (from molecular mechanics to semiempirical, to *ab initio*, to density functional) and the experimental data for the bonds in question carry large error brackets, in this case one can easily rule on the issue of accuracy in favor of theory.

The strain present in $C_{70}$ is apparent in the sums of bond angles around individual carbon atoms, which are smaller than $360°$. From the calculated bond angles [21, 31, 35] one concludes that the local strain is almost evenly distributed among the atoms, diminishing slightly in the direction from the

## Table 4.2.  Bond lengths [Å] in the C$_{70}$ fullerene molecule

| Method | Bond[a] | | | | | | | |
|---|---|---|---|---|---|---|---|---|
| | a | b | c | d | e | f | g | h |
| MM2 [13] | 1.443 | 1.395 | 1.440 | 1.386 | 1.451 | 1.415 | 1.414 | 1.452 |
| MM3 [15] | 1.452 | 1.392 | 1.448 | 1.380 | 1.459 | 1.422 | 1.416 | 1.465 |
| Tight-binding Hamilt. [32] | 1.457 | 1.397 | 1.454 | 1.389 | 1.456 | 1.443 | 1.418 | 1.452 |
| INDO/1 [33] | 1.450 | 1.400 | 1.448 | 1.391 | 1.456 | 1.427 | 1.421 | 1.453 |
| MNDO [21] | 1.473 | 1.402 | 1.469 | 1.389 | 1.478 | 1.442 | 1.430 | 1.484 |
| AM1 [34] | 1.456 | 1.380 | 1.459 | 1.383 | 1.468 | 1.447 | 1.413 | 1.461 |
| AM1 [14] | 1.464 | 1.380 | 1.465 | 1.371 | 1.465 | 1.433 | 1.415 | 1.463 |
| PM3 [22, 23] | 1.457 | 1.386 | 1.453 | 1.375 | 1.463 | 1.426 | 1.412 | 1.463 |
| HF/STO-3G [31, 35] | 1.461 | 1.379 | 1.455 | 1.364 | 1.470 | 1.417 | 1.414 | 1.486 |
| HF/3-21G [21] | 1.452 | 1.370 | 1.447 | 1.356 | 1.458 | 1.414 | 1.403 | 1.475 |
| HF/dz [35] | 1.450 | 1.370 | 1.445 | 1.357 | 1.455 | 1.413 | 1.402 | 1.474 |
| HF/DZ [35] | 1.454 | 1.382 | 1.449 | 1.369 | 1.459 | 1.421 | 1.412 | 1.478 |
| HF/DZ [25] | 1.450 | 1.377 | 1.445 | 1.365 | 1.455 | 1.418 | 1.408 | 1.474 |
| HF/dzP [35, 36] | 1.451 | 1.375 | 1.446 | 1.361 | 1.457 | 1.415 | 1.407 | 1.475 |
| HF/BLYP [25] | 1.470 | 1.413 | 1.466 | 1.406 | 1.464 | 1.455 | 1.436 | 1.487 |
| LDA [37] | 1.448 | 1.393 | 1.444 | 1.386 | 1.442 | 1.434 | 1.415 | 1.467 |
| HF/LDA [25] | 1.450 | 1.398 | 1.447 | 1.391 | 1.445 | 1.437 | 1.419 | 1.465 |
| Electron diffr. [11] | 1.464 | 1.37 | 1.47 | 1.37 | 1.46 | 1.47 | 1.39 | 1.41 |
| X-ray diffr. (C$_{70}\cdot$S$_{48}$) [12] | 1.458 | 1.380 | 1.459 | 1.370 | 1.460 | 1.430 | 1.407 | 1.476 |

[a] See Fig. 4.1.

poles of the cage to its equator.  The same trend concerning local strain is observed in the results of the POAV analysis [21].

## 4.4  Vibrational Frequencies

The $D_{5h}$ symmetry of C$_{70}$ gives rise to $12A_1' + 9A_2' + 9A_1'' + 10A_2'' + 21E_1' + 22E_2' + 19E_1'' + 20E_2''$ vibrational modes [21, 39].  Among the 204 normal modes, those belonging to the $A_2''$ and $E_1'$ irreducible representations (a total of 52

## Table 4.3. Theoretical and experimental IR vibrational frequencies [cm$^{-1}$] of the C$_{70}$ fullerene

| $A_2''$ modes | | | | $E_1'$ modes | | | |
|---|---|---|---|---|---|---|---|
| TBH[a] | QCFF/PI | MNDO | Exp. | TBH[a] | QCFF/PI | MNDO | Exp. |
| [32] | [40] | [21, 41] | [42–45] | [32] | [40] | [21, 41] | [42–45] |
| 307 | 326 | 326 | | 297 | 327 | 334 | |
| 420 | 485 | 492 | | 329 | 361 | 367 | |
| 590 | 592 | 701 | | 385 | 412 | 428 | |
| 686 | 684 | 772 | | 479 | 498 | 518 | |
| 946 | 895 | 1000 | | 547 | 560 | 627 | 578 |
| 1159 | 1168 | 1168 | | 558 | 585 | 650 | |
| 1233 | 1217 | 1314 | 1133 | 620 | 650 | 669 | |
| 1438 | 1270 | 1448 | 1461? | 661 | 711 | 740 | 674 |
| 1591 | 1389 | 1637 | | 780 | 748 | 833 | |
| 1659 | 1565 | 1742 | | 830 | 819 | 863 | |
| | | | | 892 | 931 | 929 | 795 |
| | | | | 906 | 961 | 1051 | |
| | | | | 1124 | 1118 | 1248 | |
| | | | | 1203 | 1200 | 1328 | |
| | | | | 1269 | 1245 | 1397 | |
| | | | | 1328 | 1369 | 1429 | |
| | | | | 1352 | 1383 | 1470 | |
| | | | | 1481 | 1424 | 1555 | 1414 |
| | | | | 1533 | 1499 | 1592 | 1431 |
| | | | | 1572 | 1568 | 1659 | 1461? |
| | | | | 1648 | 1640 | 1721 | 1461? |

[a] The tight-binding Hamiltonian.

modes corresponding to 31 lines) are IR-active, whereas those of $A_1'$, $E_2'$, and $E_1''$ symmetries (a total of 94 modes corresponding to 53 lines) appear in the Raman spectra. Because of the large number of lines in the IR spectra of C$_{70}$, detailed assignment of all measured frequencies is difficult if not impossible. Nevertheless, select strong lines in the IR spectrum can be ascribed to individual vibrational modes by comparing the experimental frequencies and intensities with those obtained from electronic structure calculations (Table 4.3). In particular, the very strong absorption at 1431 cm$^{-1}$ [42–45] can be unequivocally attributed to an $E_1''$ transition on the basis of the MNDO data (the calculated frequency 1592 cm$^{-1}$ [21, 41]). Similarly, the absorptions at 1133 and 1414 cm$^{-1}$ are almost certainly due to the $A_2''$ and $E_1'$ vibrational modes, respectively. Assignment of other peaks in the IR spectrum of $C_{70}$ is more tentative.

Analysis of these data leads to the conclusion that, as in the case of $C_{60}$, the MNDO method overestimates the vibrational frequencies of $C_{70}$ by about 10–15%. The overestimation is even more pronounced at the AM1 level of theory (some of the previously published AM1 frequencies [39] have been shown to be actually incorrect [41]). Despite the fact that force fields can be parameterized to reproduce the vibrational frequencies of $C_{70}$ more accurately than the semiempirical methods of quantum chemistry, they do not provide estimates of IR intensities, which makes the calculated frequencies potentially prone to misassignments [46]. The ability to compute the transition probabilities is even more crucial in the analysis of the measured Raman frequencies of $C_{70}$ [44, 45, 47], where the assignment of modes is still mostly an open issue.

## 4.5 Orbital Energy Levels, Ionization Potentials, Electron Affinities, and Positive and Negative Ions

The $C_{70}$ fullerene possesses a closed-shell electronic configuration of $16a_1' 7a_2' 23e_1' 23e_2' 5a_1'' 14a_2'' 19e_1'' 19e_2''$ [31, 35]. A closed-shell electronic structure for $C_{70}$ is predicted even by simple Hückel calculations, which yield the total $\pi$-electron energy of 108.814$\beta$, a bonding $a_2''$ HOMO with the energy of 0.529$\beta$, and a nonbonding LUMO of $a_1''$ symmetry [9, 48, 49]. Inclusion of electron–electron repulsion within the Hartree–Fock approximation slightly rearranges the orbital energy levels, resulting in an $a_2''$ HOMO and an $e_1''$ LUMO, both with negative orbital energies (Table 4.4). However, the splitting between HOMO and HOMO–1 (which has $e_1''$ symmetry) is found to be very small, amounting to less than 0.03 eV at the HF/dzP level of theory [35]. Similarly, the $a_1''$ LUMO+1 lies only 0.27 eV above the LUMO. Interestingly, LDA calculations predict an $e_1''$ HOMO located 0.07 eV above an $a_2''$ HOMO–1 [37].

In agreement with the general trend (Chapter 5), the first vertical ionization potential ($IP_1$) of $C_{70}$ is somewhat lower than that of $C_{60}$. This lowering of $IP_1$ is observed in the accurate PES [50] and single-photon ionization experiments [51], as well as in most of the theoretical predictions (Table 4.4). On the other hand, charge-transfer bracketing measurements [52, 53] appear to have resolution insufficient for distinguishing between the ionization potentials of $C_{60}$ and $C_{70}$. Inspection of the *ab initio* results listed in Table 4.4 reveals that Koopmans' theorem works reasonably well for the $C_{70}$ fullerene, provided

## Table 4.4. Theoretical and experimental values of the first vertical ionization potential and electron affinity of the $C_{70}$ fullerene

| Method | $IP_1$ [eV] | $EA_1$ [eV] |
|---|---|---|
| VEH (EHT)[a] [34] | 7.39 | n/a |
| INDO/S[a] [33] | 6.44 | 1.26 |
| MINDO/3[b] [55] | n/a | 0.79 |
| MNDO[a] [20] | 8.67 | 2.84 |
| AM1[a] [14, 34] | 9.07 | n/a |
| PM3[a] [22] | 9.01 | 3.19 |
| PM3[b] [23] | 8.82 | 3.45 |
| HF/STO-3G[a] [31, 35] | 5.14 | −2.75 |
| HF/dz[a] [35] | 7.92 | 0.79 |
| HF/DZ[a] [35] | 8.05 | 1.14 |
| HF/6-31G//MNDO[a] [24] | 7.64 | 0.80 |
| HF/dzP[a] [35] | 7.59 | 0.52 |
| HF/DZP//MNDO[a] [24] | 7.72 | 0.95 |
| HF/6-31G//MNDO[c] [24] | 7.23 | 2.87 |
| HF/DZP//MNDO[c] [24] | 7.30 | 2.82 |
| PES[d] [50] | 7.47±0.02 | n/a |
| Single-photon ion.[d] [51] | 7.3±0.2 | n/a |
| CT bracketing[d] [52] | 7.6±0.1 | n/a |
| CT bracketing[d] [53] | 7.61±0.11 | n/a |
| CT bracketing[d] [54] | n/a | 2.7±0.1 |

[a] Koopmans' theorem.

[b] $\Delta$SCF calculation.

[c] Koopmans' theorem (corrected, Chapter 5).

[d] Experimental.

a sufficiently large basis set is used. A quite accurate value of $IP_1$ is also obtained from the VEH approximation [34], which is a version of the EHT method. As in the case of $C_{60}$ (Chapter 3), approaches of the MNDO family badly overestimate the first vertical ionization potential of the $C_{70}$ fullerene.

The second vertical ionization potential ($IP_2$) of $C_{70}$ has been measured several times. Synchrotron radiation experiments have yielded $11.5 \pm 0.2$ eV for the value of $IP_2$ [56], in reasonable agreement with those of $12.2 \pm 0.5$ [57] and $12.0 \pm 0.5$ eV [58] obtained from charge-stripping measurements. The $IP_2$ of $9.2 \pm 0.1$ eV inferred from the charge-transfer bracketing data [52, 59] has been shown to be substantially underestimated and the correct value of $IP_2 \leq 11.3$ eV has been quoted [60]. In light of these experimental data, the $IP_2$ of 11.5 eV, afforded by $\Delta$SCF PM3 calculations [23], appears quite accurate. The reactivity of the $C_{70}^{2+}$ cation toward several molecules has been studied [61].

In agreement with the results of charge-transfer bracketing experiments [54], $C_{70}$ is predicted to possess a positive first vertical electron affinity ($EA_1$, Table 4.4). Relatively few investigations have been devoted to anions of $C_{70}$. The $C_{70}^-$, $C_{70}^{2-}$, $C_{70}^{3-}$, and $C_{70}^{4-}$ species have been generated in solution [43, 62–66]. In addition, properties of the $C_{70}^-$ anion in glassy matrices have been studied [64] and $C_{70}^{2-}$ has been observed in the gas phase [67].

MINDO/3 calculations on the $C_{70}^-$ anion indicate that the $D_{5h}$ symmetry of the neutral fullerene is retained upon electron attachment [55]. Compared to $C_{70}$, the anion is slightly more elongated and the alternation of bond lengths is diminished. The relaxation of geometry is accompanied by an energy lowering of 15.3 kcal/mol. Interestingly, the excess electron in $C_{70}^-$ is distributed less evenly than in $C_{60}^-$. Contradicting the MINDO/3 results, RHF calculations carried out with the PM3 semiempirical method predict the $C_{70}^-$, $C_{70}^{2-}$, $C_{70}^{3-}$, $C_{70}^{5-}$, and $C_{70}^{6-}$ anions to possess only $C_s$ symmetry [23]. In contrast to the $C_{60}^{2-}$ species, the $C_{70}^{2-}$ anion is calculated to be stable with respect to electron detachment. The computed quadrupole moments of the anions are strongly negative and increase in magnitude with the increasing negative charge. A similar trend is observed in the positive quadrupole moments of the $C_{70}^+$, $C_{70}^{2+}$, $C_{70}^{3+}$, $C_{70}^{5+}$, and $C_{70}^{6+}$ cations, all of which possess only a plane of symmetry. On the other hand, the $C_{70}^{4-}$ and $C_{70}^{4+}$ species retain full $D_{5h}$ symmetry in their triplet electronic states.


## 4.6  Excited States

There are two classes of dipole-allowed electronic excitations from the ground state of the $C_{70}$ fullerene. Transitions to the $^1A_2''$ excited states give rise to absorptions polarized along the long molecular axis of $C_{70}$, whereas those to the $^1E_1'$ states are polarized along its two short axes [68, 69]. Despite this rather severe symmetry restriction, a multitude of absorption bands is present in the experimental UV/VIS spectrum of $C_{70}$ [70–73]. The onset of absorption occurs at *ca.* 650 nm, corresponding to the singlet excitation energy of 1.9 eV. Several bands are located within the 550–620 nm range, and very strong absorptions are observed around the wavelengths of 210 and 240 nm.

Thus far, there have been no accurate electronic structure calculations on the excited states of $C_{70}$, although several semiempirical studies have been reported. Tight-binding Hamiltonian calculations [69] attribute the onset of the absorption to a $^1E_1'$ excited state located 1.9 eV [74] above the ground state. Studies employing the INDO/S method in conjunction with the CIS [33, 68] and RPA [33] approximations appear to yield excitation energies in reasonable agreement with the experimental spectrum. However, because of the large number of bands present, the definite assignment of individual transitions will remain uncertain until both polarization measurements and more sophisticated calculations become available. One should also mention that no theoretical predictions have been reported so far for the first triplet state of $C_{70}$, which is located between 1.4 [74] and 1.6 eV [75] above the ground state, although several of its photophysical properties have been measured [76–81].

The UV/VIS spectra of $C_{70}^-$, $C_{70}^{2-}$, $C_{70}^{3-}$, and $C_{70}^{4-}$ anions are very different from that of the parent fullerene [63–66]. In particular, they exhibit near-infrared absorption bands that are absent in the spectrum of $C_{70}$. Attempts have been made to explain the UV/VIS spectra of the $C_{70}^-$ and $C_{70}^+$ species with the help of CNDO/S calculations [64]. However, the CNDO/S results fail to account for all of the observed absorption bands of $C_{70}^-$ [65]. The proposed explanation of this failure, invoking a symmetry lowering of $C_{70}^-$ from $D_{5h}$ to $C_{5v}$ [65], is inconsistent with the results of MINDO/3 [55] and PM3 calculations [23] (Section 4.5).

## 4.7 Electric Polarizabilities

Due to its high symmetry, the $C_{70}$ fullerene does not possess a permanent dipole moment. However, the electric polarizability ($\alpha$) of $C_{70}$ is quite large. HF/STO-3G calculations [31] yield 1.25 for the ratio of spherically averaged electric polarizabilities of $C_{70}$ and $C_{60}$, which is about the same as the value of 1.24 obtained from PM3 calculations [22] and slightly higher than that of 1.17 predicted from a simple atomic additivity scheme. The polarizability is anisotropic, with the HF/STO-3G $\alpha_{zz}/\alpha_{xx}$ ratio amounting to 1.07.

Calculations within the VEH approximation, carried out at AM1 optimized geometries, estimate the ratio of the $C_{70}$ and $C_{60}$ average polarizabilities at 1.39 and the $\alpha_{zz}/\alpha_{xx}$ ratio at 1.13 [34]. In light of these results, the VEH approximation appears to overestimate $\alpha$ of $C_{70}$ relative to $C_{60}$. One should also note that all three approaches (HF/STO-3G, PM3, and VEH) are too approximate to yield accurate absolute values of $\alpha$.

The second-order hyperpolarizability ($\gamma$) of $C_{70}$ has been measured. As in the case of $C_{60}$ (Chapter 3), large discrepancies exist between the experimental value of $1.3\pm0.3\cdot10^{-33}$ esu [82], and the theoretical predictions of $4.5\cdot10^{-35}$ (PM3) [22] and $8.6\cdot10^{-34}$ esu (VEH) [34].

## 4.8  The $^{13}$C NMR Spectrum and Other Magnetic Properties

In contrast to $^{13}$C NMR spectra of higher fullerenes (Chapter 5), the observed peaks have been assigned to individual nuclei of the $C_{70}$ fullerene with the help of the INADEQUATE technique [10]. The nuclei of atoms $A$, $B$, $C$, and $D$ (Fig. 4.1), which are shared by one pentagon and two hexagons (Section 4.1), have larger chemical shifts than those of the atoms $E$, which are located at the junctions of three hexagons (Table 4.5). This distinction between "pyracylene" and "pyrene" carbons [86] is also observed in the theoretical chemical shifts calculated from the formula [83]

$$\delta = \delta_{ref} + \sigma_{ref} - \sigma \qquad (4.2)$$

where $\delta_{ref}$ is the $^{13}$C chemical shift in $C_{60}$, equal to 143.2 ppm [71, 86], and $\sigma_{ref}$ and $\sigma$ are the calculated absolute shieldings of the carbon nuclei in $C_{60}$ and $C_{70}$, respectively. Despite reproduction of the qualitative trends, the computed shifts are quite inaccurate, the reversal of ordering between the chemical shifts of nuclei $B$ and $C$ being particularly troublesome. Interestingly, a proper choice of geometry (MNDO versus HF/dzP) appears to be more important than the quality of the basis set (6-31G versus TZP) used in the GIAO CPHF calculations (Chapter 5) [83].

**Table 4.5.  Theoretical and experimental chemical shifts [ppm] of the $^{13}$C nuclei in the $C_{70}$ fullerene**

|  | Nucleus[a] | | | | |
|---|---|---|---|---|---|
| Method | $A$ | $B$ | $C$ | $D$ | $E$ |
| GIAO HF/6-31G//MNDO [83] | 148.9 | 146.0 | 144.9 | 142.7 | 130.9 |
| GIAO HF/DZP//HF/dzP [84] | 146.6 | 143.5 | 142.2 | 140.2 | 130.3 |
| GIAO HF/TZP//HF/dzP [84, 85] | 147.1 | 144.3 | 143.4 | 140.9 | 129.7 |
| Exp. (in $CDCl_2CDCl_2$) [86] | 150.8 | 147.8 | 148.3 | 144.4 | 130.8 |
| Exp. (in toluene-$d_8$) [10] | 150.8 | 147.8 | n/a | 144.4 | 130.8 |
| Exp. (in $CCl_4$) [43] | 150.4 | 147.2 | 147.9 | 145.2 | 130.9 |
| Exp. [70] | 150.1 | 146.8 | 147.5 | 144.8 | 130.3 |
| Exp. (in benzene-$d_8$) [71] | 150.7 | 147.4 | 148.1 | 145.4 | 130.9 |

[a] See Fig. 4.1.

HF/STO-3G calculations of the $^{13}$C NMR shifts in $C_{70}$ have also been published [31, 87]. However, because these calculations are not gauge invariant, the reported shifts are in poor agreement with the experimental data. The same is true for the calculated magnetic susceptibility of $C_{70}$, which has been experimentally determined to be 1.97 [88] and 2.12 [89] times greater than that of $C_{60}$.

Finally, a theory of ring currents in $C_{70}$ [90], based on the London approximation, and an investigation of properties possessed by the eigenfunctions of a magnetic Hamiltonian describing $C_{70}$ [91], should be mentioned.

## 4.9 Non-IPR Isomers of the $C_{70}$ Fullerene

Several non-IPR isomers of $C_{70}$ are possible. Experimental evidence for such isomers is scarce, although there has been a report of their presence in fullerene mixtures produced by sooting flames [92]. Low-energy non-IPR isomers of $C_{70}$ can be readily constructed by subjecting the IPR cage of $C_{70}$ to a single Stone–Wales transformation [93]. In $C_{70}$, there are two possible sites at which such a transformation can be carried out [94, 95]. Rotating bond $b$ (Fig. 4.1) results in structure **1** (Fig. 4.2) with two pairs of abutting pentagons. This $C_s$ structure, which is analogous to the lowest-energy non-IPR isomer of $C_{60}$ (Chapter 3), lies 50.3 (MNDO) and 56.3 kcal/mol (HF/3-21G//MNDO) above the IPR isomer of the $C_{70}$ fullerene.

Interestingly, unlike in the case of $C_{60}$, there exists a non-IPR isomer of $C_{70}$ with only one pair of adjacent pentagons [94, 95]. This unique $C_s$ isomer **2** (Fig. 4.2) is obtained through the Stone–Wales transformation around bond $d$ (Fig. 4.1). As expected, isomer **2** is lower in energy than isomer **1**, but still less stable (by 30.4 and 33.9 kcal/mol at the MNDO and HF/3-21G//MNDO levels of theory, respectively) than the IPR species.

A sequence of two Stone–Wales transformations involving both bonds $b$ and $d$ (Fig. 4.1) leads to another $C_s$ structure **3** (Fig. 4.2), which possesses two pairs of abutting pentagons [94]. With its energy 57.7 (MNDO) and 63.2 kcal/mol (HF/3-21G//MNDO) above the IPR structure, isomer **3** is somewhat less stable than isomer **1**. All three isomers are true energy minima, as indicated by the MNDO vibrational frequencies that are all real. Other $C_{70}$ carbon clusters with several pairs of adjacent pentagons have been studied and found very unfavorable energetically [20].

## 4.10 The $C_{68}$ and $C_{72}$ Fullerenes

No IPR structures are possible for fullerenes with 62, 64, 66, or 68 carbon atoms [8, 17]. Therefore, it is not surprising that the $C_{68}$ fullerene is much less stable than $C_{70}$. Tight-binding Hamiltonian calculations estimate the standard

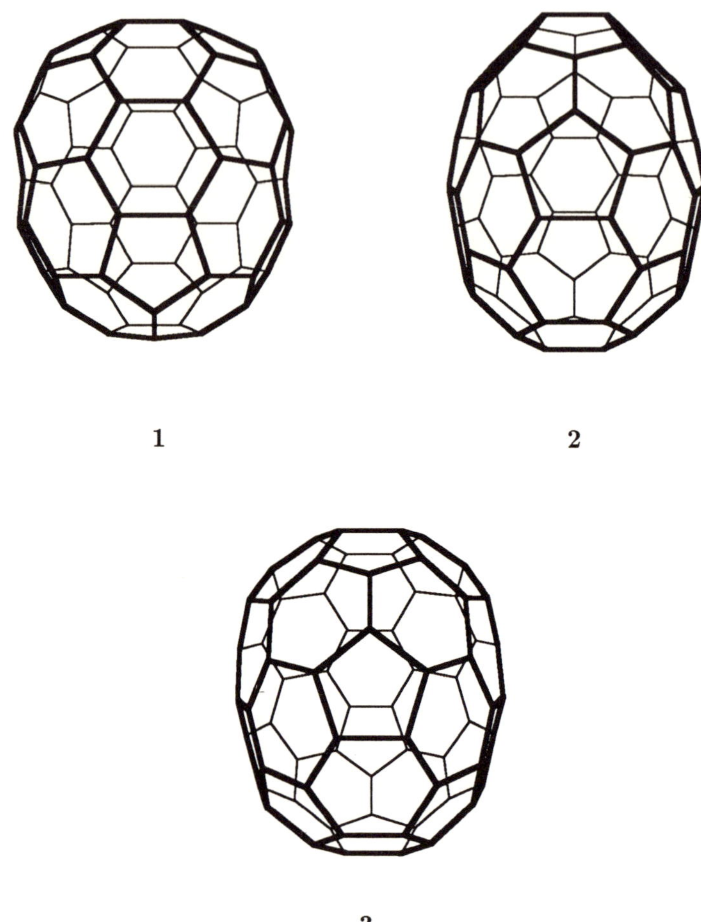

1          2

3

*Fig. 4.2. Three low-energy non-IPR isomers of the $C_{70}$ cluster.*

enthalpy of the reaction

$$\frac{70}{68} \, C_{68} \rightarrow C_{70} \tag{4.3}$$

at $-54.9$ kcal/mol [17]. The energy required to remove a $C_2$ fragment from the $C_{70}$ fullerene equals 279, 295, and 254 kcal/mol at the MNDO//MM3, HF/DZ//MM3, and BLYP/DZ//MM3 levels of theory, respectively [95].

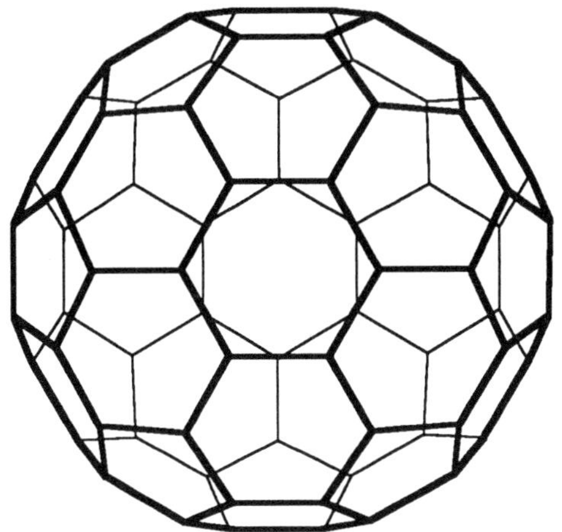

*Fig. 4.3. The IPR isomer of the $C_{72}$ fullerene.*

Although, in contrast to $C_{68}$, one IPR isomer is compatible with a fullerene cage composed of 72 carbon atoms [9, 17, 96, 97], the $C_{72}$ cluster has never been isolated. This failure appears surprising at first glance, as Hückel calculations predict the $D_{6d}$ structure (Fig. 4.3) to possess a HOMO–LUMO gap comparable to that of $C_{60}$ [9, 96]. However, more sophisticated electronic structure calculations predict the reaction

$$\frac{70}{72} \; C_{72} \rightarrow C_{70} \tag{4.4}$$

to be exothermic by 19.4 (tight-binding Hamiltonian) [17, 98], 18.4 (MNDO), and 20.0 kcal/mol (HF/3-21G) [97], contrary to the general trend of larger fullerenes being more stable on a per atom basis than the smaller ones (Chapters 5 and 6). The low stability of $C_{72}$ is most probably caused by the presence of two six-membered rings that are each surrounded by six other six-membered rings [97]. Such an arrangement of bonds strongly prefers planarity of all rings involved and introduces substantial strain in the closed fullerene cage.

Should the $C_{72}$ fullerene ever be isolated, it would be easily recognized by its simple IR (26 allowed transitions [97]) and $^{13}C$ NMR (only four peaks [8, 98]) spectra.

# References


1. W. Krätschmer, L. D. Lamb, K. Fostiropoulos, and D. R. Huffman, *Solid $C_{60}$: A New Form of Carbon*, Nature **347**, 354 (1990).

2. A. Mittelbach, W. Hönle, H. G. von Schnering, J. Carlsen, R. Janiak, and H. Quast, *Optimization of the Production and Separation of Fullerenes* Angew. Chem. Int. Ed. Engl. **31**, 1640 (1992).

3. J. B. Howard, J. T. McKinnon, M. E. Johnson, Y. Makarovsky, and A. L. Lafleur, *Production of $C_{60}$ and $C_{70}$ Fullerenes in Benzene-Oxygen Flames*, J. Phys. Chem. **96**, 6657 (1992).

4. J. B. Howard, J. T. McKinnon, Y. Makarovsky, A. L. Lafleur, and M. E. Johnson, *Fullerenes $C_{60}$ and $C_{70}$ in Flames*, Nature **352**, 139 (1991).

5. P. R. Buseck, S. J. Tsipursky, and R. Hettich, *Fullerenes from the Geological Environment*, Science **257**, 215 (1992).

6. T. K. Daly, P. R. Buseck, P. Williams, and C. F. Lewis, *Fullerenes from a Fulgurite*, Science **259**, 1599 (1993).

7. T. Wakabayashi and Y. Achiba, *A Model for the $C_{60}$ and $C_{70}$ Growth Mechanism*, Chem. Phys. Lett. **190**, 465 (1992).

8. X. Liu, T. G. Schmalz, and D. J. Klein, *Favorable Structures for Higher Fullerenes*, Chem. Phys. Lett. **188**, 550 (1992).

9. T. G. Schmalz, W. A. Seitz, D. J. Klein, and G. E. Hite, *Elemental Carbon Cages*, J. Am. Chem. Soc. **110**, 1113 (1988).

10. R. D. Johnson, G. Meijer, J. R. Salem, and D. S. Bethune, *2D Nuclear Magnetic Resonance Study of the Structure of the Fullerene $C_{70}$*, J. Am. Chem. Soc. **113**, 3619 (1991).

11. D. R. McKenzie, C. A. Davis, D. J. H. Cockayne, D. A. Muller, and A. M. Vassallo, *The Structure of the $C_{70}$ Molecule*, Nature **355**, 622 (1992).

12. G. Roth and P. Adelmann, *The Crystal Structure of $C_{70}S_{48}$: The First a Priori Structure Determination of a $C_{70}$-Containing Compound*, J. Phys. I France **2**, 1541 (1992).

13. M. Froimowitz, *Molecular Geometries and Heats of Formation of $C_{60}$ and $C_{70}$ as Computed by MM2-87*, J. Comp. Chem. **12**, 1129 (1991).

14. J. M. Rudziński, Z. Slanina, M. Togasi, E. Osawa, and T. Iizuka, *Computational Study of Relative Stabilities of $C_{60}(I_h)$ and $C_{70}(D_{5h})$ Gas-Phase Clusters*, Thermochim. Acta **125**, 155 (1988).

15. R. L. Murry, J. R. Colt, and G. E. Scuseria, *How Accurate Are Molecular Mechanics Predictions for Fullerenes? A Benchmark Comparison with Hartree-Fock Self-Consistent Field Results*, J. Phys. Chem. **97**, 4954 (1993).

16. B. L. Zhang, C. Z. Wang, K. M. Ho, C. H. Xu, and C. T. Chan, *The Geometry of Small Fullerene Cages: $C_{20}$ to $C_{70}$*, J. Chem. Phys. **97**, 5007 (1992).

17. B. L. Zhang, C. Z. Wang, and K. M. Ho, *Structures of Large Fullerenes: $C_{60}$ to $C_{94}$*, Chem. Phys. Lett. **193**, 225 (1992).
18. X.-Q. Wang, C. Z. Wang, B. L. Zhang, and K. M. Ho, *The Structural and Electronic Properties of $C_{84}$: A First-Principles Study*, Phys. Rev. Lett. **69**, 69 (1992).
19. M. D. Newton and R. E. Stanton, *Stability of Buckminsterfullerene and Related Carbon Clusters*, J. Am. Chem. Soc. **108**, 2469 (1986).
20. D. Bakowies and W. Thiel, *MNDO Study of Large Carbon Clusters*, J. Am. Chem. Soc. **113**, 3704 (1991).
21. K. Raghavachari and C. M. Rohlfing, *Structures and Vibrational Frequencies of $C_{60}$, $C_{70}$, and $C_{84}$*, J. Phys. Chem. **95**, 5768 (1991).
22. N. Matsuzawa and D. A. Dixon, *Semiempirical Calculations of the Polarizability and Second-Order Hyperpolarizability of $C_{60}$, $C_{70}$, and Model Aromatic Compounds*, J. Phys. Chem. **96**, 6241 (1992).
23. E. Roduner and I. D. Reid, *Structure and Charge Distribution of Multiply Charged $C_{70}$*, Chem. Phys. Lett. **223**, 149 (1994).
24. J. Cioslowski and K. Raghavachari, *Electrostatic Potential, Polarization, Shielding, and Charge Transfer in Endohedral Complexes of the $C_{60}$, $C_{70}$, $C_{76}$, $C_{78}$, $C_{82}$, and $C_{84}$ Clusters*, J. Chem. Phys. **98**, 8734 (1993).
25. C. van Wüllen, *An Implementation of a Kohn-Sham Density Functional Program Using a Gaussian-Type Basis Set. Application to the Equilibrium Geometry of $C_{60}$ and $C_{70}$*, Chem. Phys. Lett. **219**, 8 (1994).
26. J. Cioslowski, *Heats of Formation of Higher Fullerenes from ab Initio Hartree–Fock and Correlation Energy Functional Calculations*, Chem. Phys. Lett. **216**, 389 (1993).
27. N. Kurita, K. Kobayashi, H. Kumahora, K. Tago, and K. Ozawa, *Nonlocal Density Functional Calculations of Binding Energies of Carbon Fullerenes $C_n$, with $n = 10, 12, 16, 20, 24, 28, 32, 36, 50, 60, 70, 80, 90, 100, 110$ and $120$*, Chem. Phys. Lett. **188**, 181 (1992).
28. T. Kiyobayashi and M. Sakiyama, *Combustion Calorimetric Studies on $C_{60}$ and $C_{70}$*, Full. Sci. Tech. **1**, 269 (1993).
29. C. Pan, M. P. Sampson, Y. Chai, R. H. Hauge, and J. L. Margrave, *Heats of Sublimation from a Polycrystalline Mixture of $C_{60}$ and $C_{70}$*, J. Phys. Chem. **95**, 2944 (1991).
30. H.-D. Beckhaus, S. Verevkin, C. Rüchardt, F. Diederich, C. Thilgen, H.-U. ter Meer, H. Mohn, and W. Müller, *$C_{70}$ Is More Stable than $C_{60}$: Experimental Determination of the Heat of Formation of $C_{70}$*, Angew. Chem. Int. Ed. Engl. **33**, 996 (1994).
31. J. Baker, P. W. Fowler, P. Lazzeretti, M. Malagoli, and R. Zanasi, *Structure and Properties of $C_{70}$*, Chem. Phys. Lett. **184**, 182 (1991).
32. C. Z. Wang, C. T. Chan, and K. M. Ho, *Structure and Dynamics of $C_{60}$ and $C_{70}$ from Tight-Binding Molecular Dynamics*, Phys. Rev. B **46**, 9761 (1992).

33. R. D. Bendale, J. D. Baker, and M. C. Zerner, *Calculations on the Electronic Structure and Spectroscopy of $C_{60}$ and $C_{70}$ Cage Structures*, Int. J. Quant. Chem. Quant. Chem. Symp. **25**, 557 (1991).

34. Z. Shuai and J. L. Brédas, *Electronic Structure and Nonlinear Optical Properties of the Fullerenes $C_{60}$ and $C_{70}$: A Valence-Effective-Hamiltonian Study*, Phys. Rev. B **46**, 16135 (1992).

35. G. E. Scuseria, *The Equilibrium Structure of $C_{70}$. An ab Initio Hartree-Fock Study*, Chem. Phys. Lett. **180**, 451 (1991).

36. G. E. Scuseria, *Ab Initio Theoretical Predictions of Fullerenes*, in *Buckminsterfullerenes* (W. E. Billups and M. A. Ciufolini, eds.), VCH Publishers, New York, 1993, Ch. 5, p. 103.

37. W. Andreoni, F. Gygi, and M. Parrinello, *Structural and Electronic Properties of $C_{70}$*, Chem. Phys. Lett. **189**, 241 (1992).

38. R. Taylor, $C_{60}$, $C_{70}$, $C_{76}$, $C_{78}$ and $C_{84}$: *Numbering, $\pi$-Bond Order Calculations and Addition Pattern Considerations*, J. Chem. Soc. Perkin Trans. 2 813 (1993).

39. Z. Slanina, J. M. Rudziński, M. Togasi, and E. Ōsawa, *Quantum-Chemically Supported Vibrational Analysis of Giant Molecules: The $C_{60}$ and $C_{70}$ Clusters*, J. Mol. Struct. (Theochem) **202**, 169 (1989).

40. F. Negri, G. Orlandi, and G. Zerbetto, *QCFF/PI Vibrational Frequencies of Some Spherical Carbon Clusters*, J. Am. Chem. Soc. **113**, 6037 (1991).

41. D. Bakowies and W. Thiel, *Theoretical Infrared Spectra of Large Carbon Clusters*, Chem. Phys. **151**, 309 (1991).

42. J. P. Hare, T. J. Dennis, H. W. Kroto, R. Taylor, A. W. Allaf, S. Balm, and D. R. M. Walton, *The IR Spectra of Fullerene-60 and -70*, J. Chem. Soc. Chem. Commun. 412 (1991).

43. D. M. Cox, S. Behal, M. Disko, S. M. Gorun, M. Greaney, C. S. Hsu, E. B. Kollin, J. Millar, J. Robbins, W. Robbins, R. D. Sherwood, and P. Tindall, *Characterization of $C_{60}$ and $C_{70}$ Clusters*, J. Am. Chem. Soc. **113**, 2940 (1991).

44. B. Chase, N. Herron, and E. Holler, *Vibrational Spectroscopy of $C_{60}$ and $C_{70}$ Temperature-Dependent Studies*, J. Phys. Chem. **96**, 4262 (1992).

45. D. S. Bethune, G. Meijer, W. C. Tang, H. J. Rosen, W. G. Golden, H. Seki, C. A. Brown, and M. S. de Vries, *Vibrational Raman and Infrared Spectra of Chromatographically Separated $C_{60}$ and $C_{70}$ Fullerene Clusters*, Chem. Phys. Lett. **179**, 181 (1991).

46. P. Procacci, G. Cardini, P. R. Salvi, and V. Schettino, *Vibrational Frequencies of $C_{70}$*, Chem. Phys. Lett. **195**, 347 (1992).

47. D. S. Bethune, G. Meijer, W. C. Tang, and H. J. Rosen, *The Vibrational Raman Spectra of Purified Solid Films of $C_{60}$ and $C_{70}$*, Chem. Phys. Lett. **174**, 219 (1990).

48. P. W. Fowler and J. Woolrich, *$\pi$-Systems in Three Dimensions*, Chem. Phys. Lett. **127**, 78 (1986).

49. R. C. Haddon, L. E. Brus, and K. Raghavachari, *Rehybridization and $\pi$-Orbital Alignment: The Key to the Existence of Spheroidal Carbon Clusters*, Chem. Phys. Lett. **131**, 165 (1986).

50. D. L. Lichtenberger, M. E. Rempe, and S. B. Gogosha, *The He I Valence Photoelectron Spectrum of $C_{70}$ in the Gas Phase*, Chem. Phys. Lett. **198**, 454 (1992).

51. I. V. Hertel, H. Steger, J. de Vries, B. Weisser, C. Menzel, B. Kamke, and W. Kamke, *Giant Plasmon Excitations in $C_{60}$ and $C_{70}$ Studied by Photoionization*, Phys. Rev. Lett. **68**, 784 (1992).

52. S. W. McElvany and S. B. H. Bach, *First and Second Ionization Potentials and Reactions of $C_{60}$ and Related Clusters*, in *Proc. 39th ASMS Conf. on Mass Spectrometry and Allied Topics*, 422 (1991).

53. J. A. Zimmerman, J. R. Eyler, S. B. H. Bach, and S. W. McElvany, *"Magic Number" Carbon Clusters: Ionization Potentials and Selective Reactivity*, J. Chem. Phys. **94**, 3556 (1991).

54. S. B. H. Bach, J. E. Bruce, R. Ramanathan, C. H. Watson, J. A. Zimmerman, and J. R. Eyler, *Ionization Potentials and Electron Affinities of Semiconductor Clusters from Charge Transfer Reactions*, in *On Clusters and Clustering, from Atoms to Fractals* (P. J. Reynolds, ed.), Elsevier Science Publishers, New York, 1993, p. 59.

55. K. Tanaka, M. Okada, K. Okahara, and T. Yamabe, *Electronic Structure of $C_{70}^-$*, Chem. Phys. Lett. **202**, 394 (1993).

56. H. Steger, J. de Vries, B. Kamke, and W. Kamke, *Direct Double Ionization of $C_{60}$ and $C_{70}$ Fullerenes Using Synchrotron Radiation*, Chem. Phys. Lett. **194**, 452 (1992).

57. C. Lifshitz, M. Iraqi, T. Peres, and J. E. Fischer, *Charge Stripping $C_{60}^+$*, Rapid Commun. Mass Spectrom. **5**, 238 (1991).

58. K. A. Caldwell, D. E. Giblin, and M. L. Gross, *High-Energy Collisions of Fullerene Radical Cations with Target Gases: Capture of the Target Gas and Charge Stripping of $C_{60}^{\bullet+}$, $C_{76}^{\bullet+}$, and $C_{84}^{\bullet+}$*, J. Am. Chem. Soc. **114**, 3743 (1992).

59. S. W. McElvany, M. M. Ross, and J. H. Callahan, *First and Second Ionization Potentials, Reactions, and Surface Collisions of $C_{60}$ and Related Clusters*, Mat. Res. Soc. Symp. Proc. **206**, 697 (1991).

60. S. Petrie, G. Javahery, J. Wang, and D. K. Bohme, *Selected-Ion Flow Tube Study of Charge Transfer from Fullerene Dications: "Bracketing" the Second Ionization Energies of $C_{60}$ and $C_{70}$*, J. Phys. Chem. **96**, 6121 (1992).

61. S. Petrie, G. Javahery, J. Wang, and D. K. Bohme, *Derivatization of the Fullerene Dications $C_{60}^{2+}$ and $C_{70}^{2+}$ by Ion-Molecule Reactions in the Gas Phase*, J. Am. Chem. Soc. **114**, 9177 (1992).

62. P.-M. Allemand, A. Koch, F. Wudl, Y. Rubin, F. Diederich, M. M. Alvarez, S. J. Anz, and R. L. Whetten, *Two Different Fullerenes Have the Same Cyclic Voltammetry*, J. Am. Chem. Soc. **113**, 1050 (1991).

63. D. Dubois, K. M. Kadish, S. Flanagan, R. E. Haufler, L. P. F. Chibante, and L. J. Wilson, *Spectroelectrochemical Study of the $C_{60}$ and $C_{70}$ Fullerenes and Their Mono-, Di-, Tri- and Tetraanions*, J. Am. Chem. Soc. **113**, 4364 (1991).

64. T. Kato, T. Kodama, T. Shida, T. Nakagawa, Y. Matsui, S. Suzuki, H. Shiromaru, K. Yamauchi, and Y. Achiba, *Electronic Absorption Spectra of the Radical Anions and Cations of Fullerenes: $C_{60}$ and $C_{70}$*, Chem. Phys. Lett. **180**, 446 (1991)

65. D. R. Lawson, D. L. Feldheim, C. A. Foss, P. K. Dorhout, C. M. Elliott, C. R. Martin, and B. Parkinson, *Near-IR Absorption Spectra for the $C_{70}$ Fullerene Anions*, J. Phys. Chem. **96**, 7175 (1992).

66. M. A. Greaney and S. M. Gorun, *Production, Spectroscopy, and Electronic Structure of Soluble Fullerene Ions*, J. Phys. Chem. **95**, 7142 (1991).

67. P. A. Limbach, L. Schweikhard, K. A. Cowen, M. T. McDermott, A. G. Marshall, and J. V. Coe, *Observation of the Doubly Charged, Gas-Phase Fullerene Anions $C_{60}^{2-}$ and $C_{70}^{2-}$*, J. Am. Chem. Soc. **113**, 6795 (1991).

68. J. Feng, J. Li, Z. Li, and M. C. Zerner, *Quantum Chemical Calculations on Buckminsterfullerene and Related Structures. II. The Electronic Structure and Spectra of Some $C_n$ and $C_n Ca^{2+}$ Cages*, Int. J. Quant. Chem. **39**, 331 (1991).

69. J. Shumway and S. Satpathy, *Polarization-Dependent Optical Properties of $C_{70}$*, Chem. Phys. Lett. **211**, 595 (1993).

70. R. Taylor, J. P. Hare, A. K. Abdul-Sada, and H. W. Kroto, *Isolation, Separation and Characterization of the Fullerenes $C_{60}$ and $C_{70}$: The Third Form of Carbon*, J. Chem. Soc. Chem. Commun. 1423 (1990).

71. H. Ajie, M. M. Alvarez, S. J. Anz, R. D. Beck, F. Diederich, K. Fostiropoulos, D. R. Huffman, W. Krätschmer, Y. Rubin, K. E. Schriver, D. Sensharma, and R. L. Whetten, *Characterization of the Soluble All-Carbon Molecules $C_{60}$ and $C_{70}$*, J. Phys. Chem. **94**, 8630 (1990).

72. J. P. Hare, H. W. Kroto, and R. Taylor, *Preparation and UV/Visible Spectra of Fullerenes $C_{60}$ and $C_{70}$*, Chem. Phys. Lett. **177**, 394 (1991).

73. R. L. Whetten, M. M. Alvarez, S. J. Anz, K. E. Schriver, R. D. Beck, F. N. Diederich, Y. Rubin, R. Ettl, C. S. Foote, A. P. Darmanyan, and J. W. Arbogast, *Spectroscopic and Photophysical Properties of the Soluble $C_n$ Molecules, $n = 60, 70, 76/78, 84$*, Mat. Res. Soc. Symp. Proc. **206**, 639 (1991).

74. J. W. Arbogast and C. S. Foote, *Photophysical Properties of $C_{70}$*, J. Am. Chem. Soc. **113**, 8886 (1991).

75. R. E. Haufler, L.-S. Wang, L. P. F. Chibante, C. Jin, J. Conceicao, Y. Chai, and R. E. Smalley, *Fullerene Triplet State Production and Decay: R2PI Probes of $C_{60}$ and $C_{70}$ in a Supersonic Beam*, Chem. Phys. Lett. **179**, 449 (1991).

76. Y. Zeng, L. Biczok, and H. Linschitz, *External Heavy Atom Induced Phosphorescence Emission of Fullerenes: The Energy of Triplet $C_{60}$*, J. Phys. Chem. **96**, 5237 (1992).

77. D. K. Palit, A. V. Sapre, J. P. Mittal, and C. N. R. Rao, *Photophysical Properties of the Fullerenes, $C_{60}$ and $C_{70}$*, Chem. Phys. Lett. **195**, 1 (1992).

78. M. R. Wasielewski, M. P. O'Neil, K. R. Lykke, M. J. Pellin, and D. M. Gruen, *Triplet States of Fullerenes $C_{60}$ and $C_{70}$: Electron Paramagnetic Resonance Spectra, Photophysics, and Electronic Structures*, J. Am. Chem. Soc. **113**, 2774 (1991).

79. M. Terazima, N. Hirota, H. Shinohara, and Y. Saito, *Time-Resolved EPR Investigation of the Triplet States of $C_{60}$ and $C_{70}$*, Chem. Phys. Lett. **195**, 333 (1992).

80. H. Levanon, V. Meiklyar, S. Michaeli, and D. Gamliel, *Triplet State and Dynamics of Photoexcited $C_{70}$*, J. Am. Chem. Soc. **115**, 8722 (1993).

81. K. Tanigaki, T. W. Ebbesen, and S. Kuroshima, *Picosecond and Nanosecond Studies of the Excited State Properties of $C_{70}$*, Chem. Phys. Lett. **185**, 189 (1991).

82. Y. Wang and L.-T. Cheng, *Nonlinear Optical Properties of Fullerenes and Charge-Transfer Complexes of Fullerenes*, J. Phys. Chem. **96**, 1530 (1992).

83. J. Cioslowski, unpublished.

84. M. Häser, R. Ahlrichs, H. P. Baron, P. Weis, and H. Horn, *Direct Computation of Second-Order SCF Properties of Large Molecules on Workstation Computers with an Application to Large Carbon Clusters*, Theor. Chim. Acta **83**, 455 (1992).

85. M. Häser, personal communication (1994).

86. F. Diederich and R. L. Whetten, *Beyond $C_{60}$: The Higher Fullerenes*, Acc. Chem. Res. **25**, 119 (1992).

87. P. W. Fowler, P. Lazzeretti, M. Malagoli, and R. Zanasi, *Magnetic Properties of $C_{60}$ and $C_{70}$*, Chem. Phys. Lett. **179**, 174 (1991).

88. R. S. Ruoff, D. Beach, J. Cuomo, T. McGuire, R. L. Whetten, and F. Diederich, *Confirmation of a Vanishingly Small Ring-Current Magnetic Susceptibility of Icosahedral $C_{60}$*, J. Phys. Chem. **95**, 3457 (1991).

89. R. C. Haddon, L. F. Schneemeyer, J. V. Waszczak, S. H. Glarum, R. Tycko, G. Dabbagh, A. R. Kortan, A. J. Muller, A. M. Mujsce, M. J. Rosseinsky, S. M. Zahurak, A. V. Makhija, F. A. Thiel, K. Raghavachari, E. Cockayne, and V. Elser, *Experimental and Theoretical Determination of the Magnetic Susceptibility of $C_{60}$ and $C_{70}$*, Nature **350**, 46 (1991).

90. A. Pasquarello, M. A. Schlüter, and R. C.Haddon, *Ring Currents in Topologically Complex Molecules: Application to $C_{60}$, $C_{70}$, and Their Hexa-Anions*, Phys. Rev. A **47**, 1783 (1993).

91. D. Coffey and S. A. Trugman, *Magnetic Properties of Undoped $C_{60}$*, Phys. Rev. Lett. **69**, 176 (1992).

92. J. F. Anacleto, H. Perreault, R. K. Boyd, S. Pleasance, M. A. Quilliam, P. G. Sim, J. B. Howard, Y. Makarovsky, and A. L. Lafleur, *$C_{60}$ and $C_{70}$ Fullerene Isomers Generated in Flames — Detection and Verification by Liquid Chromatography Mass Spectrometry Analyses*, Rapid Commun. Mass Spectrom. **6**, 214 (1992).

93. A. J. Stone and D. J. Wales, *Theoretical Studies of Icosahedral $C_{60}$ and Some Related Species*, Chem. Phys. Lett. **128**, 501 (1986).

94. K. Raghavachari, *Structural Isomers of $C_{70}$*, Mat. Res. Soc. Symp. Proc. **270**, 287 (1992).

95. W. C. Eckhoff and G. E. Scuseria, *A Theoretical Study of the $C_2$ Fragmentation Energy of $C_{60}$ and $C_{70}$*, Chem. Phys. Lett. **216**, 399 (1993).

96. P. W. Fowler and D. E. Manolopoulos, *Magic Numbers and Stable Structures for Fullerenes, Fullerides, and Fullerenium Ions*, Nature **355**, 428 (1992).

97. K. Raghavachari, *Electronic and Geometric Structure of $C_{72}$*, Z. Phys. D **26**, S261 (1993).

98. B. L. Zhang, C. Z. Wang, K. M. Ho, C. H. Xu, and C. T. Chan, *The Geometry of Large Fullerene Cages: $C_{72}$ to $C_{102}$*, J. Chem. Phys. **98**, 3095 (1993).

# Chapter 5

# $C_{76}$, $C_{78}$, $C_{82}$, and $C_{84}$: The Medium-Size Fullerenes

Extracts of the soot obtained by resistive heating or contact-arc vaporization of graphite are a rich source of not only the $C_{60}$ and $C_{70}$ clusters, but also medium-size fullerenes [1–7]. Among those species, which have also been detected in benzene–oxygen flames [8] and products of plasma vaporization of graphite [9], isomers of the $C_{76}$, $C_{78}$, $C_{82}$, and $C_{84}$ clusters are of most interest.

## 5.1 Isomerism

Unlike $C_{60}$ and $C_{70}$, the higher fullerenes have several isomers that satisfy the isolated pentagon rule (IPR) [10]. In particular, there are two IPR isomers of $C_{76}$ [11], five of $C_{78}$ [12], nine of $C_{82}$ [13], and 24 of $C_{84}$ [14]. Not all of these isomers, however, are observed experimentally.

The $C_{76}$ fullerene has one isomer with $T_d$ symmetry and one with $D_2$ symmetry (Fig. 5.1). However, the $T_d$ isomer, which has been considered in the early literature [15], has an open-shell electronic configuration with a triply degenerate, doubly occupied ($t_1^2$) HOMO [11, 16], and is therefore expected to undergo a Jahn–Teller distortion. On the other hand, the $D_2$ structure has a closed-shell configuration. Its $^{13}$C NMR spectrum is predicted to consist of 19 lines of identical intensity, which is exactly what is measured [4] for the $C_{76}$ species isolated from the soot extract [1, 2]. Interestingly, the $D_2$ symmetry implies chirality of the fullerene cage. The chirality has been confirmed experimentally by resolution (using asymmetric osmylation) of the $C_{76}$ cluster into enantiomers with very high specific rotation [17]. In light of these facts, the $D_2$ structure of the experimentally observed $C_{76}$ fullerene is well established.

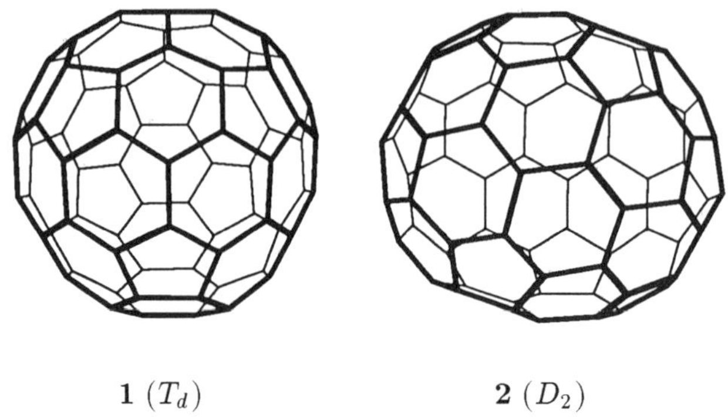

$\mathbf{1}\ (T_d)$  $\qquad\qquad\qquad$  $\mathbf{2}\ (D_2)$

Fig. 5.1.  The two IPR isomers of the $C_{76}$ fullerene.

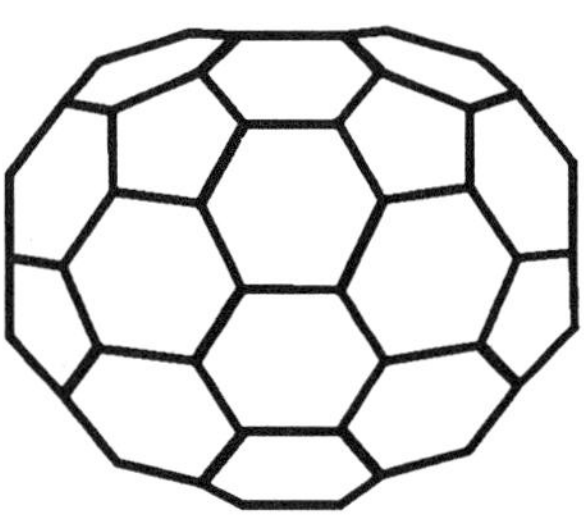

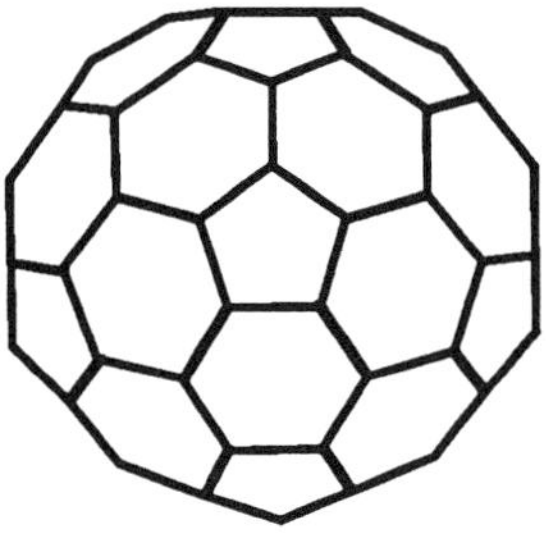

$\mathbf{1}\ (D_{3h})$  $\qquad\qquad\qquad$  $\mathbf{2}\ (D_{3h})$

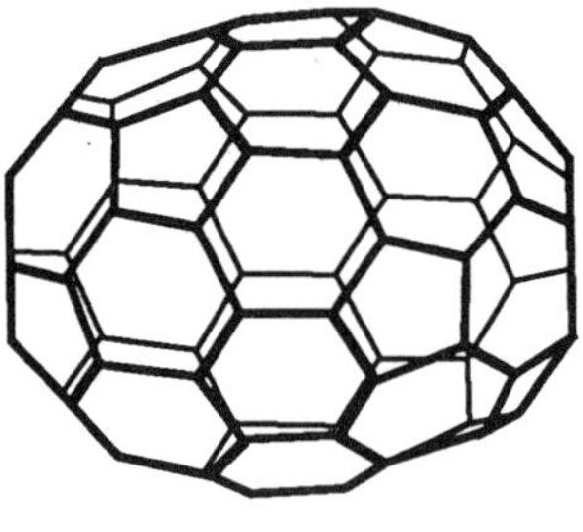

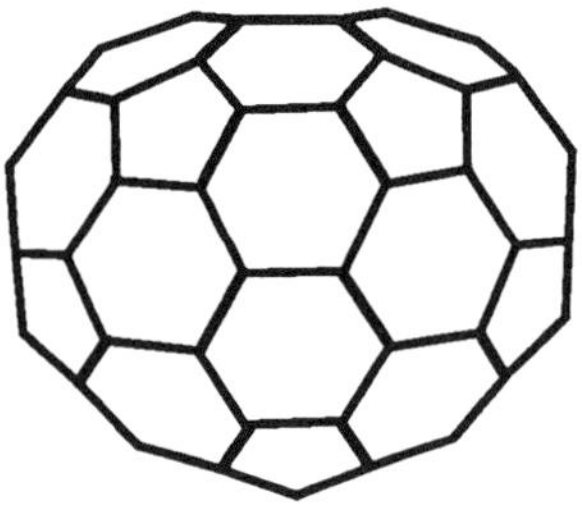

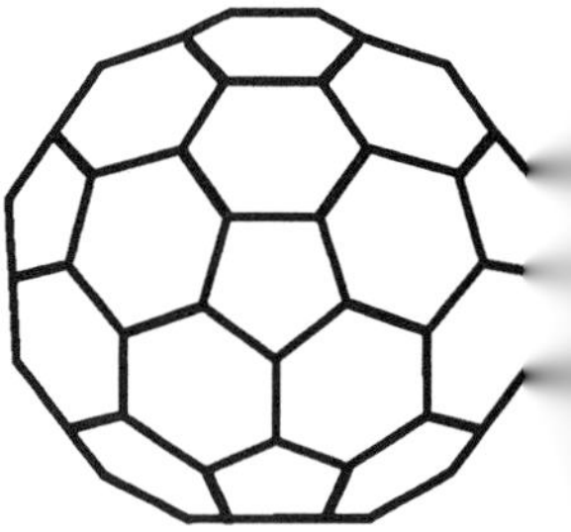

$\mathbf{3}\ (D_3)$  $\qquad\qquad$  $\mathbf{4}\ (C_{2v})$  $\qquad\qquad$  $\mathbf{5}\ (C_{2v})$

Fig. 5.2.  The five IPR isomers of the $C_{78}$ fullerene.

There are five IPR isomers of $C_{78}$ (Fig. 5.2). In addition to a structure with $D_3$ symmetry, there are two pairs of $D_{3h}$ and $C_{2v}$ isomers, which are all related [6] through a series of Stone–Wales transformations [18]. All five structures possess closed-shell electronic configurations. Because within the Hückel approximation only structure **1** has a nonbonding LUMO and a large HOMO–LUMO gap, it was originally designated as the isomer most likely to be observed experimentally [12]. However, the NMR spectra of the $C_{78}$ fullerene fraction isolated by HPLC from the soot extracts are in variance with such a designation. The 56 distinct lines observed in one of these spectra have been ascribed to a 5:2:2 mixture of structures **5**, **4**, and **3** [3]. A 5:1 mixture of **4** and **3** has also been described [6] and its constituents separated. The major component has been assigned the $C_{2v}$ structure **4** on the basis of its NMR spectrum consisting of 18 peaks of full intensity and three peaks of half intensity. With its 13 equal-intensity peaks, the spectrum of the minor component has been found consistent with the $D_3$ structure **3**. The remaining 22 peaks reported in [3] correspond to the $C_{2v}$ isomer **5**, which is predicted to give rise to 17 full-intensity and five half-intensity lines. However, in neither of these two experiments have the eight-line spectra of the $D_{3h}$ isomers been observed.

There are three IPR isomers of the $C_{82}$ fullerene with $C_2$ symmetry, three with $C_s$ symmetry, two with $C_{3v}$ symmetry, and one with $C_{2v}$ symmetry (Fig. 5.3). All of them have closed-shell electronic configurations. In contrast to $C_{76}$, $C_{78}$, and $C_{84}$, all nine isomers of $C_{82}$ can be interconverted [13] by a single series of Stone–Wales transformations [18]. The $C_2$ isomers are predicted to give rise to 41 lines of equal intensity in the NMR spectrum, whereas the spectra of the $C_s$ and $C_{2v}$ isomers are expected to consist of 38 strong and six weak, and of 17 strong and seven weak lines, respectively. Finally, the $C_{3v}$ isomers are predicted to afford 11 (or 12) full-intensity and five (or three) half-intensity lines, in addition to one line of one-sixth intensity. These predictions [13] can be compared with the measured spectrum of the $C_{82}$ fraction. The 41 strong and 29 moderate-intensity peaks can be accounted for by the presence of an 8:1:1 mixture of the $C_2$, $C_{2v}(\mathbf{9})$, and $C_{3v}(\mathbf{7})$ isomers [3]. Unfortunately, the information provided by the NMR spectrum alone is insufficient to distinguish between the $C_2$ isomers **1**, **3**, and **5**, although theoretical predictions (Section 5.2) strongly favor structure **3** as the predominant $C_{82}$ fullerene.

There are 24 isomers of $C_{84}$ that satisfy the IPR requirement (Fig. 5.4). These structures have molecular symmetries varying from $C_1$ to $T_d$ and closed-shell electronic configurations. The 24 isomers belong to two disjoint families [19], each of them consisting of species related by a series of Stone–Wales transformations [18]. The smaller family has structures **1**, **2**, and **5**, all of them chiral, as its members. Their chirality is preserved in the course of the transformations.

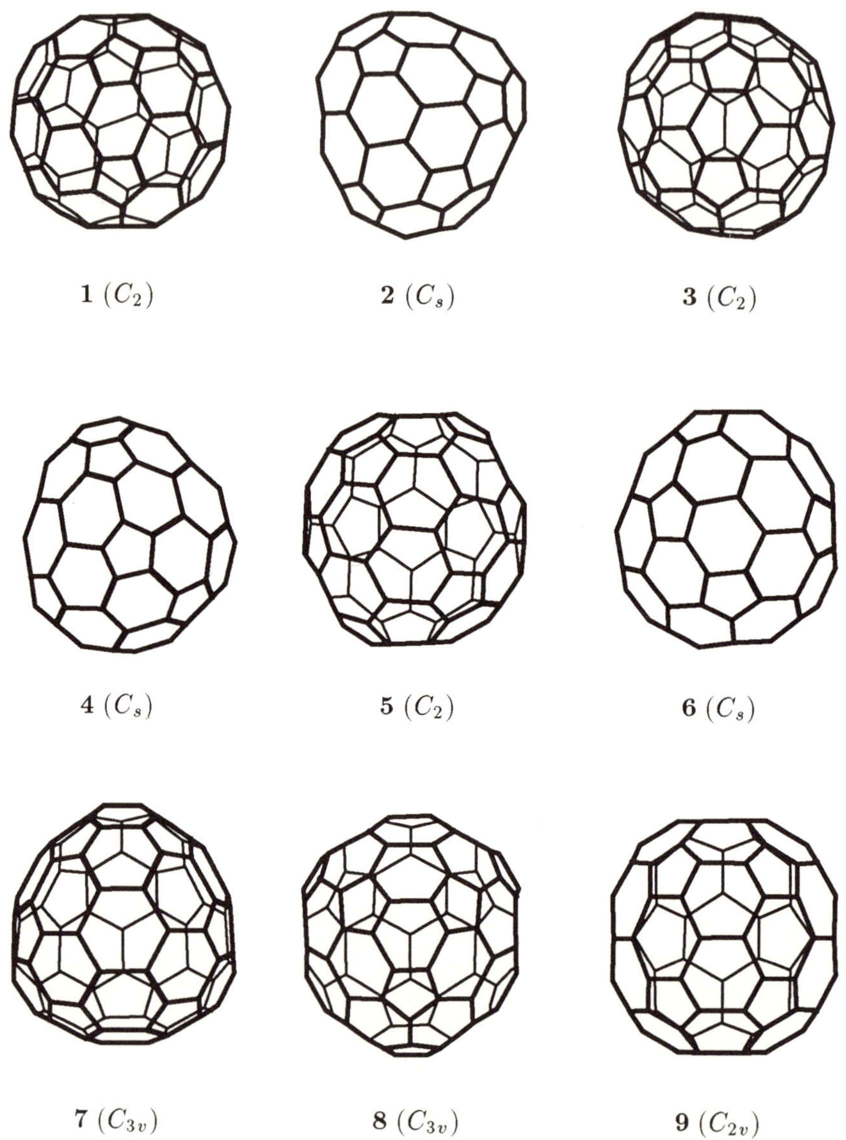

Fig. 5.3. The nine IPR isomers of the $C_{82}$ fullerene.

On the basis of their nonnegative Hückel LUMO energies, structures **1**, **20**, and **24** were originally thought [14, 20] to be the likely candidates for the experimentally observed fullerenes. However, NMR measurements contradict these predictions. The measured NMR spectrum of the $C_{84}$ fraction is far simpler than that of its $C_{82}$ counterpart. Only 31 full-intensity and one half-intensity lines are observed [3, 7], consistent with a 2:1 mixture of the $D_2$ and $D_{2d}$ isomers. There is strong evidence (Sections 5.2 and 5.6) that these isomers correspond to structures **22** and **23**, respectively.

Finally, it should be mentioned that some non-IPR isomers of the medium-size fullerenes have been considered in the chemical literature. Several non-IPR structures of the $C_{76}$ cluster that would give rise to the experimentally observed 19 NMR lines have been constructed [21]. A $C_{3v}$ structure of $C_{76}$, corresponding to a spectrum with 17 lines, has been described [5]. AM1 and MM3 calculations on non-IPR $D_7$, $D_{6h}$, $D_{3h}$, and $D_{2h}$ isomers of $C_{84}$ have also been reported [22]. A $C_i$ isomer of $C_{84}$ has been the subject of an MNDO study [23].

## 5.2  Relative Energies

The available experimental evidence demonstrates that, although the isolated pentagon rule (IPR) is never violated, approaches based solely on the $\pi$-electron Hückel approximation are of only very limited use when more detailed predictions of fullerene isomers formed in the process of graphite vaporization are sought. In order to reconcile the predictions concerning the $C_{78}$ cluster with the experimental facts, a hypothesis has been put forward [6] that the observed isomers are those of the lowest energy *within* each family of species related by the Stone–Wales transformations. Should this hypothesis be correct, one would expect the formation of only one $C_{82}$ isomer [13]. Moreover, two isomers of $C_{84}$ should be observed, one of them corresponding to structure **1**, **2**, or **5** [19]. These predictions are clearly wrong, emphasizing the need for more sophisticated approaches, preferably those based on *ab initio* calculations of relative energies.

In agreement with the expectations based on qualitative arguments [11], the $T_d$ isomer **1** of $C_{76}$ is found to undergo the first-order Jahn–Teller distortion to a $^1A_1$ state with $D_{2d}$ symmetry and the $a_2^2$ electronic configuration [16]. Despite the stabilization gained by this distortion, the $D_{2d}$ structure is found to lie 45.2 kcal/mol above the $D_2$ isomer **2** at the HF/STO-3G level of theory. This large energy difference is slightly reduced to 42.7 kcal/mol when single-point HF/dz calculations are carried out at the HF/STO-3G optimized geometries. The $D_2$ isomer, which has the $b_1^2$ configuration, is also predicted to be more stable by 5.3 and 24.2 kcal/mol by tight-binding Hamiltonian [24] and MM3 [25] calculations, respectively. From these results, it is obvious that the tight-binding Hamiltonian method drastically underestimates the energy difference in this case.

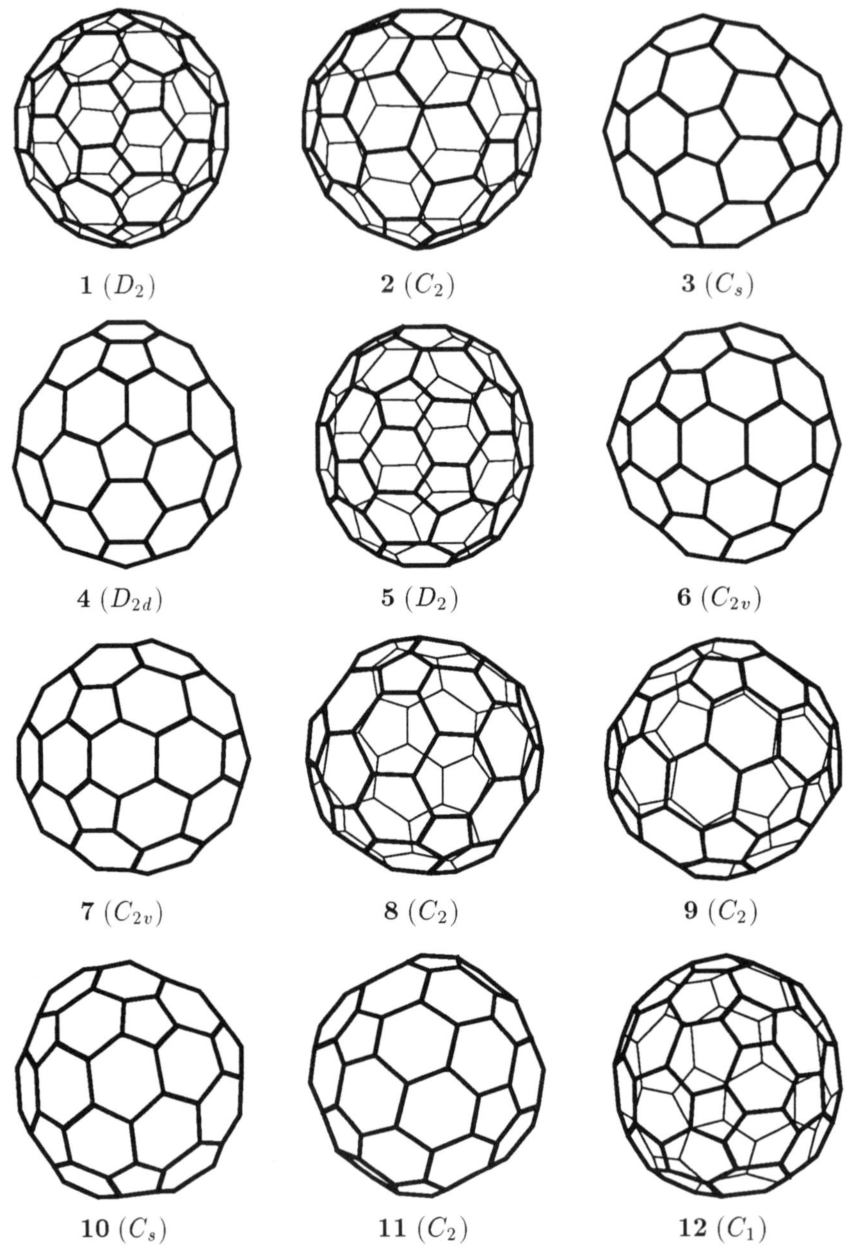


Fig. 5.4. The 24 IPR isomers of the $C_{84}$ fullerene.

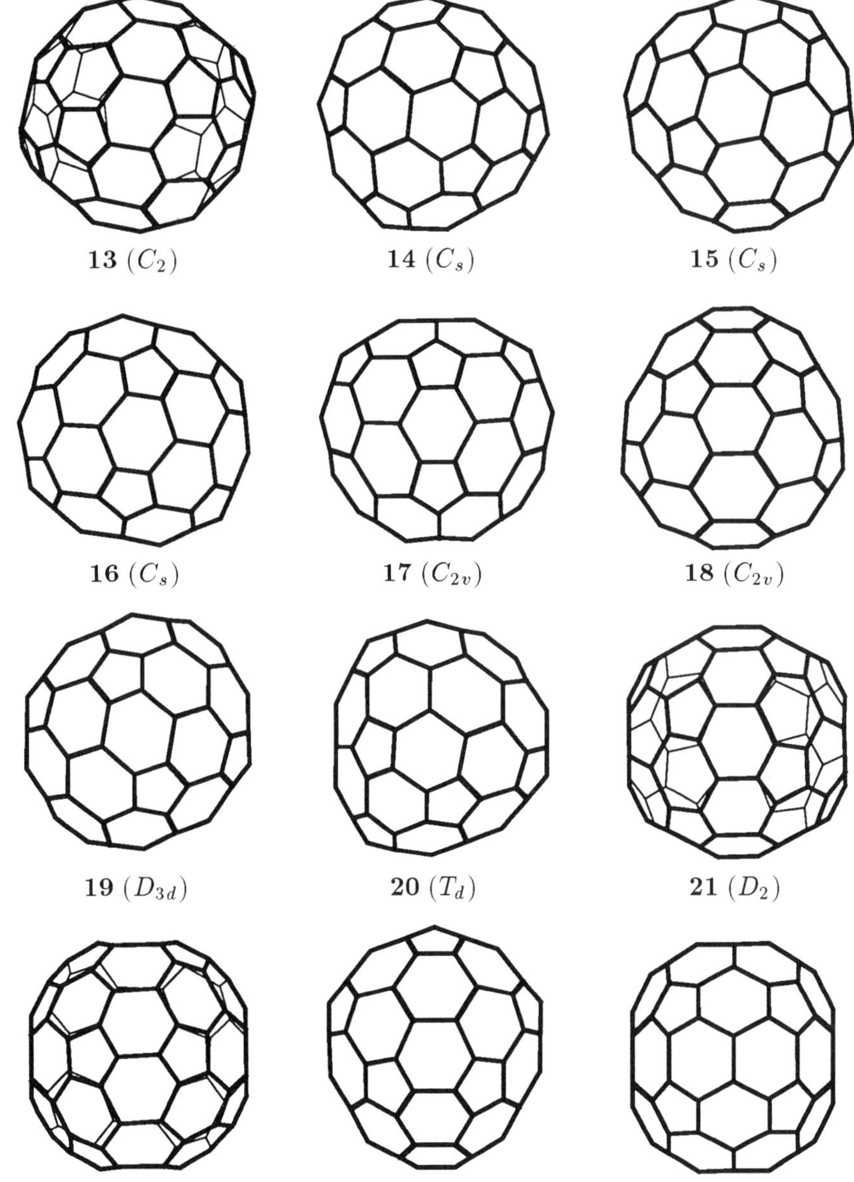

**13** $(C_2)$　　**14** $(C_s)$　　**15** $(C_s)$

**16** $(C_s)$　　**17** $(C_{2v})$　　**18** $(C_{2v})$

**19** $(D_{3d})$　　**20** $(T_d)$　　**21** $(D_2)$

**22** $(D_2)$　　**23** $(D_{2d})$　　**24** $(D_{6h})$

## Table 5.1. Relative heats of formation and relative energies [kcal/mol] of the IPR isomers of the $C_{78}$ cluster

| | Isomer[a] | | | | |
|---|---|---|---|---|---|
| Method | $D_{3h}$ (1) | $D_{3h}$ (2) | $D_3$ (3) | $C_{2v}$ (4) | $C_{2v}$ (5) |
| MM3 [25, 31] | 7.4 | 3.7 | 3.2 | 1.1 | 0.0 |
| MM3 [6] | 8.0 | 3.5 | 3.5 | 1.4 | 0.0 |
| Tight-binding Hamiltonian [24] | 21.1 | 2.0 | 7.5 | 6.5 | 0.0 |
| Curvature corrected HMO [26] | 9.9 | 2.5 | 8.0 | 2.7 | 0.0 |
| MNDO [26] | 23.4 | 1.6 | 10.5 | 7.1 | 0.0 |
| AM1 [26] | 21.3 | 5.1 | 7.1 | 5.3 | 0.0 |
| PM3 [26] | 17.5 | 5.6 | 6.3 | 3.8 | 0.0 |
| MNDOC(SCF) [26] | 20.5 | 3.8 | 9.2 | 5.4 | 0.0 |
| MNDOC(BWEN)//MNDOC(SCF) [26] | 23.4 | 3.7 | 9.5 | 6.5 | 0.0 |
| HF/STO-3G [25, 31] | 17.2 | 10.7 | 8.4 | 1.8 | 0.0 |
| HF/3-21G//MNDO [26] | 13.9 | 11.1 | 4.7 | 0.5 | 0.0 |
| HF/3-21G [29] | 12.8 | 11.1 | 4.1 | 0.2 | 0.0 |
| HF/dz//HF/STO-3G [31] | 16.4 | 9.8 | 6.0 | 1.6 | 0.0 |
| HF/6-31G*(5d)//HF/3-21G [29] | 20.2 | 7.2 | 3.4 | 4.0 | 0.0 |
| HF/DZP//MNDO [30] | n/a | n/a | n/a | 3.9 | 0.0 |
| LYP/DZP//MNDO [30] | n/a | n/a | n/a | 3.5 | 0.0 |
| LDA [27] | 25.7 | 3.4 | 7.7 | 4.5 | 0.0 |

[a] See Fig. 5.2.

A broad spectrum of theoretical approaches has been used to compute the relative energies of IPR isomers of the $C_{78}$ cluster. The results of these calculations are summarized in Table 5.1. Several observations are in order. First of all, contrary to the arguments based on the Hückel approximation [12], there is a consensus among all methods employed that the $C_{2v}$ isomer **5** is the most stable one, whereas the $D_{3h}$ isomer **1** is in fact the least stable of the five structures! However, the failure of the Hückel method can be averted [26] by correcting the resonance integrals $\beta$ for curvature [23]. Second,

## Table 5.2. Relative heats of formation and relative energies [kcal/mol] of the IPR isomers of the $C_{82}$ cluster

| Method | Isomer[a] | | | | | | | | |
|---|---|---|---|---|---|---|---|---|---|
| | $C_2$ | $C_s$ | $C_2$ | $C_s$ | $C_2$ | $C_s$ | $C_{3v}$ | $C_{3v}$ | $C_{2v}$ |
| | (1) | (2) | (3) | (4) | (5) | (6) | (7) | (8) | (9) |
| Tight-binding Hamiltonian [32] | 4.4 | 1.8 | 0.0 | 5.2 | 3.8 | 5.3 | 13.1 | 16.9 | 6.6 |
| QCFF/PI [33] | 5.7 | 6.1 | 0.0 | 3.8 | 6.9 | 9.7 | 25.1 | 23.7 | 13.0 |
| MNDO [34] | 5.3 | 6.7 | 0.0 | 4.3 | 7.9 | 10.6 | 25.7 | 27.0 | 14.2 |
| AM1 [35] | 4.0 | 6.5 | 0.0 | 6.0 | 12.0 | n/a | n/a | 31.0 | 21.8 |
| PM3 [33] | 4.1 | 5.2 | 0.0 | 6.4 | 11.8 | 16.6 | 29.4 | 35.2 | 22.1 |
| HF/3-21G//MNDO [34] | 5.3 | 4.2 | 0.0 | 8.8 | 17.6 | 26.6 | 39.5 | 54.7 | 36.0 |
| HF/6-31G*(5d)//MNDO [34] | 7.9 | 6.4 | 0.0 | 8.0 | 16.6 | 24.0 | 40.0 | 49.1 | 27.8 |
| LDA [36] | 9.6 | 2.8 | 0.0 | 8.9 | 7.2 | 10.8 | 23.6 | 29.7 | 15.2 |

[a] See Fig. 5.3.

the MM3 parameterization of molecular mechanics severely underestimates the energy differences [25, 31], while semiempirical approaches, including the tight-binding Hamiltonian approximation [24], the curvature-corrected HMO method [26], and the approaches of the MNDO family (except PM3) [26], tend to place the unobserved $D_{3h}$ isomer **2** as the second most stable structure. The same is true about the LDA method [27]. On the other hand, results of *ab initio* calculations clearly suggest the ordering of stabilities **5** > **4** > **3** or **5** > **4** ≈ **3**. The latter ordering, which is in perfect agreement with the experimental results [3, 28], is obtained when better basis sets are used [29], as the inclusion of polarization functions appears to be essential for obtaining accurate energy separation between isomers **4** and **5** [29, 30]. Third, electron correlation effects affect the computed relative energies only marginally, at least at the MNDOC/BWEN [26] and LYP/DZP [30] levels of theory. Moreover, only small energy changes are observed when MNDO optimized geometries are used in single-point *ab initio* calculations instead of their fully optimized counterparts. Finally, it should be mentioned that differences in the zero-point energies between the isomers are small, being estimated [31] with the MM3 method at $-0.3$, $-0.3$, $0.1$, and $-0.7$ kcal/mol for structures **4**, **3**, **2**, and **1**, respectively (values relative to isomer **5**).

The relative energies of the $C_{82}$ IPR isomers have been the subject of several investigations (Table 5.2). Despite some differences in the calculated

## Table 5.3. Relative heats of formation and relative energies [kcal/mol] of selected IPR isomers of the $C_{84}$ cluster

| | Isomer[a] | | | | |
|---|---|---|---|---|---|
| | $C_{2v}$ | $T_d$ | $D_2$ | $D_{2d}$ | $D_{6h}$ |
| Method | (**18**) | (**20**) | (**22**) | (**23**) | (**24**) |
| MM3 [39] | 6.4 | 13.6 | 0.3 | 0.0 | 2.4 |
| Tight-binding Hamilt. [37] | 14.0 | 27.2 | 0.8 | 0.0 | 7.1 |
| MNDO [39] | 15.6 | 31.3 | 0.4 | 0.0 | 8.0 |
| AM1 [39] | 15.2 | 30.4 | 0.4 | 0.0 | 8.1 |
| PM3 [39] | 12.9 | 25.8 | 0.4 | 0.0 | 6.8 |
| HF/sv7s4p [39] | 13.8 | 27.7 | 0.4 | 0.0 | 5.9 |
| HF/sv7s4p//MNDO [39] | 15.5 | 30.3 | 1.4 | 0.0 | 7.2 |
| HF/DZP//MNDO [30] | n/a | n/a | 0.9 | 0.0 | n/a |
| LYP/DZP//MNDO [30] | n/a | n/a | 0.5 | 0.0 | n/a |
| LDA [38] | n/a | n/a | −0.6 | 0.0 | n/a |

[a] See Fig. 5.4.

relative energies, all methods predict the $C_2$ isomer **3** to have the lowest energy, in agreement with the observed NMR spectrum of the $C_{82}$ fraction [3]. However, contrary to the published interpretation of the NMR spectrum, the other $C_2$ (**1** and **5**) and $C_s$ (**2** and **4**) structures are predicted to be much more stable than their $C_{2v}$ (**9**) and $C_{3v}$ (**7** and **8**) counterparts. Further doubts about the proposed NMR assignments are raised by the fact that the $C_{3v}$ structures have their symmetries lowered to $C_s$ upon geometry optimization [32, 34–36]. Similarly, the geometry of the $C_{2v}$ isomer **9** distorts to either $C_1$ [35] or $C_2$ [32, 34, 36] symmetry.

In the case of the $C_{84}$ cluster, the two lowest-energy isomers are found to be almost isoenergetic (Table 5.3). The $D_2$ structure **22** is calculated to have a slightly higher energy than the $D_{2d}$ structure **23** by all methods except LDA [38]. In addition, isomers **11**, **16**, **19**, and **24** are predicted [39] within the MNDO, AM1, and PM3 approaches to be almost isoenergetic and lie at least 7.0 kcal/mol above the **22/23** pair. This energy ordering is also confirmed by limited *ab initio* calculations [39]. Interestingly, the $D_{6h}$ isomer **24** (a "carbon cylinder") is one of the three structures originally considered [14, 20] to be the most stable on the basis of qualitative $\pi$-electron arguments. The other two

## Table 5.4.  Relative heats of formation and relative energies [kcal/mol] of selected high-energy IPR isomers of the $C_{84}$ cluster

| Method | Isomer[a] | | | |
| --- | --- | --- | --- | --- |
| | $D_2$ | $D_2$ | $T_d$ | $D_{6h}$ |
| | (**1**) | (**5**) | (**20**) | (**24**) |
| MM3 [39] | 21.0 | 9.9 | 11.2 | 0.0 |
| Tight-binding Hamiltonian [37] | 36.9 | 8.5 | 20.2 | 0.0 |
| QCFF/PI [46, 47] | 55.6 | n/a | 23.6 | 0.0 |
| MNDO [39, 48] | 50.1 | 15.5 | 23.3 | 0.0 |
| AM1 [39] | 43.6 | 12.8 | 22.3 | 0.0 |
| PM3 [39] | 39.4 | 12.4 | 19.0 | 0.0 |
| HF/3-21G [48] | n/a | n/a | 18.2 | 0.0 |

[a] See Fig. 5.4.

structures, $D_2$ (**1**) and $T_d$ (**20**) ("leapfrogs"), are very high in energy (Table 5.4). In fact, isomer **1** is the least stable of the 24 IPR structures and much less stable than some non-IPR structures of $C_{84}$ [39].

Among the members of the smaller family of $C_{84}$ isomers related by the Stone–Wales transformations (Section 5.1), the $D_2$ structure **5** has the lowest energy. This structure has also been proposed [40] as the likely candidate for the observed $D_2$ species [3, 7]. However, taking into account the relatively high energy of structure **5** (Table 5.4), it is clear that the $D_2$ species is in fact isomer **22**. This conclusion is further supported by the calculations on equilibrium compositions of the $C_{84}$ vapor [41] that predict the **22/23** ratio close to 2:1 within a wide range of temperatures (1000–5000 K), in perfect agreement with the observed abundances [3, 7].

There are some general trends observed in the aforedescribed relative energy calculations that are worth mentioning. The energies of fullerene isomers are determined by an intricate interplay between the $\pi$-electron conjugation effects and the strain introduced by the local curvature [29, 42]. As these two effects are mostly uncorrelated, approaches that invoke only $\pi$ electrons (such as the total $\pi$-electron energies or the $\pi$-electron HOMO–LUMO gaps) are completely useless for even qualitative predictions of the relative stabilities of fullerene isomers. However, the curvature-corrected HMO approach [23] affords qualitatively correct energy orderings for both $C_{78}$ [26] and $C_{84}$ [39] isomers. The most reliable estimates of relative stabilities are provided by Hartree–Fock *ab initio* calculations that include polarization functions. Such calculations can be performed at MNDO optimized geometries without adversely affecting the accuracy of the computed relative energies. The relative energies obtained from the methods of the MNDO family usually decrease in the order MNDO > AM1 > PM3 and are quite reliable, at least as far as the lowest-energy isomers are concerned. On the other hand, both MM3 and tight-binding Hamiltonian calculations often seriously underestimate the energy differences, although the former method usually yields correct stability ordering.

## 5.3  Heats of Formation

Total energies of fullerenes, calculated with reliable *ab initio* methods, can be readily converted into the standard enthalpies of fullerene interconversion reactions

$$\frac{N}{60}\, C_{60} \rightarrow C_N \tag{5.1}$$

by assuming negligible differences in the zero-point energies (ZPEs) and the entropic contributions. Justification for this assumption is based upon the fact that such contributions are proportional to the number of carbon atoms and therefore cancel to a large extent for the reactions, Eq. (5.1).

**Table 5.5.  Estimated standard enthalpies of the fullerene interconversion reactions, Eq. (5.1), and standard enthalpies of formation for selected medium-size fullerenes**

| | $\Delta H^0_r$ [Eq. (5.1)] | | $\Delta H^0_f$(gas) |
|---|---|---|---|
| | [kcal/mol] | | [kcal/mol] |
| Cluster | HF/DZP | LYP/DZP | LYP/DZP |
| $C_{76}$ ($D_2$ isomer **2**) | −82.9 | −88.4 | 671.6 |
| $C_{78}$ ($C_{2v}$ isomer **4**) | −96.7 | −103.5 | 676.5 |
| $C_{78}$ ($C_{2v}$ isomer **5**) | −100.6 | −107.0 | 673.0 |
| $C_{82}$ ($C_2$ isomer **3**) | −129.6 | −138.0 | 682.0 |
| $C_{84}$ ($D_2$ isomer **22**) | −157.5 | −166.9 | 673.1 |
| $C_{84}$ ($D_{2d}$ isomer **23**) | −158.4 | −167.4 | 672.6 |

Standard enthalpies computed according to the foregoing prescription within the HF/DZP and LYP/DZP levels of theory (at MNDO optimized geometries) [30] are listed in Table 5.5. The data clearly show that the higher fullerenes are significantly more stable than $C_{60}$ on a per-atom basis. Inclusion of electron correlation effects through the approximate LYP density functional additionally lowers the standard enthalpies of the interconversion reactions, Eq. (5.1), by 0.35–0.39 $(N - 60)$ kcal/mol.

Using the estimated value of 600.0 kcal/mol for $\Delta H_f^0(C_{60}, \text{gas})$ [30] makes it possible to compute the standard enthalpies of formation for higher fullerenes. The enthalpies listed in Table 5.5 are in qualitative agreement with the previously published HF/STO-3G and MM3 estimates [25], which are expected to be less reliable because of the lower levels of theory employed. The calculated heats of formation converge to a limit of 670–680 kcal/mol with the increasing cluster sizes. This limit is approximately equal to the strain energy acquired by the introduction of 12 five-membered rings into a graphite sheet.

## 5.4 Geometries and Vibrational Frequencies

Because most of the medium-size fullerenes possess rather low symmetries, optimization of their geometries often presents a formidable task. For example, the $D_2$ isomer **2** of the $C_{76}$ cluster has 57 internal degrees of freedom, 30 of them being the first-neighbor stretches [16]. Geometries of the $C_{78}$ isomers are characterized by between 21 and 60 independent parameters, including between 13 and 34 stretches [31]. Nevertheless, geometries of all isomers of $C_{78}$ [26], $C_{82}$ [33, 34], and $C_{84}$ [39] have been optimized within semiempirical methods of the MNDO family. The HF/STO-3G optimized geometries are available for all isomers of the $C_{76}$ [16] and $C_{78}$ [31] clusters. Geometries of the latter fullerenes have also been optimized at the HF/3-21G level of theory [29]. As demonstrated by the close agreement between the HF/STO-3G bond lengths in the $D_2$ isomer **2** of $C_{76}$ and their experimental counterparts determined from X-ray measurements on the $C_{76} \cdot (S_8)_6$ solid-state adduct [43], *ab initio* electronic structure calculations are capable of affording reasonably accurate geometries of higher fullerenes.

The varying shapes and sizes of the fullerene cages notwithstanding, some interesting trends emerge from the optimized geometries. In particular, despite substantial bond alternation (of the order of 0.1 Å), the average bond lengths are almost constant for all medium-size fullerenes. For example, at the MNDO level of theory the average of all bond lengths in the $C_{78}$ family of clusters is 1.4475 Å, with the variation between isomers amounting to only 0.0003 Å [26]. The analogous average for the $C_{84}$ clusters is 1.4471 Å, with a variation of 0.0005 Å [39]. There is a possible explanation for this unexpected regularity, which is also observed [26, 39] in the AM1 and PM3 optimized geometries: For each three bonds connected to a particular carbon atom, the sum of the Pauling bond orders is constant and equal to 1.0. If one assumes a linear relationship between the Pauling bond orders and the corresponding bond lengths, one concludes that the sum of bond lengths for each atom (and thus the average bond length) should also be constant.

In general, bond lengths calculated within the MNDO family of methods follow the order MNDO > AM1 > PM3 and possess less alternation than those optimized at the HF/STO-3G or HF/3-21G level. Taking into account the fact that geometries of fullerenes optimized within the Hartree–Fock approximation exhibit exaggerated bond alternation [44], the MNDO geometries are expected to be somewhat more accurate.

Because of their low symmetries, some of the fullerenes are expected to possess readily measurable dipole moments. In particular, the HF/DZP calculations (at the MNDO optimized geometries) [45] predict dipole moments of 0.16 and 0.43 D for the $C_{2v}$ isomers **4** and **5** of the $C_{78}$ cluster. For the $C_2$ isomer **3** of the $C_{82}$ species, the computed dipole moments are 0.36 [34] and 0.34 D [45] at the HF/3-21G//MNDO and HF/DZP//MNDO levels of theory, respectively. The former calculations predict the other isomers of the $C_{82}$ cluster to have dipole moments as large as 1.91 D. On the other hand, dipole moments calculated with the PM3 method are much smaller, amounting to only 0.03 D for the $C_2$ isomer **3** [33].

In general, the medium-size fullerenes with closed-shell electronic configurations retain their topological symmetries upon geometry optimization. However, at the MNDO, AM1, and PM3 levels of theory, the $C_{3v}$ (**7**) and $C_{2v}$ (**9**) isomers of $C_{82}$ have been reported [35] to deform to $C_s$ and $C_1$ symmetries, respectively, the deformation energies for the latter isomer being as large as 7.8 kcal/mol. Similar symmetry breaking has been observed for isomers **7**, **8**, and **9** in LDA geometry optimizations and attributed to the smallness of the HOMO–LUMO gaps that persists even after symmetry lowering [36]. Breaking of symmetry has also been reported [39] for the $C_s$ isomer **10** of the $C_{84}$ cluster, which has its symmetry lowered to $C_1$ at the MNDO, PM3, and MM3 levels of theory, but not at the AM1 level. All other isomers of the $C_{78}$ and $C_{84}$ clusters have been confirmed [26, 35] to be local minima by an MNDO vibrational analysis.

Several semiempirical calculations of vibrational frequencies in higher fullerenes have been published. The IR and Raman spectra of the $C_{76}$ cluster, which consist of 165 and 222 modes, respectively, have been investigated [46] with the help of the QCFF/PI method, allowing a partial assignment of the experimentally observed IR absorptions [4]. The IR spectra of all five isomers of the $C_{78}$ cluster have been calculated [26] with the MNDO method. There are 58, 58, 113, 172, and 173 IR-active modes for isomers **1–5**, respectively. The spectrum of the $D_{3h}$ isomer **1** has been found to possess a characteristic pair of strong bands in the vicinity of 600 cm$^{-1}$, which is missing in the spectra of the remaining four species. However, isomers **2–5** are all predicted to absorb strongly around 900 $cm^{-1}$, whereas structure **1** is not. QCFF/PI calculations of vibrational frequencies have also been carried out for all nine isomers of the $C_{82}$ fullerene [33]. Although the predicted IR spectra do not vary much from one structure to another, differences in absorptions within the 800–900 and 1400–1500 $cm^{-1}$ ranges have been proposed as the means for identification of individual isomers.

Motivated by the early (incorrect) predictions concerning stabilities of the $C_{84}$ species (Sections 5.1 and 5.2), several theoretical studies on vibrational frequencies of the $D_2$ (**1**), $T_d$ (**20**), and $D_{6h}$ (**24**) isomers have been carried out [47–50]. There are 183 IR-active and 246 Raman-active modes for isomer **1**, 31 and 62 modes for isomer **20**, and 31 and 53 modes for isomer **24**. Within the QCFF/PI approximation, both the IR and the Raman spectra of isomer **1** have been found [48] to span broader frequency regions than those of isomers **20** and **24** [47]. Results of MNDO calculations on IR spectra of all the 24 IPR isomers of the $C_{84}$ cluster have been published [39]. These calculations predict both of the most stable isomers **22** and **23** to absorb strongly around 1550 $cm^{-1}$.

## 5.5 Ionization Potentials, Electron Affinities, and Excited States

The large sizes of higher fullerenes presently limit calculations of their ionization potentials ($IPs$) and electron affinities ($EAs$) to those invoking either the $\Delta$SCF approximation or Koopmans' theorem (KT). The effects of electron correlation and orbital relaxation can be approximately accounted for by correcting the KT-based estimates according to the equations [45]

$$IP_{corr} = IP_{Koopmans} + IP(C_{60}) - IP_{Koopmans}(C_{60}), \qquad (5.2)$$

and

$$EA_{corr} = EA_{Koopmans} + EA(C_{60}) - EA_{Koopmans}(C_{60}). \qquad (5.3)$$

In Eqs. (5.2) and (5.3), $IP_{Koopmans}$ and $EA_{Koopmans}$ stand for the estimates obtained from Koopmans' theorem, whereas $IP(C_{60})$ and $EA(C_{60})$ refer to the experimental values for the $C_{60}$ cluster.

The corrected estimates for ionization potentials and electron affinities of selected medium-size fullerenes, calculated at the HF/6-31G and HF/DZP levels of theory (at MNDO optimized geometries) [45], are compiled in Table 5.6. The two sets of values agree within less than 0.1 eV. The ionization potentials are lower than those of either $C_{60}$ or $C_{70}$, whereas the electron affinities are consistently higher (and therefore always positive). In other words, the higher fullerenes are better electron donors as well as acceptors than either $C_{60}$ or

## Table 5.6. Estimated ionization potentials and electron affinities of selected medium-size fullerenes

| Cluster | IP [eV] | | EA [eV] | |
|---|---|---|---|---|
| | HF/6-31G | HF/DZP | HF/6-31G | HF/DZP |
| $C_{76}$ ($D_2$ isomer **2**) | 6.66 | 6.71 | 3.20 | 3.13 |
| $C_{78}$ ($C_{2v}$ isomer **4**) | 6.98 | 7.04 | 3.36 | 3.29 |
| $C_{78}$ ($C_{2v}$ isomer **5**) | 6.65 | 6.72 | 3.54 | 3.44 |
| $C_{82}$ ($C_2$ isomer **3**) | 6.56 | 6.61 | 3.55 | 3.49 |
| $C_{84}$ ($D_2$ isomer **22**) | 6.99 | 7.04 | 3.49 | 3.43 |
| $C_{84}$ ($D_{2d}$ isomer **23**) | 7.02 | 7.07 | 3.37 | 3.30 |

$C_{70}$. These trends can be explained by the fact that the larger fullerenes have a greater proportion of six-membered rings and hence possess electronic structures that are closer to that of graphite. One can also conclude from the data in Table 5.6 that the ionization potentials and electron affinities are sensitive to local curvatures of the fullerene cages, as the $IPs$ and $EAs$ of the two $C_{78}$ isomers differ by as much as 0.32 and 0.15 eV, respectively. These observations about trends in $IPs$ and $EAs$ are also borne out by the HOMO and LUMO orbital energies of various fullerenes calculated at the HF/STO-3G, HF/dz [16, 31], QCFF/PI [46, 48], and MINDO/3 [51] levels of theory.

Despite the fact that the higher fullerenes are better electron acceptors than $C_{60}$, they are less likely to form highly charged negative ions. This is so because, in contrast to $C_{60}$, they usually possess nondegenerate LUMOs. The monoanions of $C_{76}$ (the $D_2$ isomer **2**), $C_{78}$ (the $C_{2v}$ isomer **5**), $C_{82}$ (an unidentified $C_2$ isomer), and $C_{84}$ (the $D_2$ and $D_{2d}$ isomers **22** and **23**) have been studied [51] with the UHF variant of the semiempirical MINDO/3 method. Full geometry optimizations reveal that bond lengths in these fullerenes become more equalized upon electron attachment. As expected from the lack of degenerate LUMOs, a Jahn–Teller symmetry breaking is not observed, except for the $D_{2d}$ isomer **23** of the $C_{84}^-$ anion, which has its symmetry lowered to $C_2$.

An interesting case is presented by the high-energy $C_{3v}$ isomer **7** of the $C_{82}$ cluster, for which LDA calculations [36, 52] predict a HOMO–LUMO gap of only 0.1 eV. On the other hand, the LUMO–(LUMO+1) gap of 1.0 eV is calculated, suggesting a particularly stable $C_{82}^{2-}$ dianion. The predicted stability of $C_{82}^{2-}$ is in line with the conclusions based on Hückel calculations [53] carried out for this and other fullerene dianions.

Finally, it should be mentioned that the UV spectrum of the $D_2$ isomer **2** of the $C_{76}$ cluster has been investigated with the LDA [54] and QCFF/PI [46] methods. The spectrum predicted by the latter calculations commences with the low-energy transitions to the $^1B_3$, $^1B_2$, and $^1B_1$ excited states that are located at 1.73, 2.23, and 2.28 eV, respectively, in rough agreement with the experimental [4] excitation energies of 1.37, 1.63, and 1.73 eV.

## 5.6 $^{13}$C NMR Spectra

The $^{13}$C NMR spectroscopy has repeatedly proved itself to be an indispensable tool for the determination of structures of medium-size carbon clusters. Unfortunately, because of the small amounts of higher fullerenes available, measurements of their two-dimentional $^{13}$C NMR spectra are not presently feasible. This limits the usefulness of the NMR data to the determination of molecular symmetries, which are directly related to the number of peaks observed in the NMR spectra. Although the knowledge of molecular symmetry is often sufficient to rule out certain isomers, it provides only limited information in the case of larger clusters, such as $C_{84}$.

With the advent of fast RISC-based workstations and efficient computer implementations of the GIAO CPHF method, reliable *ab initio* electronic structure calculations of the NMR chemical shifts are now a reality. However, as discussed in Chapters 3 and 4, the shifts calculated within the Hartree–Fock approximation are not accurate enough to assign individual nuclei to the experimentally observed lines in the NMR spectra of carbon clusters.

Despite this lack of accuracy, critically important insights can be gained from the theoretical predictions of the NMR spectra of fullerenes. The case of the $C_{84}$ cluster well illustrates this point. As mentioned in Section 5.1, the experimental NMR spectrum of the $C_{84}$ fraction is consistent with a 2:1 mixture of the $D_2$ and $D_{2d}$ isomers. Among the 24 IPR structures of $C_{84}$, four (**1, 5, 21,** and **22**) possess $D_2$ symmetry. Although isomer **22** is consistently predicted to be the most stable one by a wide spectrum of theoretical methods (Section 5.2), arguments for structure **5** being the experimentally observed species have been put forward [40]. In order to resolve this controversy, GIAO HF/TZP calculations (at HF/sv7s4p optimized geometries) on the NMR chemical shifts in isomers **5, 22,** and **23** of $C_{84}$ have been carried out [55]. Although the predicted positions of NMR peaks do not match their experimental counterparts on a nucleus-by-nucleus basis, the calculated ranges of 132.8–142.1 ppm for **22** and 133.4–142.5 ppm for **23** agree very well with that of 133.8–144.6 ppm [3] observed for the $C_{84}$ fraction. At the same time, isomer **5** can be readily rejected on the basis of its computed range of 127.8–151.8 ppm.

Quite importantly, the aforementioned chemical shifts can in fact be reproduced with much less sophisticated basis sets. In particular, the GIAO HF/6-31G calculations (at MNDO optimized geometries) [56] yield the ranges of 131.0–142.2 ppm for the $D_2$ isomer **22** and 131.2–141.4 ppm for the $D_{2d}$ isomer **23**. The half-intensity peak in the spectrum of **23** is predicted to occur at 137.5 ppm with the TZP basis set [55] and at 138.9 ppm with the 6-31G basis set [56], whereas the experimental value is 139.8 ppm [3]. The overall features of the NMR spectrum of the $D_2$ isomer **2** of $C_{76}$ are also well reproduced, as shown by the comparison between the GIAO HF/6-31G computed range of 127.8–149.7 ppm and its experimental [57] counterpart of 129.6–150.0 ppm.

# References


1. F. Diederich, R. Ettl, Y. Rubin, R. L. Whetten, R. Beck, M. Alvarez, S. Anz, D. Sensharma, F. Wudl, K. C. Khemani, and A. Koch, *The Higher Fullerenes: Isolation and Characterization of $C_{76}$, $C_{84}$, $C_{90}$, $C_{94}$, and $C_{70}O$, an Oxide of $D_{5h}$-$C_{70}$*, Science **252**, 548 (1991).
2. K. Kikuchi, N. Nakahara, T. Wakabayashi, M. Honda, H. Matsumiya, T. Moriwaki, S. Suzuki, H. Shiromaru, K. Saito, K. Yamauchi, I. Ikemoto, and Y. Achiba, *Isolation and Identification of Fullerene Family: $C_{76}$, $C_{78}$, $C_{82}$, $C_{84}$, $C_{90}$, and $C_{96}$*, Chem. Phys. Lett. **188**, 177 (1992).
3. K. Kikuchi, N. Nakahara, T. Wakabayashi, S. Suzuki, H. Shiromaru, Y. Miyake, K. Saito, I. Ikemoto, M. Kainosho, and Y. Achiba, *NMR Characterization of Isomers of $C_{78}$, $C_{82}$ and $C_{84}$ Fullerenes*, Nature **357**, 142 (1992).
4. R. Ettl, I. Chao, F. Diederich, and R. L. Whetten, *Isolation of $C_{76}$, a Chiral ($D_2$) Allotrope of Carbon*, Nature **353**, 149 (1991).
5. D. Ben-Amotz, R. G. Cooks, L. Dejarme, J. C. Gunderson, S. H. Hoke II, B. Kahr, G. L. Payne, and J. M. Wood, *Occurrence and Fragmentation of High-Mass Fullerenes*, Chem. Phys. Lett. **183**, 149 (1991).
6. F. Diederich, R. L. Whetten, C. Thilgen, R. Ettl, I. Chao, and M. M. Alvarez, *Fullerene Isomerism: Isolation of $C_{2v}$-$C_{78}$ and $D_3$-$C_{78}$*, Science **254**, 1768 (1991).
7. D. E. Manolopoulos, P. W. Fowler, R. Taylor, H. W. Kroto, and D. R. M. Walton, *An End to the Search for the Ground State of $C_{84}$?*, J. Chem. Soc. Faraday Trans. **88**, 3117 (1992).
8. J. B. Howard, J. T. McKinnon, M. E. Johnson, Y. Makarovsky, and A. L. Lafleur, *Production of $C_{60}$ and $C_{70}$ Fullerenes in Benzene–Oxygen Flames*, J. Phys. Chem. **96**, 6657 (1992).
9. D. H. Parker, P. Wurz, K. Chatterjee, K. R. Lykke, J. E. Hunt, M. J. Pellin, J. C. Hemminger, D. M. Gruen, and L. M. Stock, *High-Yield Synthesis, Separation, and Mass-Spectrometric Characterization of Fullerenes $C_{60}$ to $C_{266}$*, J. Am. Chem. Soc. **113**, 7499 (1991).
10. X. Liu, T. G. Schmalz, and D. J. Klein, *Favorable Structures for Higher Fullerenes*, Chem. Phys. Lett. **188**, 550 (1992).

11. D. E. Manolopoulos, *Proposal of a Chiral Structure for the Fullerene $C_{76}$*, J. Chem. Soc. Faraday Trans. **87**, 2861 (1991).

12. P. W. Fowler, R. C. Batten, and D. E. Manolopoulos, *The Higher Fullerenes: A Candidate for the Structure of $C_{78}$*, J. Chem. Soc. Faraday Trans. **87**, 3103 (1991).

13. D. E. Manolopoulos, P. W. Fowler, and R. P. Ryan, *Hypothetical Isomerisations of $LaC_{82}$*, J. Chem. Soc. Faraday Trans. **88**, 1225 (1992).

14. D. E. Manolopoulos and P. W. Fowler, *Molecular Graphs, Point Groups, and Fullerenes*, J. Chem. Phys. **96**, 7603 (1992).

15. P. W. Fowler, J. E. Cremona, and J. I. Steer, *Systematics of Bonding in Non-icosahedral Carbon Clusters*, Theor. Chim. Acta **73**, 1 (1988).

16. J. R. Colt and G. E. Scuseria, *An ab Initio Study of the $C_{76}$ Fullerene Isomers*, J. Phys. Chem. **96**, 10265 (1992).

17. J. M. Hawkins and A. Meyer, *Optically Active Carbon: Kinetic Resolution of $C_{76}$ by Asymmetric Osmylation*, Science **260**, 1918 (1993).

18. A. J. Stone and D. J. Wales, *Theoretical Studies of Icosahedral $C_{60}$ and Some Related Species*, Chem. Phys. Lett. **128**, 501 (1986).

19. P. W. Fowler, D. E. Manolopoulos, and R. P. Ryan, *Stone–Wales Pyracylene Transformations of the Isomers of $C_{84}$*, J. Chem. Soc. Chem. Commun. 408 (1992).

20. P. W. Fowler, *Three Candidates for the Structure of $C_{84}$*, J. Chem. Soc. Faraday Trans. **87**, 1945 (1991).

21. O. Ori and M. D'Mello, *A Topological Study of the Structure of the $C_{76}$ Fullerene*, Chem. Phys. Lett. **197**, 49 (1992).

22. S. Tsuzuki and K. Tanabe, *Relative Stability of Nine Candidate Structures of $C_{84}$ Calculated with $AM1$ and $MM3$*, Chem. Phys. Lett. **195**, 352 (1992).

23. D. Bakowies and W. Thiel, *MNDO Study of Large Carbon Clusters*, J. Am. Chem. Soc. **113**, 3704 (1991).

24. B. L. Zhang, C. Z. Wang, and K. M. Ho, *Structures of Large Fullerenes: $C_{60}$ to $C_{94}$*, Chem. Phys. Lett. **193**, 225 (1992).

25. R. L. Murry, J. R. Colt, and G. E. Scuseria, *How Accurate Are Molecular Mechanics Predictions for Fullerenes? A Benchmark Comparison with Hartree–Fock Self-Consistent Field Results*, J. Phys. Chem. **97**, 4954 (1993).

26. D. Bakowies, A. Gelessus, and W. Thiel, *Quantum-Chemical Study of $C_{78}$ Fullerene Isomers*, Chem. Phys. Lett. **197**, 324 (1992).

27. X.-Q. Wang, C. Z. Wang, B. L. Zhang, and K. M. Ho, *Relative Stability of $C_{78}$ Isomers*, Chem. Phys. Lett. **200**, 35 (1992).

28. T. Wakabayashi, K. Kikuchi, S. Suzuki, H. Shiromaru, and Y. Achiba, *Pressure-Controlled Selective Isomer Formation of Fullerene $C_{78}$*, J. Phys. Chem. **98**, 3090 (1994).

29. K. Raghavachari and C. M. Rohlfing, *Isomers of $C_{78}$. Competition between Electronic and Steric Factors*, Chem. Phys. Lett. **208**, 436 (1993).

30. J. Cioslowski, *Heats of Formation of Higher Fullerenes from ab Initio Hartree-Fock and Correlation Energy Functional Calculations*, Chem. Phys. Lett. **216**, 389 (1993).

31. J. R. Colt and G. E. Scuseria, *An ab Initio Study of the $C_{78}$ Fullerene Isomers*, Chem. Phys. Lett. **199**, 505 (1992).

32. B. L. Zhang, C. Z. Wang, K. M. Ho, C. H. Xu, and C. T. Chan, *The Geometry of Large Fullerene Cages: $C_{72}$ to $C_{102}$*, J. Chem. Phys. **98**, 3095 (1993).

33. G. Orlandi, F. Zerbetto, and P. W. Fowler, *Infrared Fingerprints of Nine $C_{82}$ Isomers: A Semiempirical Prediction*, J. Phys. Chem. **97**, 13575 (1993).

34. K. Raghavachari, personal communication (1993).

35. S. Nagase, K. Kobayashi, T. Kato, and Y. Achiba, *A Theoretical Approach to $C_{82}$ and $LaC_{82}$*, Chem. Phys. Lett. **201**, 475 (1993).

36. X. Q. Wang, C. Z. Wang, B. L. Zhang, and K. M. Ho, *Electronic Structures of $C_{82}$ Fullerene Isomers*, Chem. Phys. Lett. **217**, 199 (1994).

37. B. L. Zhang, C. Z. Wang, and K. M. Ho, *Search for the Ground-State Structure of $C_{84}$*, J. Chem. Phys. **96**, 7183 (1992).

38. X.-Q. Wang, C. Z. Wang, B. L. Zhang, and K. M. Ho, *Structural and Electronic Properties of $C_{84}$: A First-Principles Study*, Phys. Rev. Lett. **69**, 69 (1992).

39. D. Bakowies, M. Kolb, W. Thiel, S. Richard, R. Ahlrichs, and M. M. Kappes, *Quantum-Chemical Study of $C_{84}$ Fullerene Isomers*, Chem. Phys. Lett. **200**, 411 (1992).

40. T. Wakabayashi, H. Shiromaru, K. Kikuchi, and Y. Achiba, *A Selective Isomer Growth of Fullerenes*, Chem. Phys. Lett. **201**, 470 (1993).

41. Z. Slanina, J.-P. François, M. Kolb, D. Bakowies, and W. Thiel, *Calculated Relative Stabilities of $C_{84}$*, Full. Sci. Tech. **1**, 221 (1993).

42. K. Raghavachari, *Ground State of $C_{84}$: Two Almost Isoenergetic Isomers*, Chem. Phys. Lett. **190**, 397 (1992).

43. M. M. Kappes, personal communication (1994).

44. M. Häser, J. Almlöf, and G. E. Scuseria, *The Equilibrium Geometry of $C_{60}$ as Predicted by Second-Order (MP2) Perturbation Theory*, Chem. Phys. Lett. **181**, 497 (1991).

45. J. Cioslowski and K. Raghavachari, *Electrostatic Potential, Polarization, Shielding, and Charge Transfer in Endohedral Complexes of the $C_{60}$, $C_{70}$, $C_{76}$, $C_{78}$, $C_{82}$, and $C_{84}$ Clusters*, J. Chem. Phys. **98**, 8734 (1993).

46. G. Orlandi, F. Zerbetto, P. W. Fowler, and D. E. Manolopoulos, *The Electronic Structure and Vibrational Frequencies of the Stable $C_{76}$ Isomer of $D_2$ Symmetry*, Chem. Phys. Lett. **208**, 441 (1993).

47. F. Negri, G. Orlandi, and F. Zerbetto, *QCFF/PI Vibrational Frequencies of Some Spherical Carbon Clusters*, J. Am. Chem. Soc. **113**, 6037 (1991).

48. F. Negri, G. Orlandi, and F. Zerbetto, *Prediction of the Structure and the Vibrational Frequencies of a $C_{84}$ Isomer of $D_2$ Symmetry*, Chem. Phys. Lett. **189**, 495 (1992).

49. K. Raghavachari and C. M. Rohlfing, *Structures and Vibrational Frequencies of $C_{60}$, $C_{70}$, and $C_{84}$*, J. Phys. Chem. **95**, 5768 (1991).

50. B. L. Zhang, C. Z. Wang, and K. M. Ho, *Vibrational Spectra of $C_{84}$ Isomers*, Phys. Rev. B **47**, 1643 (1993).

51. M. Okada, K. Okahara, K. Tanaka, and T. Yamabe, *Electronic Structures of Mono-anionized Higher Fullerenes. $C_{76}^-$, $C_{78}^-$, $C_{82}^-$ and $C_{84}^-$*, Chem. Phys. Lett. **209**, 91 (1993).

52. K. Laasonen, W. Andreoni, and M. Parrinello, *Structural and Electronic Properties of $La@C_{82}$*, Science **258**, 1916 (1992).

53. D. E. Manolopoulos and P. W. Fowler, *Structural Proposals for Endohedral Metal-Fullerene Complexes*, Chem. Phys. Lett. **187**, 1 (1991).

54. H.-P. Cheng and R. L. Whetten, *Electronic States and Structure of $D_2$ $C_{76}$*, Chem. Phys. Lett. **197**, 44 (1992).

55. U. Schneider, S. Richard, M. M. Kappes, and R. Ahlrichs, *Ab Initio $^{13}C$ NMR Shifts of Several $C_{84}$ Isomers*, Chem. Phys. Lett. **210**, 165 (1993).

56. J. Cioslowski, unpublished.

57. F. Diederich and R. L. Whetten, *Beyond $C_{60}$: The Higher Fullerenes*, Acc. Chem. Res. **25**, 119 (1992).

# Chapter 6

# Large Spheroidal and Tubular Fullerenes, Graphitic Microtubules, and Hypothetical Polymeric Allotropes of Carbon

The discovery of $C_{60}$ and related fullerenes rekindled interest in structures of various hypothetical allotropes of carbon that are composed of individual large clusters, linear polymers, two-dimensional networks, or three-dimensional lattices. In many instances, these allotropes have energies that are only slightly higher than those of the known forms of elemental carbon.

## 6.1  Large Spheroidal Fullerenes

The soot produced by resistive heating of graphite contains carbon clusters with a wide spectrum of sizes. In addition to the $C_{60}$, $C_{70}$, $C_{76}$, $C_{78}$, $C_{82}$, and $C_{84}$ species, fullerenes with between 90 and 330 atoms can be extracted from the soot with a variety of solvents. In particular, $C_{90}$ and $C_{96}$ fractions have been isolated from the $CS_2$ extracts [1], whereas giant fullerenes with up to 330 carbons have been observed in the material extracted with toluene under high pressure [2]. The yield of fullerenes is affected by the way the soot is prepared, with the plasma-discharge vaporization reported to produce solids that are up to 44% extractable with high-boiling-point solvents, such as 1,2,3,5-tetramethylbenzene [3]. Large clusters (presumably fullerenes) that possess up to 400 carbon atoms are also formed from higher carbon oxides, such as $C_{24}O_6$, $C_{32}O_8$, and $C_{40}O_{10}$, upon laser desorption [4].

The mechanism of formation of large fullerenes is presently unknown, although the coalescence of smaller cages that is observed experimentally in hot dense vapors of $C_{60}$ and $C_{70}$ [5] constitutes one obvious pathway. There is also some evidence [6] pointing to graphitic microtubules as possible precursors of fullerenes. In either case, preferential formation of a particular class of large fullerenes appears very unlikely. Thus, the prospects of isolating individual isomers of large fullerenes from the soot remain remote. Moreover, because of the explosive increase in the number of IPR structures that is already evident for the medium-size clusters, the use of reliable quantum-chemical methods in exhaustive searches for the lowest-energy isomers of fullerenes with more than about 100 atoms is not expected to become feasible in the near future. However, combined molecular mechanics, semiempirical, and *ab initio* studies, in which only a few species are selected with low-level calculations and then investigated with more sophisticated methods, show more promise, as exemplified by the recent successful identification of the lowest-energy isomers among 46 IPR structures of $C_{90}$ and 187 IPR structures of $C_{96}$ [7].

Enumeration of the isomers of large fullerenes poses potential difficulties that extend beyond the computational cost arising from the sheer number of possible structures. By cutting along selected edges, many fullerene cages can be mapped into at least one continuous spiral strip of edge-sharing five- and six-membered rings [8]. Such a mapping assigns a string of fives and sixes to a cage, greatly facilitating the generation, enumeration, classification, and identification of fullerene isomers. Unfortunately, the existence of cages (the smallest one having 380 atoms) that cannot be mapped into spirals has been proven [9]. However, a different enumeration scheme, based on the fact that an arbitrary fullerene can be uniquely specified by the positions of 12 pentagonal defects on a hexagonal honeycomb lattice [10], may provide a possible viable alternative to the spiral algorithm.

In light of these limitations, it is not surprising that for large fullerenes theoretical predictions of electronic properties remain mostly the domain of qualitative methods of quantum chemistry. Among those approaches, the $\pi$-electron Hückel Hamiltonian is often used because of its direct relationship to the topology of chemical bonds. Topological considerations afford general conclusions about many important characteristics of carbon clusters, such as their possible symmetries. It has been rigorously proven [11] that a fullerene cage must belong to one of the following 28 point groups: $I_h$, $I$, $T_d$, $T_h$, $T$, $D_{2h}$, $D_{3h}$, $D_{5h}$, $D_{6h}$, $D_{2d}$, $D_{3d}$, $D_{5d}$, $D_{6d}$, $D_2$, $D_3$, $D_5$, $D_6$, $S_4$, $S_6$, $C_{2h}$, $C_{3h}$, $C_{2v}$, $C_{3v}$, $C_2$, $C_3$, $C_s$, $C_i$, and $C_1$. Some of these symmetries are realizable only for quite large fullerenes. For example, one of the $C_{140}$ isomers is the smallest cluster belonging to the $I$ point group. Even more interestingly, it has been shown [11] that peaks in the $^{13}C$ NMR spectrum of any single fullerene isomer can have at most three different intensities. Some of the point groups ($C_1$, $C_i$, $C_2$, $S_4$, $D_2$, $D_5$, and $D_6$) admit only spectra in which all peaks have equal intensity.

Identification of infinite homologous series of stable fullerenes constitutes another important application of the Hückel method. One interesting series of this kind is formed by the icosahedral cages which have the number of vertices given by [12, 13]

$$v = 20 \left( h^2 + hk + k^2 \right) ,  \qquad (6.1)$$

where $0 \leq k \leq h$ are integers. Such clusters have $I_h$ symmetry when $k$ is equal to either 0 or $h$, otherwise they belong to the $I$ point group, thus lacking the center of symmetry [14]. The smallest members of the $I_h$ family, which includes the prototypical $C_{60}$ cluster, possess 20, 60, 80, 180, and 240 carbon atoms. The other family commences with the $C_{140}$ and $C_{260}$ fullerenes. An electron-counting rule for the icosahedral clusters has been formulated on the basis of a limited number of calculations within the Hückel approximation [12, 13]. It states that, if $h - k$ in Eq. (6.1) is divisible by 3, the cluster has a multiple of 60 atoms and a closed-shell electronic configuration, otherwise it has $60n + 20$ atoms and an open-shell configuration [13]. Such a prescription, which originates from the symmetry considerations independent of the approximations to the electronic Hamiltonian, leads to closed-shell icosahedral clusters with 60, 180, 240, 420, 540, ... carbon atoms.

Two members of the aforementioned series have been studied [15] with the LDA method. The icosahedral $C_{180}$ and $C_{240}$ fullerenes have been found more stable than $C_{60}$ on a per-atom basis. In continuation of the trends that are already apparent in the medium-size fullerenes (Section 5.5), the calculations predict ionization potentials that decrease and electron affinities that increase with the increasing cluster size. Both clusters have $h_u$ HOMOs, $t_{1u}$ LUMOs, and $t_{1g}$ LUMO+1s. These orbital symmetries are the same as in $C_{60}$.

Results of MNDO calculations on the icosahedral $C_{140}$, $C_{180}$, $C_{240}$, and $C_{540}$ fullerenes [16] corroborate the predictions based on the Hückel method. The $C_{140}$ molecule, which is expected to be open-shell, undergoes a Jahn–Teller distortion that lowers its symmetry all the way down to $C_1$. On the other hand, the closed-shell $C_{180}$, $C_{240}$, and $C_{540}$ clusters retain their $I_h$ symmetries and are local minima on the corresponding energy hypersurfaces, as evidenced by the calculated vibrational frequencies [16, 17]. Interestingly, the optimized geometry of $C_{540}$ corresponds to a shape that is icosahedral rather than spherical. Thus, the $C_{540}$ cluster can be described as a collection of graphite lattice fragments joined in a symmetrical fashion by 12 pentagons. The MNDO data also reproduce the trends in ionization potentials and electron affinities observed in the results of the aforediscussed LDA calculations.

Other homologous series of stable fullerenes have been described [14]. In fact, it can be shown [14, 18] that for all clusters with $60 + 6k$ ($k = 2, 3, \ldots$) atoms, there is at least one isomer with a closed-shell electronic configuration. The proof of this statement is based on the so-called leapfrog principle [13, 14, 18]. The leapfrog transformation, which is a very useful tool for constructing infinite series of closed-shell fullerenes, proceeds as follows (Fig. 6.1). Let **A** be a polyhedron composed of only five- and six-membered rings. First, additional vertices are placed at each of the ring centers of **A**. This produces an intermediate structure **B**. In the second step, each of the triangular centers in **B** becomes a vertex of a new polyhedron **C**. The resulting polyhedron has again only five- and six-membered rings, but the number of its vertices is three times that of **A**. The usefulness of the leapfrog transformation stems from the fact that the new polyhedron belongs to the same point group as the original one [13, 18]. Even more importantly, clusters with structures obtained through the leapfrog transformation are guaranteed to possess closed-shell electronic configurations [18]. The transformation can be iterated, affording infinite homologous series.

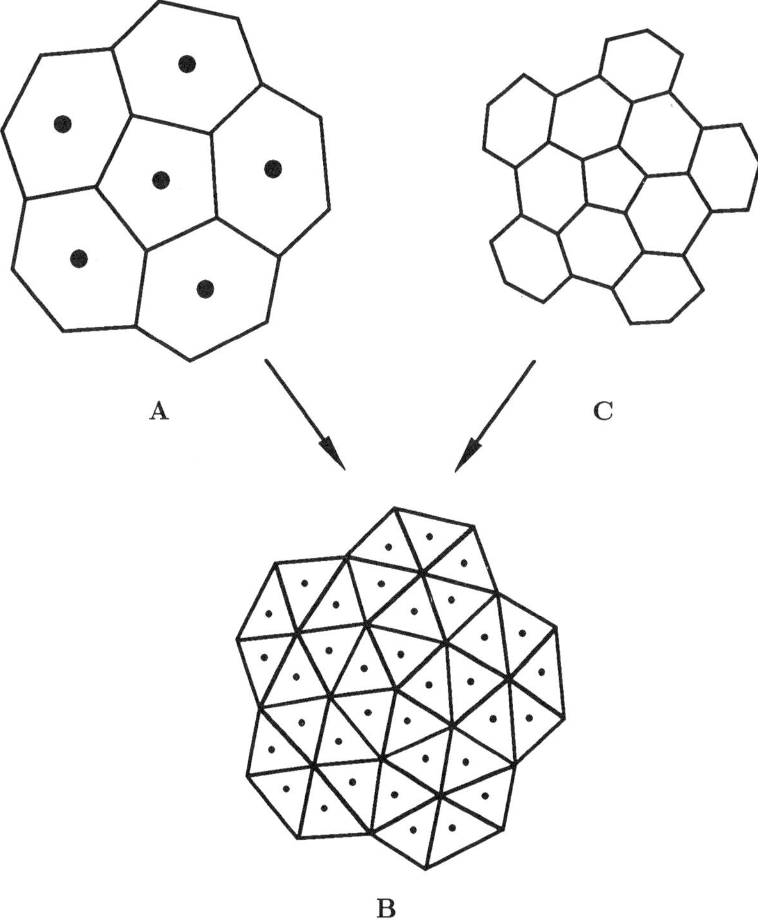

*Fig. 6.1.  The effect of the leapfrog*
*transformation on the corannulene fragment.*

There exist, of course, homologous series of stable fullerenes that are not related through the leapfrog transformation. Closed-shell "carbon cylinders" have $70+30k$ or $84+36k$ ($k = 0,\ 1,\ \ldots$) atoms and $D_{5h}$ or $D_{5d}$ symmetries [19]. Tetrahedral clusters are composed of

$$v = 4\left[h^2 + hk + k^2 + l^2 + lm + m^2 + 3(hm - kl)\right] \qquad (6.2)$$

atoms and have $T$, $T_d$, or $T_h$ symmetries [14]. In Eq. (6.2), $h, k, l$, and $m$ are integers satisfying the conditions $h > 0$, $k \geq 0$, and $hm - kl > 0$. Several other series of fullerenes have also been described [20].

An even greater variety of cage-like carbon clusters can be obtained from all of the aforementioned structures by replacing some or all of their C–C bonds by the C–C≡C–C chains. The resulting "fullereneynes" [21] are less stable on a per-atom basis than the parent fullerenes.

Because of the large number of possible structures, it is important to understand how the stabilities of large fullerenes are affected by their cage sizes and shapes. The available theoretical evidence [16, 22, 23] indicates that the heats of formation of higher spheroidal and icosahedral fullerenes are greater than that of $C_{60}$ (but lower on a per-atom basis) and approach a constant for large cage sizes. The limiting value of the heat of formation is estimated at 670–680 kcal/mol from LYP/DZP calculations on the most stable isomers of medium-size fullerenes [22]. For icosahedral fullerenes, the estimates of 965 and 850–960 kcal/mol are obtained from the energies computed within the tight-binding Hamiltonian [23] and MNDO approximations [16], respectively. The limit appears to be approached like $N^{-1}$ [23], where $N$ is the number of carbon atoms in a cluster. The alternative asymptotics of $\ln N$ [24], deduced from elasticity theory, is in variance with the available data.

The important issue of the shape dependence of energies has also been addressed. HF/STO-3G calculations show that spheroidal fullerenes are energetically preferable to long fullerene capsules with the same number of atoms. For example, the $C_{120}$ capsular fullerenes with $D_{6d}$ and $D_{5d}$ symmetries are less stable than their pear-shaped $T_d$ counterpart by 32 and 108 kcal/mol, respectively [25, 26]. The corresponding energy differences computed with the MM3 method amount to 20 and 59 kcal/mol [25], whereas those obtained within the tight-binding Hamiltonian approximation equal 21 and 61 kcal/mol [23]. The MNDO calculations place the $D_{5d}$ structure 99 kcal/mol above the $T_d$ one [16]. The energy differences become even more pronounced for larger fullerenes, with the capsular $D_{6h}$ $C_{180}$ and $D_{5d}$ $C_{240}$ fullerenes predicted at the HF/STO-3G level of theory to be less stable than their spheroidal $I_h$ counterparts by 208 and 750 kcal/mol, respectively [26]. The higher stability of spheroidal fullerenes can be directly attributed to the diminished strain, which is minimized when the five-membered rings are distributed over the cage as evenly as possible [16, 23]. It should also be mentioned here that the "archimedene" structure originally proposed [27] for the $C_{120}$ cluster possesses four-, six-, and ten-membered rings and is expected to be far less stable than any of the three structures discussed earlier. The recently proposed $C_{120}$ torus with $D_{5d}$ symmetry [28] and the $C_{120}$ species obtained by fusing two $C_{60}$ cages through four-membered rings or single bonds [29, 30] are also predicted to be relatively unstable.

The existence of concentric aggregates of spheroidal fullerenes has been proposed [31] on the basis of early photomicrographs showing onion-like, multilayered graphitic particles [32, 33]. MM3 calculations on such systems ("hyperfullerenes"), composed of the $C_{60}$, $C_{180}$, $C_{240}$, and $C_{540}$ icosahedral clusters, have yielded interesting results [34]. The formation of hyperfullerenes from their constituting carbon clusters is an exothermic process, provided the spacing between the fullerene cages is large enough. The calculated stabilization energies are substantial, amounting to as much as 412 kcal/mol for the hyperfullerene containing the $C_{60}$, $C_{240}$, and $C_{540}$ layers. In this particular hyperfullerene, the innermost $C_{60}$ shell can rotate almost freely, but the rotation of the middle $C_{240}$ shell is hindered by a barrier of *ca.* 30 kcal/mol. As expected, at the equilibrium geometry the angular orientations of the five-membered rings are matched among all three shells.

## 6.2 Large Tubular Fullerenes and Graphitic Microtubules

Although, as mentioned in Section 6.1, tubular (capsular) fullerenes are less stable than their spheroidal counterparts, experimental conditions can be devised in which their formation is preferred [35]. At present, graphitic microtubules can be produced in gram quantities by a direct-current arc vaporization of graphite under an atmosphere of helium [36]. The resulting material is highly crystalline and different from graphite, as revealed by electron diffraction [35] and Raman spectroscopy [37]. Its transmission electron microscope (TEM) images show the presence of structures consisting of concentric tubes that are up to 10 000 Å long [35]. The separation between the tubes amounts to *ca.* 3.4 Å, and their number varies between 2 and 50. The tubes are usually capped by polyhedral structures. There is experimental evidence [38] that the absence of caps is necessary for tube growth, suggesting a mechanism of formation in which carbon atoms are added to the open ends of the tubes. The alternative mode of growth by a direct insertion into a graphite network [39] is therefore less likely.

In order to close a tube, its terminating caps must contain a total of 12 five-membered rings. The observed morphologies of graphitic microtubules [6, 40] indicate the lack of preference for a particular distribution of these pentagons [41]. Most of the observed tips are asymmetric. Therefore, as the topology of the tips determines the arrangement of hexagons in the tube, most tubes are both helical and chiral [6, 35, 41].

Of particular interest are graphitic microtubules consisting of only one tube with a diameter commensurate with that of $C_{60}$. Such structures are occasionally observed in the TEM images [42]. These simple microtubules can be capped by two halves of the $C_{60}$ cluster. The equatorial cut that produces the halves can be normal to either the fivefold or the threefold axis of $C_{60}$ [43]. The former cut affords caps that are compatible with tubes consisting of hexagons with long axes perpendicular to the tube axis [Fig. 6.2($a$)]. The resulting $C_{60+10k}$ ($k = 1, 2, \ldots$) tubules are called armchair [43], serpentine [44], crenelated [45], or $T_\perp$ [23]. Conversely, the latter cut results in the $C_{60+18k}$ ($k = 1, 2, \ldots$) zigzag tubules [43] (also called sawtooth [44], or $T_\parallel$ [23]), in which the long axes of hexagons are parallel to the tube axis [Fig. 6.2($b$)]. The symmetry characteristics of the armchair and zigzag tubules have been studied [43].

The aforementioned two classes of tubules are special cases of more general structures. Classification of graphitic microtubules is facilitated by the observation [43, 44, 46–49] that any tubule is uniquely defined by the vector connecting two centers of hexagons on a single graphite layer (graphene) lattice (Fig. 6.3). The infinite tube is constructed by rolling the graphene sheet in such a way that the end of the vector (point $B$) is superimposed on its origin (point $A$). The vector, which becomes a circumference of the tube upon rolling, can be conveniently represented by a pair of integers $n_1$ and $n_2$ such that

$$\overrightarrow{AB} = n_1\vec{R}_1 + n_2\vec{R}_2 \ , \tag{6.3}$$

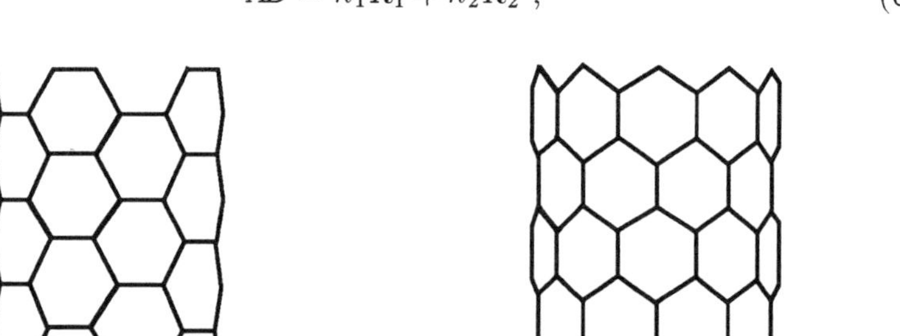

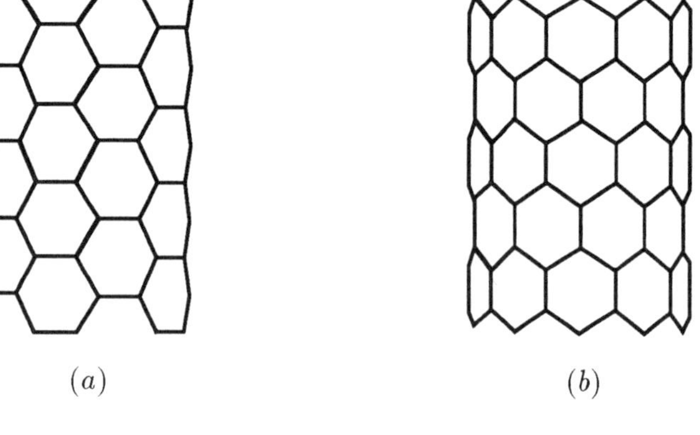

($a$)                                          ($b$)

*Fig. 6.2. The building blocks of the (a) armchair and (b) zigzag graphitic microtubules.*

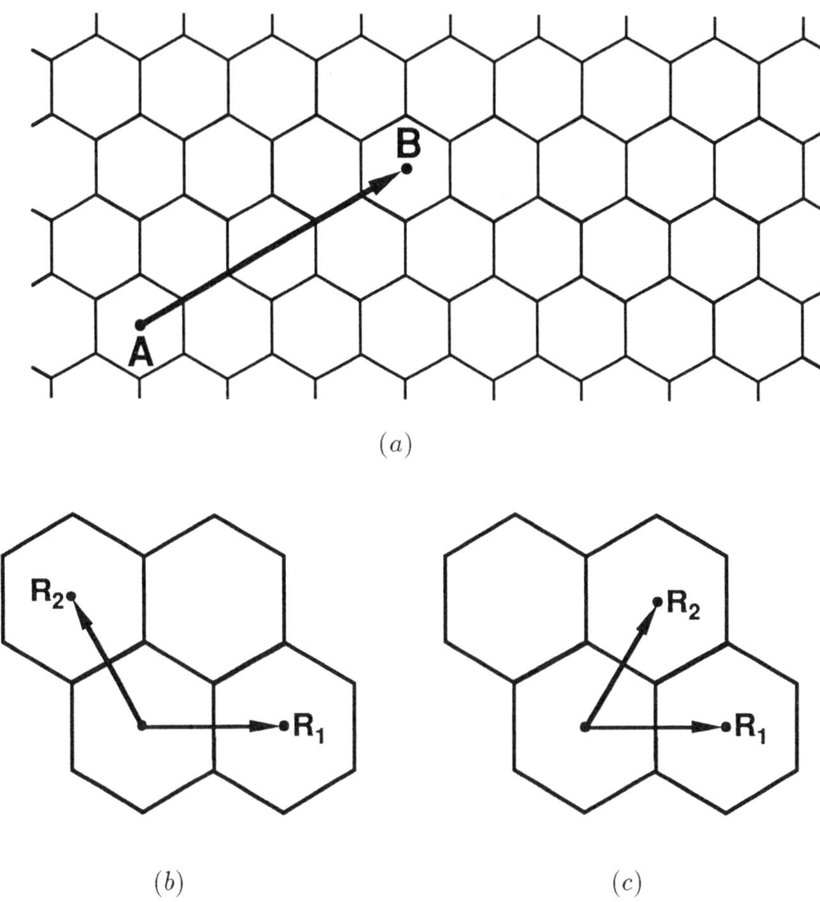



Fig. 6.3. (a) The graphene lattice with the vector $\overrightarrow{AB}$ and the pairs of unit vectors from (b) [46] and (c) [44].

where $\vec{R}_1$ and $\vec{R}_2$ are two unit vectors. Two choices of the unit vectors have been proposed (Fig. 6.3), leading to the notations $A(n_1, n_2)$ $(n_1 \geq 2n_2 \geq 0)$ [46] and $[n_1, n_2]$ $(n_1 \geq n_2 \geq 0)$ [44, 47], respectively. These two notations can be easily interconverted by observing that $A(n_1, n_2) \equiv [n_1 - n_2, n_2]$. In particular, the armchair tubules are denoted by $A(2n, n) \equiv [n, n]$, whereas the zigzag tubules correspond to the $A(n, 0) \equiv [n, 0]$ structures. It should be noted that tubules other than armchair or zigzag are chiral.

As first demonstrated in [46], electronic properties of open-ended graphitic microtubules are easily deduced from the band structure of the underlying graphene sheet. While the boundary conditions along the tube axis are the same as in graphene, the presence of finite periodicity along the circumference selects Bloch functions with particular reciprocal vectors. A rule, which states that a tubule $A(n_1, n_2)$ has a vanishing gap if $n_1 - 2n_2 = 0$, a narrow gap if $n_1 - 2n_2 = 3k$ ($k = 1, 2, \ldots$), and a moderate gap otherwise [46], can be derived by carefully examining the symmetry of the first Brillouin zone of the graphene sheet and assuming that its band structure is only weakly perturbed by the curvature of the tube. Thus, all armchair tubules are predicted to be metallic. The same conclusion has been reached by other researchers [48–50].

More detailed consideration of the helical and rotational symmetries of open-ended graphitic microtubules results in the prediction that band gaps of the $A(n_1, n_2)$ tubules with $n_1 - 2n_2 \neq 3k$ are accurately approximated by the expression [47]

$$\Delta = \frac{d_0}{R} |\beta| , \qquad (6.4)$$

where $\beta$ is the curvature-corrected resonance integral between the $p_z$ orbitals of carbon, $d_0$ is the C–C distance along the tubule surface, and $R$ is the tubule radius. By considering the magnitudes of $\beta$ and $d_0$, one concludes that such tubules do not have band gaps smaller than the ambient thermal energy ($kT$) for diameters smaller than *ca.* 300 Å.

The prediction that all armchair tubules are metallic does not address the possibility of Peierls distortions that would open the gap and result in semiconducting rather than metallic character [51]. LDA calculations carried out for the $A(10, 5)$ tubule confirm the presence of a crossing between the $a_1$ and $a_2$ bands at the Fermi level and the vanishing band gap [52]. Moreover, they yield *ca.* 1 K as the estimate of the transition temperature above which the metallic state exists. Thus, the $A(10, 5)$ tubule is predicted to be metallic at room temperature. Even more interestingly, the calculations find the density of carriers close to that of metals. This means that, unlike graphite (which is not a good conductor because of the low density of carriers), a material composed of such tubules should exhibit high electrical conductivity.

Simple considerations based on elasticity theory [53] predict the strain energy per carbon atom in graphitic microtubules to be inversely proportional to the square of the tubule radius $R$. The results of LDA [44], molecular mechanics [44, 54], and tight-binding Hamiltonian [23] calculations are in agreement with this prediction. As mentioned in Section 6.1, capped tubules are less stable than the spheroidal fullerenes with the same number of atoms. However, as the strain is concentrated mostly within the caps, open-ended tubules actually have energies lower on a per-atom basis than both spheroidal fullerenes of the same radius [44] and single graphite sheets of width equal to the tube circumference [54]. The proportionality constants between the strain energy and $R^{-2}$ are almost identical for the armchair and zigzag tubules [23, 44].

TEM images showing microtubules of different diameters, connected through conical segments, have been published [55]. The joints connecting tubular and conical segments or tubules at an angle possess pairs of five- and seven-membered rings. Two such joints have been studied in detail [45]. Because the joints introduce a 30° kink, 12 of them are sufficient to form toroidal structures with 540 or 576 carbon atoms. LDA calculations show that such tori have binding energies comparable on a per-atom basis to that of the $C_{60}$ cluster. Other joints of this type ("elbows"), built from rings of various sizes, and the resulting $C_{340}$ and $C_{440}$ tori have also been described [56]. In addition, L-, Y-, T-, and X-shaped tubule connectors, related to the so-called schwarzons (Section 6.4), have been discussed [57].

Like the $C_{60}$ cluster, the graphitic microtubules are rich in $\pi$ electrons. Therefore, they are expected to be highly polarizable. This high polarizability means that polar molecules can be trapped inside the tubules by capillarity-like forces originating from the induced dipole–dipole interactions. LDA calculations, in which an $A(12,6)$-type tubule consisting of 120 carbon atoms and terminated with 24 hydrogens at both ends has been investigated [58], confirm these expectations. The tubule is 12.8 Å long and has a diameter of 8.2 Å. When two hydrogen fluoride molecules are placed inside the tubule, they are found to interact strongly, with a stabilization energy as large as 5 kcal/mol. This "nanocapillarity" has been observed in experiments in which liquid lead is seen filling graphitic microtubules by capillary suction [59]. Finally, it should be mentioned that the large polarizabilities of graphitic microtubules are expected to give rise to relatively strong van der Waals interactions between individual tubes. LDA calculations, carried out for concentric armchair tubules, show that the barriers to rotation and translation arising from such interactions are low [60]. Experimental evidence for radial deformations of tubules due to these interactions has been recently reported [61].

## 6.3 Hypothetical Polymeric Allotropes of Carbon

Carbon clusters form molecular crystals in which individual fullerene molecules are held together by van der Waals interactions. In addition to these solids, which are discussed in Chapter 10, several hypothetical polymeric allotropes consisting of one-, two-, and three-dimensional carbon networks have been described in the literature. The properties of some of these substances and the possible routes to their synthesis have been reviewed in recent articles [62, 63].

There have been some speculative reports on the "carbyne" or "chaoite" form of carbon, in which long chains of $=C=C=$ or $-C\equiv C-$ units are reportedly present [62, 64]. The carbyne is supposed to be stable between 2600 and 3800 K, and at pressures below 6 GPa. At present, these reports still await definitive confirmation. Hypothetical allotropes of carbons consisting of two-dimensional lattices have been the subject of recent theoretical studies

[62, 65, 66]. In particular, properties of several new forms of carbon called graphynes have been calculated with the MNDO method [66]. The graphynes, which are composed of benzene rings and $-C\equiv C-$ units, can be obtained by replacing some or all of the C–C bonds in graphite by $C-C\equiv C-C$ chains (compare the relationship between fullerenes and fullereneynes; Section 6.1). They are predicted to be less stable than graphite by 15–26 kcal/mol of C and refractory to thermal graphitization. A representative member of the graphyne family, 6,6,6-graphyne (Fig. 6.4), is expected to crystallize in the *p6m* space group (with the in-plane unit cell dimensions: $a = b = 6.86$ Å) and be a semiconductor with a band gap of 1.2 eV. The heat of formation of 6,6,6-graphyne is estimated at 14.9 kcal/mol of C, which is much lower than that of the $(-C\equiv C-)_\infty$ carbyne (25.4 kcal/mol of C).

The accuracy of the MNDO method used in the aforementioned calculations has been assessed [67] by comparing the calculated heats of formation of graphite and diamond with the experimental data. The computed difference between the standard enthalpies of diamond and graphite has been found too high by *ca.* 6 kcal/mol of C. One possible source of this discrepancy is the difference in correlation energies of these two forms of carbon, estimated at *ca.* 4 kcal/mol of C [68]. Interestingly, the MNDO method predicts several hypothetical forms of carbon, composed of triptycyl moieties and containing both $sp^2$ and $sp^3$ carbon atoms, to be more stable than diamond [67].

There are at least two allotropic forms of the $sp^3$-hybridized carbon, namely the cubic and the hexagonal diamond, that are well characterized experimentally. The cubic modification is the constituent of the common diamond. The hexagonal form, called lonsdaleite, has been synthesized from

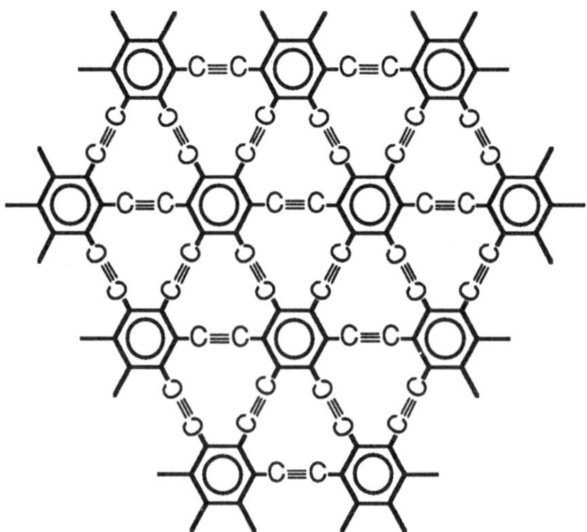

*Fig. 6.4.   The 6,6,6-graphyne lattice.*

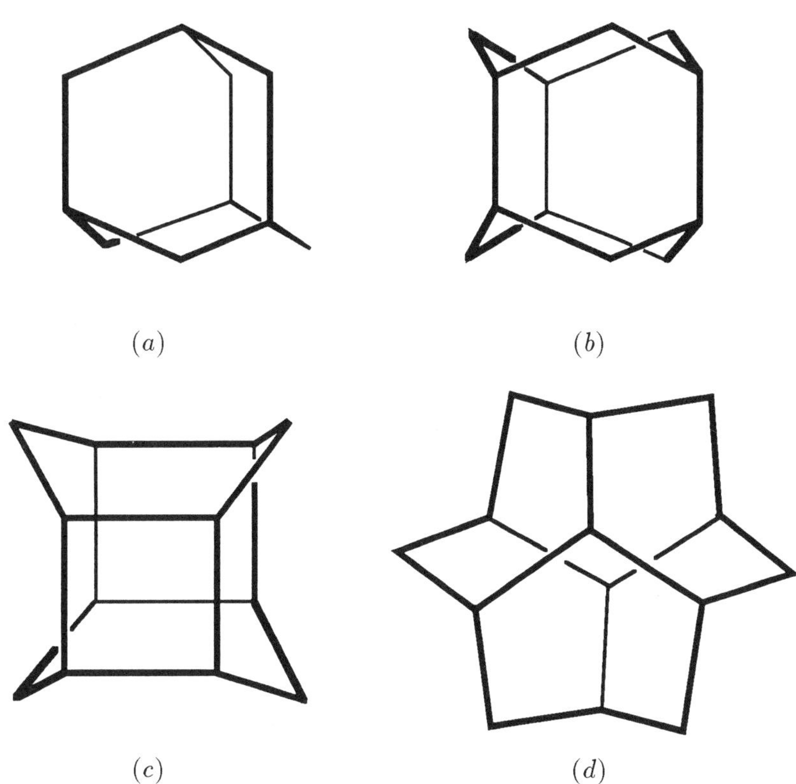

Fig. 6.5. The building blocks of some $sp^3$-hybridized
carbon allotropes: (a) adamantane, (b) iceane,
(c) pentacyclo[3.3.3.1$^{2,11}$.1$^{4,8}$.1$^{6,9}$]tetradecane, (d) tetraasterane.

graphite by applying high temperature and pressure [69]. Both modifications possess infinite, three-dimensional networks of condensed cyclohexane rings. However, all rings in the cubic diamond are in the chair conformation, whereas in the hexagonal diamond, rings in both chair and boat conformations are present. In other words, the cubic diamond consists of carbon networks derived from adamantane [Fig. 6.5(a)], whereas the lattice of the hexagonal diamond employs iceane [Fig. 6.5(b)] as a building block [70].

At atmospheric pressure, lonsdaleite is a metastable form of diamond. The MM2 method predicts lonsdaleite to be less stable than cubic diamond by 0.48 kcal/mol of C [70], in good agreement with LDA calculations that yield the energy difference of 0.69 kcal/mol of C [71]. The MM2 calculations involve extrapolation from small clusters to the bulk limit. Properties of some of these clusters have also been computed with several semiempirical and *ab initio* approaches [72]. The extrapolations yield bond lengths in both forms of diamond equal to 1.537 Å, which compares well with the experimental value of 1.545 Å for the cubic modification. The calculated density equals 3.57 g/cm$^3$.

There have been reports [73] of the "$C_8$" or "superdense" form of carbon. The originally proposed structure composed of cubane units has been found incorrect and a lattice, which not only is consistent with all experimental data but also has lower energy, has been suggested [74]. As in the case of the two forms of diamond, this BC-8 or $\gamma$-Si lattice consists of condensed cyclohexane rings. However, all of these rings are in the twist-boat conformation. With the building blocks of pentacyclo[3.3.3.1$^{2,11}$.1$^{4,8}$.1$^{6,9}$]tetradecane [Fig. 6.5($c$)], the resulting structure is both denser and less stable than those of diamond. The energy difference between the BC-8 carbon and cubic diamond amounts to 11.2 [71] and 15.9 kcal/mol of C [70] at the MM2 and LDA levels of theory, respectively. According to the results of the MM2 calculations, each carbon atom in the BC-8 lattice is connected to its neighbors through one shorter (1.49 Å) and three longer (1.60 Å) bonds. The bond angles of 100.0° and 117.2° deviate significantly from the ideal value of 109.47°. The calculated density of 3.68 g/cm$^3$ means that the BC-8 carbon is indeed denser than diamond.

LDA calculations predict that BC-8 carbon becomes more stable than diamond under pressures greater than *ca.* 1100 GPa [71]. Other, even denser forms of carbon, in which each atom is sixfold coordinated, have energies as much as 58 kcal/mol of C higher than diamond and are expected to be stable only under extremely high pressures.

X-ray absorption spectroscopy (XAS) provides evidence that "fullerene black", which is the insoluble fraction of the fullerene soot, differs from conventional soot [75]. It is possible that fullerene black is composed of three-dimensional carbon lattices derived from fullerene cages. Much interest in such lattices has been engendered recently by reports on polymerization of the $C_{60}$ cluster that is brought about by either high pressure or irradiation with UV light. Pressure-induced changes in Raman spectra of solid $C_{60}$ have been observed [76]. The experimental data indicate that $C_{60}$ transforms reversibly to a semitransparent phase at pressures between 15 and 25 GPa, but the transformation becomes irreversible above 27 GPa. The transformed solid is metastable and diamond-like at ambient conditions. Its Raman spectrum reveals the presence of $sp^3$-hybridized carbon atoms and indicates covalent bonding between the $C_{60}$ cages in the new high-pressure phase. Similar conclusions have been reached from experiments in which optical absorption of solid $C_{60}$ has been monitored [77]. The depressurized samples have shown no traces of $C_{60}$, diamond, or graphite. It should be mentioned, however, that transformation of $C_{60}$ to amorphous carbon [78] and diamond [79] has been observed under high pressures, contradicting the aforementioned results.

Crosslinking of $C_{60}$ has also been observed upon irradiation with UV light [80–83]. Mass spectra of the irradiated material show the presence of $C_{60k}$ clusters, with $k$ as high as 20 [81]. A low-frequency peak, indicative of the intercage mode, appears in the Raman spectrum upon irradiation, and a contraction of the $C_{60}$ lattice is evidenced by the changes in the X-ray diffraction pattern [80]. The experimental data are consistent with polymerization of $C_{60}$ through a series of $[2+2]$ cycloadditions involving most probably the "double" bonds (Section 3.1) of the clusters [82]. Recent MNDO, HF/dz, BLYP/dz [29], AM1, PM3, and LDA calculations [30] have shown that the $C_{120}$ cycloaddition products lie within 50 kcal/mol of two isolated $C_{60}$ molecules, the cycloaddition involving two "double" bonds being energetically preferable.

Hypothetical allotropes of the $sp^3$-hybridized carbon, obtained by crosslinking graphitic microtubules and called tubulanes, have been studied with molecular mechanics [84]. One of these allotropes, consisting of the $A(2,2)$ armchair tubules and denoted as 8-tetra(2,2)tubulane, is predicted to be only slightly less stable than graphite ($\Delta H_f^0 = 5.9$ kcal/mol of C) and have a density of 3.43 g/cm$^3$. In the lattice of this tubulane, which consists of the tetraasterane building blocks [Fig. 6.5($d$)], each carbon atom belongs to one four-membered and five six-membered rings. 8-tetra(2,2)tubulane is expected to crystallize in the $I4/mmm$ space group (with the unit cell dimensions: $a = 4.28$ Å and $c = 2.54$ Å) and possess a bulk modulus comparable to that of diamond.

Structures of other tubulanes can also be derived from the lattices of known zeolites. Electronic structures of six hypothetical carbon allotropes, obtained by replacing heteroatoms in the lattices of zeolites ZSM-39, melanophlogite, bikitaite, ZSM-23, Theta-1, and ferrierite with $sp^3$-hybridized carbon atoms, have been investigated with the LDA method [85]. The calculations have found the energies of the $C_{MTN}$ and $C_{MEP}$ carbon lattices, which are based on the ZSM-39 zeolite and melanophlogite, respectively, to be only 1.6 and 2.1 kcal/mol of C higher than that of diamond. The surprising stability of these carbon allotropes is directly attributable to their structures consisting of space-filling $C_{20}/C_{24}$ ($C_{MEP}$) and $C_{20}/C_{28}$ ($C_{MTN}$) polyhedra [85, 86] that give rise to bond angles deviating only slightly from the ideal 109.47°. The $C_{MEP}$ and $C_{MNT}$ solids are insulators with band gaps predicted to be only slightly smaller than that of diamond.

## 6.4 Negative-Curvature Structures

The presence of 12 five-membered rings in fullerenes introduces the positive curvature necessary for the formation of closed polyhedral cages. As exemplified by the flat graphene lattice, the six-membered rings impart neither positive nor negative curvature. On the other hand, rings with more than six vertices confer negative curvatures, resulting in either finite or infinite structures.

The infinite negative-curvature structures are called schwarzites [87]. The schwarzite lattices are obtained by tiling periodic surfaces with rings of $sp^2$-hybridized carbon atoms. The surfaces, which are periodic in all three directions and divide Cartesian space into two domains that are mutually exclusive and congruent, are topologically equivalent to the so-called minimal surfaces [88]. Shapes of these minimal surfaces can be easily pictured by imagining periodic soap films subject to equal pressures on both sides.

Construction and characterization of schwarzite lattices are greatly facilitated by the fact that they are related to only three relevant classes of minimal surfaces, namely D (diamond, also called F), P (primitive; Fig. 6.6), and G (gyroid). The D216, P216, D168, and P192 schwarzites have been the first proposed negative-curvature carbon lattices. The D216 and P216 schwarzites [87] have 80 six-membered and 24 seven-membered rings per unit cell. The D216 structure consists of repeated tetrahedral confluences of pores built around the cubic diamond lattice, whereas the confluences of six pores wrapped around a simple cubic lattice are the repeating motif of the P216 structure. Because of this "molecular porosity", allotropes of carbon based on such schwarzite

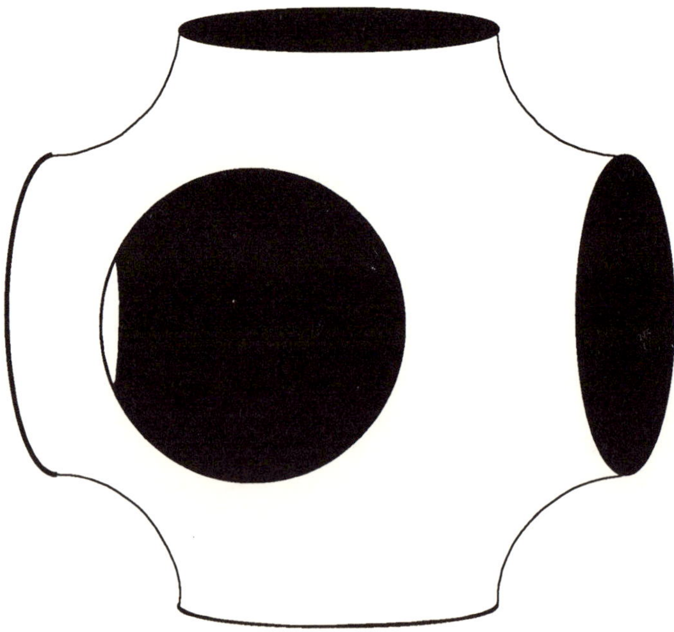

*Fig. 6.6. The P (primitive) minimal surface.*

lattices are expected to have quite low densities. Indeed, molecular mechanics calculations [87] predict densities of 1.15 and 1.02 $g/cm^3$ for the D216 and P216 structures, respectively. The corresponding calculated heats of formation, relative to the graphene lattice, are 4.2 and 4.6 kcal/mol of C.

The D168 structure [89], also known as "buckygym" or "plumber's nightmare", possesses 168 atoms per unit cell. As all schwarzites of the D type, its consists of surfaces wrapped around the fcc lattice of cubic diamond. The structure is built from six- and seven-membered rings, with each seven-membered ring surrounded only by hexagons, and each six-membered ring surrounded by three hexagons and three heptagons. In other words, each of the carbon atoms of the D168 schwarzite belongs to one seven-membered and two six-membered rings. Molecular mechanics calculations predict the D168 schwarzite to be less stable than the graphene lattice by between 2.5 [89] and 5.0 kcal/mol of C [90]. This high stability is attributed to the fact that the bond angles in the D168 carbon networks are very close to $120°$. This remarkably low strain can be attained thanks to the presence of the seven-membered rings that, in contrast to their five-membered counterparts in fullerenes, are quite flexible.

The P192 schwarzite [91], called a "graphite foam", has 80 six-membered and 12 eight-membered rings per simple cubic unit cell. For this structure, molecular mechanics calculations [87] predict the density of 1.16 $g/cm^3$ and the heat of formation equal to 4.4 kcal/mol of C.

Less porous negative-curvature P24 and D24 structures have also been described [92]. Interestingly, either structure contains only one set of equivalent carbon atoms. Each of these carbon atoms belongs to one six-membered and two eight-membered rings. The P24 or $6.8^2P$ lattice consists of three-dimensional linkages of eight-membered rings, whereas the D24 or $6.8^2D$ lattice, in which all of the bond angles are equal to $120°$, can be best described as a linkage of benzene rings. In fact, the latter structure has been originally suggested for the poorly characterized "polybenzene" obtained by pyrolysis of $C_6I_6$ or the reaction of $C_6Cl_6$ with sodium amalgam [93]. The P24 schwarzite is predicted to be metallic by tight-binding Hamiltonian calculations [92], which yield 2.04 $g/cm^3$ for its density and 11.3 kcal/mol of C for its heat of formation relative to the graphene lattice. The D24 schwarzite, with the predicted density of 2.19 $g/cm^3$, the heat of formation estimated at 4.8 kcal/mol of C, and the band gap of 3.4 eV, is expected to be much more stable than its P24 counterpart.

As in the case of fullerenes, several isomeric structures of schwarzites that possess the same number of carbon atoms per unit cell are possible. Enumeration of such isomers is facilitated by the recently described [94] computational algorithm for generating lattices of the $sp^2$-hybridized carbon atoms on arbitrary smooth surfaces. Using this algorithm, alternative structures for the D216 and P216 schwarzites have been identified [94]. The new structures give rise to solids that are slightly denser and more stable than their previously proposed [87] counterparts. Other new structures generated with the aforementioned algorithm include D200, P56, P168, G56, and G216 schwarzites.

The electronic structures of schwarzites have been the subject of several calculations. The D168 lattice [89] has been investigated [95] with the LDA method. The D168 schwarzite is predicted to be a semiconductor with an indirect band gap of 0.47 eV and an electronic dielectric constant of 3.99. For comparison, non-self-consistent LDA calculations involving full geometry optimizations yield 1.34 eV for the band gap and 5.7 kcal/mol of C for the heat of formation relative to the graphene lattice [90]. With the same method, band gaps of the P216, D216 [87], and P192 [91] schwarzites are estimated at 1.35, 1.24, and 1.29 eV, respectively. The corresponding heats of formation

**Table 6.1.  Results of approximate
LDA calculations on schwarzites [96]**

| Structure | Rings | Lattice constant [Å] | Density [g/cm$^3$] | Band gap [eV] |
|---|---|---|---|---|
| P24 [92] | 6,8 | 7.770 (bcc) | 2.03 | 0.00 |
| D24 [92] | 6,8 | 6.033 (sc) | 2.17 | 2.96 |
| G48 | 6,8 | 9.520 (bcc) | 2.21 | 0.00 |
| P56 | 7 | 8.163 (sc) | 2.05 | 0.00 |
| G56 | 7 | 10.145 (bcc) | 2.13 | 0.00 |
| P168 | 6,7 | 13.875 (sc) | 1.28 | 0.00 |
| D168 [89][a] | 6,7 | 21.878 (fcc) | 1.28 | 0.59 |
| D168 [89][a] | 6,7 | 22.438 (fcc) | 1.20 | 0.47 |
| P192 [91] | 6,8 | 14.869 (sc) | 1.16 | 0.63 |
| D200 [94] | 6,7 | 23.640 (fcc) | 1.22 | 0.00 |
| D216 [87][a] | 6,7 | 24.672 (fcc) | 1.16 | 0.29 |
| D216 [87][a] | 6,7 | 24.597 (fcc) | 1.16 | 0.73 |
| P216 [87][a] | 6,7 | 16.186 (sc) | 1.02 | 0.17 |
| P216 [87][a] | 6,7 | 15.718 (sc) | 1.12 | 0.00 |
| G216 [94] | 6,7 | 19.373 (bcc) | 1.18 | 0.00 |

[a] Two different geometries.

are calculated at 4.9, 5.0, and 4.5 kcal/mol of C. Taking into account the fact that for the $C_{60}$ cluster a very reasonable value of $\Delta H_f^0$ equal to 606 kcal/mol is predicted at the same level of theory, one expects the above energy estimates to be quite accurate.

A comprehensive study of eight previously proposed and four new schwarzites has been published [96]. The results of approximate LDA calculations, compiled in Table 6.1, demonstrate that most of the schwarzites are metals. The computed band gaps are quite sensitive to the lattice geometries (see the double D168, D216, and P216 entries). In general, except for the P216 structure, lattices with isolated seven- and eight-membered rings give rise to insulating or semiconducting solids, whereas in the lattices of metallic schwarzites these rings abut. The large band gap of 2.96 eV predicted for the "polybenzene" D24 schwarzite is in good agreement with that obtained from tight-binding Hamiltonian calculations [92].

Finally, it should be mentioned that it is possible to construct finite structures from fragments of the schwarzite lattices by either terminating them with suitable fullerene caps, or connecting them in an appropriate manner. Such finite structures can be viewed as assemblies of negative-curvature units called schwarzons [26, 97]. One of these schwarzons contains 132 carbon atoms distributed among 20 six-membered and 12 seven-membered rings. Its $T_d$ symmetry makes it an equivalent of the $sp^3$-hybridized carbon atom. Another schwarzon, with 114 carbon atoms and $D_{3h}$ symmetry, is an equivalent of the $sp^2$-hybridized carbon. The $sp^2$ schwarzon has six seven-membered rings and is a junction of three carbon tubes intersecting at $120^o$.

Using these schwarzons, one can assemble hyperstructures analogous to those of the known and hypothetical allotropes of carbon. Lattices of "hypergraphite" and "hyperdiamond" are obtained from the $sp^2$ and $sp^3$ schwarzons, respectively. One can imagine 60 $sp^2$ schwarzons connected through short graphitic tubules to form a hyperfullerene with 10 080 carbon atoms [26, 97]. The entire process of constructing schwarzons from carbon atoms can be iterated, leading to hyperschwarzons, and so on. One may only wonder whether such structures will ever be synthesized ...

# References

1. K. Kikuchi, N. Nakahara, T. Wakabayashi, M. Honda, H. Matsumiya, T. Moriwaki, S. Suzuki, H. Shiromaru, K. Saito, K. Yamauchi, I. Ikemoto, and Y. Achiba, *Isolation and Identification of Fullerene Family:* $C_{76}$, $C_{78}$, $C_{82}$, $C_{84}$, $C_{90}$ *and* $C_{96}$, Chem. Phys. Lett. **188**, 177 (1992).

2. L. D. Lamb, D. R. Huffman, R. K. Workman, S. Howells, T. Chen, D. Sarid, and R. F. Ziolo, *Extraction and STM Imaging of Spherical Giant Fullerenes*, Science **255**, 1413 (1992).

3. D. H. Parker, P. Wurz, K. Chatterjee, K. R. Lykke, J. E. Hunt, M. J. Pellin, J. C. Hemminger, D. M. Gruen, and L. M. Stock, *High-Yield Synthesis, Separation, and Mass-Spectrometric Characterization of Fullerenes $C_{60}$ to $C_{266}$*, J. Am. Chem. Soc. **113**, 7499 (1991).

4. Y. Rubin, M. Kahr, C. B. Knobler, F. Diederich, and C. L. Wilkins, *The Higher Oxides of Carbon $C_{8n}O_{2n}$ (n= 3 − 5): Synthesis, Characterization, and X-ray Crystal Structure. Formation of Cyclo[n]carbon Ions $C_n^+$ (n= 18, 24), $C_n^-$ (n= 18, 24, 30), and Higher Carbon Ions Including $C_{60}^+$ in Laser Desorption Fourier Transform Mass Spectrometric Experiments*, J. Am. Chem. Soc. **113**, 495 (1991).

5. C. Yeretzian, K. Hansen, F. Diederich, and R. L. Whetten, *Coalescence Reactions of Fullerenes*, Nature **359**, 44 (1992).

6. V. P. Dravid, X. Lin, Y. Wang, X. K. Wang, A. Yee, J. B. Ketterson, and R. P. H. Chang, *Buckytubes and Derivatives: Their Growth and Implications for Buckyball Formation*, Science **259**, 1601 (1993).

7. R. L. Murry and G. E. Scuseria, *Theoretical Study of $C_{90}$ and $C_{96}$ Fullerene Isomers*, J. Phys. Chem. **98**, 4212 (1994).

8. D. E. Manolopoulos, J. C. May, and S. E. Down, *Theoretical Studies of the Fullerenes: $C_{34}$ to $C_{70}$*, Chem. Phys. Lett. **181**, 105 (1991).

9. D. E. Manolopoulos and P. W. Fowler, *A Fullerene without a Spiral*, Chem. Phys. Lett. **204**, 1 (1993).

10. M. Fujita, R. Saito, G. Dresselhaus, and M. S. Dresselhaus, *Formation of General Fullerenes by Their Projection on a Honeycomb Lattice*, Phys. Rev. B **45**, 13834 (1992).

11. P. W. Fowler, D. E. Manolopoulos, D. B. Redmond, and R. P. Ryan, *Possible Symmetries of Fullerene Structures*, Chem. Phys. Lett. **202**, 371 (1993).

12. D. J. Klein, W. A. Seitz, and T. G. Schmalz, *Icosahedral Symmetry Carbon Cage Molecules*, Nature **323**, 703 (1986).

13. P. W. Fowler, *How Unusual is $C_{60}$? Magic Numbers for Carbon Clusters*, Chem. Phys. Lett. **131**, 444 (1986).

14. P. W. Fowler, J. E. Cremona, and J. I. Steer, *Systematics of Bonding in Non-icosahedral Carbon Clusters*, Theor. Chim. Acta **73**, 1 (1988).

15. B. I. Dunlap, D. W. Brenner, J. W. Mintmire, R. C. Mowrey, and C. T. White, *Local Density Functional Electronic Structures of Three Stable Icosahedral Fullerenes*, J. Phys. Chem. **95**, 8737 (1991).

16. D. Bakowies and W. Thiel, *MNDO Study of Large Carbon Clusters*, J. Am. Chem. Soc. **113**, 3704 (1991).

17. D. Bakowies and W. Thiel, *Theoretical Infrared Spectra of Large Carbon Clusters*, Chem. Phys. **151**, 309 (1991).

18. P. W. Fowler and J. I. Steer, *The Leapfrog Principle: A Rule for Electron Counts of Carbon Clusters*, J. Chem. Soc. Chem. Commun. 1403 (1987).

19. P. W. Fowler, *Carbon Cylinders: A Class of Closed-Shell Clusters*, J. Chem. Soc. Faraday Trans. **86**, 2073 (1990).

20. L. A. Chernozatonskii, *Barrelenes/Tubelenes — A New Class of Cage Carbon Molecules and Its Solids*, Phys. Lett. A **166**, 55 (1992).

21. R. H. Baughman, D. S. Galvão, C. Cui, Y. Wang, and D. Tománek, *Fullereneynes: A New Family of Porous Fullerenes*, Chem. Phys. Lett. **204**, 8 (1993).

22. J. Cioslowski, *Heats of Formation of Higher Fullerenes from ab Initio Hartree-Fock and Correlation Energy Functional Calculations*, Chem. Phys. Lett. **216**, 389 (1993).

23. G. B. Adams, O. F. Sankey, J. B. Page, M. O'Keeffe, and D. A. Drabold, *Energetics of Large Fullerenes: Balls, Tubes, and Capsules*, Science **256**, 1792 (1992).

24. J. Tersoff, *Energies of Fullerenes*, Phys. Rev. B **46**, 15546 (1992).

25. R. L. Murry, J. R. Colt, and G. E. Scuseria, *How Accurate Are Molecular Mechanics Predictions for Fullerenes? A Benchmark Comparison with Hartree-Fock Self-Consistent Field Results*, J. Phys. Chem. **97**, 4954 (1993).

26. G. E. Scuseria, *Ab Initio Theoretical Predictions of Fullerenes*, in *Buckminsterfullerenes* (W.E. Billups and M.A. Ciufolini, eds.), VCH Publishers, New York, 1993, Ch. 5, p. 103.

27. A. D. J. Haymet, $C_{120}$ and $C_{60}$: *Archimedean Solids Constructed from $sp^2$ Hybridized Carbon Atoms*, Chem. Phys. Lett. **122**, 421 (1985).

28. J. C. Greer, S. Itoh, and S. Ihara, *Ab Initio Geometry and Stability of a* $C_{120}$ *Torus*, Chem. Phys. Lett. **222**, 621 (1994).

29. D. L. Strout, R. L. Murry, C. Xu, W. C. Eckhoff, G. K. Odom, and G. E. Scuseria, *A Theoretical Study of Buckminsterfullerene Reaction Products:* $C_{60}+C_{60}$, Chem. Phys. Lett. **214**, 576 (1993).

30. N. Matsuzawa, M. Ata, D. A. Dixon, and G. Fitzgerald, *Dimerization of* $C_{60}$: *The Formation of Dumbbell-Shaped* $C_{120}$, J. Phys. Chem. **98**, 2555 (1994).

31. H. W. Kroto and K. McKay, *The Formation of Quasi-icosahedral Spiral Shell Carbon Particles*, Nature **331**, 328 (1988).

32. S. Iijima, *The 60-Carbon Cluster Has Been Revealed!*, J. Phys. Chem. **91**, 3466 (1987).

33. D. Ugarte, *Canonical Structure of Large Carbon Clusters:* $C_n$, $n > 100$, Europhys. Lett. **22**, 45 (1993).

34. M. Yoshida and E. Ōsawa, *Molecular Mechanics Calculations of Giant- and Hyperfullerenes with Eicosahedral Symmetry*, Full. Sci. Tech. **1**, 55 (1993).

35. S. Iijima, *Helical Microtubules of Graphitic Carbon*, Nature **354**, 56 (1991).

36. T. W. Ebbesen and P. M. Ajayan, *Large-Scale Synthesis of Carbon Nanotubes*, Nature **358**, 220 (1992).

37. H. Hiura, T. W. Ebbesen, K. Tanigaki, and H. Takahashi, *Raman Studies of Carbon Nanotubes*, Chem. Phys. Lett. **202**, 509 (1993).

38. S. Iijima, P. M. Ajayan, and T. Ichihashi, *Growth Model for Carbon Nanotubes*, Phys. Rev. Lett. **69**, 3100 (1992).

39. M. Endo and H. W. Kroto, *Formation of Carbon Nanofibers*, J. Phys. Chem. **96**, 6941 (1992).

40. Y. Saito, T. Yoshikawa, M. Inagaki, M. Tomita, and T. Hayashi, *Growth and Structure of Graphitic Tubules and Polyhedral Particles in Arc-Discharge*, Chem. Phys. Lett. **204**, 277 (1993).

41. P. M. Ajayan, T. Ichihashi, and S. Iijima, *Distribution of Pentagons and Shapes in Carbon Nano-Tubes and Nano-Particles*, Chem. Phys. Lett. **202**, 384 (1993).

42. P. M. Ajayan and S. Iijima, *Smallest Carbon Nanotube*, Nature **358**, 23 (1992).

43. M. S. Dresselhaus, G. Dresselhaus, and R. Saito, *Carbon Fibers Based on $C_{60}$ and Their Symmetry*, Phys. Rev. B **45**, 6234 (1992).

44. D. H. Robertson, D. W. Brenner, and J. W. Mintmire, *Energetics of Nanoscale Graphitic Tubules*, Phys. Rev. B **45**, 12592 (1992).

45. B. I. Dunlap, *Connecting Carbon Tubules*, Phys. Rev. B **46**, 1933 (1992).

46. N. Hamada, S. Sawada, and A. Oshiyama, *New One-Dimensional Conductors: Graphitic Microtubules*, Phys. Rev. Lett. **68**, 1579 (1992).

47. C. T. White, D. H. Robertson, and J. W. Mintmire, *Helical and Rotational Symmetries of Nanoscale Graphitic Tubules*, Phys. Rev. B **47**, 5485 (1993).

48. R. Saito, M. Fujita, G. Dresselhaus, and M. S. Dresselhaus, *Electronic Structure of Graphene Tubules Based on $C_{60}$*, Phys. Rev. B **46**, 1804 (1992).

49. R. Saito, M. Fujita, G. Dresselhaus, and M. S. Dresselhaus, *Electronic Structure of Chiral Graphene Tubules*, Appl. Phys. Lett. **60**, 2204 (1992).

50. K. Tanaka, K. Okahara, M. Okada, and T. Yamabe, *Electronic Properties of Bucky-Tube Model*, Chem. Phys. Lett. **191**, 469 (1992).

51. K. Okahara, K. Tanaka, H. Aoki, T. Sato, and T. Yamabe, *Band Structures of Carbon Nanotubes with Bond-Alternation Patterns*, Chem. Phys. Lett. **219**, 462 (1994).

52. J. W. Mintmire, B. I. Dunlap, and C. T. White, *Are Fullerene Tubules Metallic?*, Phys. Rev. Lett. **68**, 631 (1992).

53. G. G. Tibbetts, *Why are Carbon Filaments Tubular?*, J. Cryst. Growth **66**, 632 (1984).

54. S. Sawada and N. Hamada, *Energetics of Carbon Nano-Tubes*, Solid State Commun. **83**, 917 (1992).

55. S. Iijima, T. Ichihashi, and Y. Ando, *Pentagons, Heptagons, and Negative Curvature in Graphite Microtubule Growth*, Nature **356**, 776 (1992).

56. L. A. Chernozatonskii, *Carbon Nanotube Elbow Connections and Tori*, Phys. Lett. A **170**, 37 (1992).

57. L. A. Chernozatonskii, *Carbon Nanotube Connectors and Planar Jungle Gyms*, Phys. Lett. A **172**, 173 (1992).

58. M. R. Pederson and J. Q. Broughton, *Nanocapillarity in Fullerene Tubules*, Phys. Rev. Lett. **69**, 2689 (1992).

59. P. M. Ajayan and S. Iijima, *Capillarity-Induced Filling of Carbon Nanotubes*, Nature **361**, 333 (1993).

60. J.-C. Charlier and J.-P. Michenaud, *Energetics of Multilayered Carbon Tubules*, Phys. Rev. Lett. **70**, 1858 (1993).

61. R. S. Ruoff, J. Tersoff, D. C. Lorents, S. Subramoney, and B. Chan, *Radial Deformation of Carbon Nanotubes by van der Waals Forces*, Nature **364**, 514 (1993).

62. I. V. Stankevich, M. V. Nikerov, and D. A. Bochvar, *The Structural Chemistry of Crystalline Carbon: Geometry, Stability, and Electronic Spectrum*, Russ. Chem. Rev. **53**, 640 (1984).

63. F. Diederich and Y. Rubin, *Synthetic Approaches Toward Molecular and Polymeric Carbon Allotropes*, Angew. Chem. Int. Ed. Engl. **31**, 1101 (1992).

64. A. G. Whittaker, *Carbon: A New View of Its High-Temperature Behavior*, Science **200**, 763 (1978).

65. K. M. Merz, Jr., R. Hoffmann, and A. T. Balaban, $3,4$-*Connected Carbon Nets: Through-Space and Through-Bond Interactions in the Solid State*, J. Am. Chem. Soc. **109**, 6742 (1987).

66. R. H. Baughman, H. Eckhardt, and M. Kertesz, *Structure-Property Predictions for New Planar Forms of Carbon: Layered Phases Containing $sp^2$ and $sp$ Atoms*, J. Chem. Phys. **87**, 6687 (1987).

67. H. R. Karfunkel and T. Dressler, *New Hypothetical Carbon Allotropes of Remarkable Stability Estimated by Modified Neglect of Diatomic Overlap Solid-State Self-Consistent Field Computations*, J. Am. Chem. Soc. **114**, 2285 (1992).

68. H. Stoll, *On the Correlation Energy of Graphite*, J. Chem. Phys. **97**, 8449 (1992).

69. F. P. Bundy and J. S. Kasper, *Hexagonal Diamond — A New Form of Carbon*, J. Chem. Phys. **46**, 3437 (1967).

70. G. Laqua, H. Musso, W. Boland, and R. Ahlrichs, *Force Field Calculations (MM2) of Carbon Lattices*, J. Am. Chem. Soc. **112**, 7391 (1990).

71. S. Fahy and S. G. Louie, *High-Pressure Structural and Electronic Properties of Carbon*, Phys. Rev. B **36**, 3373 (1987).

72. M. Shen, H. F. Schaefer III, C. Liang, J.-H. Lii, N. L. Allinger, and P.v.R. Schleyer, *Finite $T_d$ Symmetry Models for Diamond: From Adamantane to Superadamantane ($C_{35}H_{36}$)*, J. Am. Chem. Soc. **114**, 497 (1992).

73. N. N. Matyushenko, V. E. Strel'nitskiĭ, and V. A. Gusev, *A Dense New Version of Crystalline Carbon $C_8$*, JETP Lett. **30**, 199 (1979).

74. R. L. Johnston and R. Hoffmann, *Superdense Carbon, $C_8$: Supercubane or Analogue of $\gamma$-Si?*, J. Am. Chem. Soc. **111**, 810 (1989).

75. H. Werner, D. Herein, J. Blöcker, B. Henschke, U. Tegtmeyer, Th. Schedel-Niedrig, M. Keil, A. M. Bradshaw, and R. Schlögl, *Spectroscopic*

*and Chemical Characterisation of "Fullerene Black"*, Chem. Phys. Lett. **194**, 62 (1992).

76. C. S. Yoo and W. J. Nellis, *Phase Transition from $C_{60}$ Molecules to Strongly Interacting $C_{60}$ Agglomerates at Hydrostatic High Pressures*, Chem. Phys. Lett. **198**, 379 (1992).

77. F. Moshary, N. H. Chen, I. F. Silvera, C. A. Brown, H. C. Dorn, M. S. de Vries, and D. S. Bethune, *Gap Reduction and the Collapse of Solid $C_{60}$ to a New Phase of Carbon under Pressure*, Phys. Rev. Lett. **69**, 466 (1992).

78. D. W. Snoke, K. Syassen, and A. Mittelbach, *Optical Absorption Spectrum of $C_{60}$ at High Pressure*, Phys. Rev. B **47**, 4146 (1993).

79. M. N. Regueiro, P. Monceau, and J.-L. Hodeau, *Crushing $C_{60}$ to Diamond at Room Temperature*, Nature **355**, 237 (1992).

80. A. M. Rao, P. Zhou, K.-A. Wang, G. T. Hager, J. M. Holden, Y. Wang, W.-T. Lee, X.-X. Bi, P. C. Eklund, D. S. Cornett, M. A. Duncan, and I. J. Amster, *Photoinduced Polymerization of Solid $C_{60}$ Films*, Science **259**, 955 (1993).

81. D. S. Cornett, I. J. Amster, M. A. Duncan, A. M. Rao, and P. C. Eklund, *Laser Desorption Mass Spectrometry of Photopolymerized $C_{60}$ Films*, J. Phys. Chem. **97**, 5036 (1993).

82. P. Zhou, Z.-H. Dong, A. M. Rao, and P. C. Eklund, *Reaction Mechanism for the Photopolymerization of Solid Fullerene $C_{60}$*, Chem. Phys. Lett. **211**, 337 (1993).

83. Y. Wang, J. M. Holden, Z.-H. Dong, X.-X. Bi, and P. C. Eklund, *Photo-Dimerization Kinetics in Solid $C_{60}$ Films*, Chem. Phys. Lett. **211**, 341 (1993).

84. R. H. Baughman and D. S. Galvão, *Tubulanes: Carbon Phases Based on Cross-Linked Fullerene Tubules*, Chem. Phys. Lett. **211**, 110 (1993).

85. R. Nesper, K. Vogel, and P. E. Blöchl, *Hypothetical Carbon Modifications Derived from Zeolite Frameworks*, Angew. Chem. Int. Ed. Engl. **32**, 701 (1993).

86. K. Vogel and R. Nesper, *Novel Hypothetical Carbon Modifications as Host Frameworks for Dopants*, J. Mol. Catal. **82**, 467 (1993).

87. T. Lenosky, X. Gonze, M. Teter, and V. Elser, *Energetics of Negatively Curved Graphitic Carbon*, Nature **355**, 333 (1992).

88. A. L. Mackay, *Periodic Minimal Surfaces*, Nature **314**, 604 (1985).

89. D. Vanderbilt and J. Tersoff, *Negative-Curvature Fullerene Analog of $C_{60}$*, Phys. Rev. Lett. **68**, 511 (1992).

90. R. Phillips, D. A. Drabold, T. Lenosky, G. B. Adams, and O. F. Sankey, *Electronic Structure of Schwarzite*, Phys. Rev. B **46**, 1941 (1992).

91. A. L. Mackay and H. Terrones, *Diamond from Graphite*, Nature **352**, 762 (1991).

92. M. O'Keeffe, G. B. Adams, and O. F. Sankey, *Predicted New Low Energy Forms of Carbon*, Phys. Rev. Lett. **68**, 2325 (1992).

93. J. Gibson, M. Holohan, and H. L. Riley, *Amorphous Carbon*, J. Chem. Soc. 456 (1946).

94. S. J. Townsend, T. J. Lenosky, D. A. Muller, C. S. Nichols, and V. Elser, *Negatively Curved Graphitic Sheet Model of Amorphous Carbon*, Phys. Rev. Lett. **69**, 921 (1992).

95. W. Y. Ching, M.-Z. Huang, and Y.-N. Xu, *Electronic and Optical Properties of the Vanderbilt-Tersoff Model of Negative-Curvature Fullerene*, Phys. Rev. B **46**, 9910 (1992).

96. M.-Z. Huang, W. Y. Ching, and T. Lenosky, *Electronic Properties of Negative-Curvature Periodic Graphitic Carbon Surfaces*, Phys. Rev. B **47**, 1593 (1993).

97. G. E. Scuseria, *Negative Curvature and Hyperfullerenes*, Chem. Phys. Lett. **195**, 534 (1992).

# Chapter 7

# Fullerenes with Fewer than 60 Carbon Atoms

Carbon vapors abound in small clusters as well as in large ones [1]. Theoretical work on small carbon clusters began with the early EHT [2] and MINDO/2 [3] calculations and progressed over the last 30 years to sophisticated *ab initio* treatments that incorporate electron correlation effects. Reliable theoretical predictions for these clusters are often hindered by the fact that their isomers possess very similar energies and the computed energy differences are quite sensitive to the level of theory used in electronic structure calculations. For example, the rhombic isomer of $C_4$ is predicted to be more stable than the linear species by only 1.6 kcal/mol at the $CCSD(T)/5s4p3d2f1g$ level of theory, but the energy differences obtained with less sophisticated basis sets and electron correlation treatments vary between $-7$ and 15 kcal/mol [4]. Despite these difficulties, the consensus reached by most calculations [5–9] is that the $C_2$–$C_{10}$ clusters definitely prefer linear structures for odd numbers of carbon atoms, whereas for even numbers of atoms the cyclic and linear structures are close in energy.

The powerful technique of "ion chromatography" [10] makes it possible to separate isomers of both positive and negative carbon ions, and elucidate their structures. The experimental setup consists of an ion source, a mass spectrometer, a drift cell, and a detector. The mass-selected ions enter the drift cell and are subjected to a homogeneous electric field of 2–10 V/cm. As the ions move through the cell, they collide with atoms of the helium buffer gas present at the pressure of 2–5 Torr. Under such conditions, the ion mobilities, which are of the order of 1–20 $cm^2/V \cdot s$, are inversely proportional to the spherically averaged cross-sections. Individual species can be assigned to the recorded peaks in the arrival time distribution plots ("ion chromatograms") by calculating their cross-sections and comparing the predicted mobilities with the experimental ones [11].

For each of the positive carbon ions composed of between three and six atoms only a single peak is observed, corresponding to the linear isomer [10, 11]. Both the linear and the cyclic isomers of the $C_7^+$, $C_8^+$, $C_9^+$, and $C_{10}^+$ species are detectable, whereas the clusters consisting of between 11 and 20 atoms are present exclusively in the cyclic form. However, beginning with $C_{21}^+$, peaks due to planar bicyclic structures appear. They are joined by the peaks corresponding to nonplanar cyclic isomers at $C_{29}^+$ and then by peaks due to other ring structures. The carbon cages emerge at $C_{30}^+$ and dominate the observed abundances for clusters with more than about 45 atoms.

Simple topological arguments [12] place a lower bound on the number of carbon atoms in the fullerene cages. Let $N$ be the number of atoms, $M$ the number of bonds, and $P_k$ the number of $k$-membered rings. Since each carbon atom is linked to three neighbors,

$$2M = 3N \qquad \text{and} \qquad 2M = \sum_k kP_k \; . \qquad (7.1)$$

The Euler theorem, which relates the number of atoms, bonds, and rings for cage structures, yields

$$N + \sum_k P_k = M + 2 \; . \qquad (7.2)$$

Combining Eqs. (7.1) and (7.2) leads to

$$\sum_k (6 - k)P_k = 12 \qquad \text{and} \qquad N = 2\sum_k P_k - 4 \; . \qquad (7.3)$$

Since only five- and six-membered rings are present in fullerenes, Eq. (7.3) shows that these molecules possess exactly 12 pentagons and $(N - 20)/2$ hexagons. This means that $C_{20}$ is the smallest realizable fullerene. In fact, it is well known that at least one fullerene isomer exists for every even $N \geq 20$, except for $N = 22$ [13]. However, one should keep in mind that the IPR structures are possible only for $N \geq 60$ [12].

In light of the above considerations, the low stability of the $C_{20}$ fullerene must be the reason for its failure to appear in the ion chromatography experiments. Electronic structure calculations are thus needed to predict and explain the relative energies of various $C_{20}$ species.

## 7.1 The $C_{20}$ Clusters

The $C_{20}$ fullerene cage **1** (Fig. 7.1) consists of 12 pentagons. In principle, the cage could have the $I_h$ symmetry of a regular dodecahedron, as its fully hydrogenated analog, a $C_{20}H_{20}$ hydrocarbon called dodecahedrane [14, 15], does. However, the $I_h$ structure of **1** is subject to a Jahn–Teller distortion because of its open-shell $g_u^2$ electronic configuration [16]. At the HF/6-31G* level of theory, the singlet ground state with $C_2$ symmetry is a true minimum on the potential energy hypersurface [17]. The same symmetry is found with the tight-binding Hamiltonian method [18], but MNDO [19, 20] and AM1 [21, 22] calculations predict a $C_1$ minimum. A $D_{5d}$ geometry has been suggested for the $^3A_{2g}$ triplet state of **1** [16]. Finally, it should be mentioned that the $C_{20}^{2+}$ cation does not undergo a Jahn–Teller distortion and is believed to retain the full $I_h$ symmetry [23].

The standard enthalpy of formation for the $C_{20}$ fullerene, calculated at the MNDO level of theory, is 825.2 kcal/mol [19]. This is equivalent to 41.3 kcal/mol per carbon atom, which is much higher than the analogous value of 14.5 kcal/mol for the $C_{60}$ fullerene. The sharply lowered stability of the $C_{20}$ cluster stems from both the increased strain and the presence of abutting five-membered rings that diminish the stabilization due to resonance.

There are many other possible isomers of the $C_{20}$ cluster. Among them, the bowl structure **2** (Fig. 7.1), reminiscent of corannulene, and the cyclic isomer **3** (Fig. 7.1) have been the subjects of several studies. At the HF/6-31G* level of theory, the bowl structure **2** is found to possess $C_{5v}$ symmetry and lie 61.8 kcal/mol below **1** [17]. Inclusion of electron correlation effects through the BLYP functional increases the energy difference to 73.8 kcal/mol when the correlation energy is computed from the Hartree–Fock electron density, but lowers it to 54.7 kcal/mol when the electron density is calculated self-consistently. AM1 calculations [21, 22] also predict $C_{5v}$ geometry for **2**, but place it 49.8 kcal/mol above **1**. The LDA method also favors the fullerene cage **1** over the carbon bowl **2** by 17.3 [24] and 23.3 kcal/mol [17].

The cyclic structure **3** can exist in either a cumulenic or an acetylenic form. At the HF/6-31G* level of theory, the $C_{10h}$ acetylenic structure with bond lengths of 1.196 and 1.381 Å is the energy minimum and lies 92.9 kcal/mol below the fullerene isomer **1** [17]. This energy difference is somewhat reduced to 87.9 and 79.3 kcal/mol at the two variants of the BLYP/6-311G** method mentioned above. The AM1 method [22] also finds the $C_{10h}$ structure as a minimum and estimates it to be more stable than **1** by as much as 289 kcal/mol. Hartree–Fock calculations involving a large basis set place the species with an acetylenic ring 107 kcal/mol below the fullerene isomer, but inclusion of electron correlation effects through the MP2 approximation reverses the order of stabilities, yielding the energy difference of −85.9 kcal/mol [16]. The LDA method also favors the fullerene structure over **3** by as much as 88.6 [17] and 61.1 kcal/mol [24].

Although troublesome, the large discrepancies among the aforementioned theoretical predictions are not surprising. The interconversion reactions between different isomers of carbon clusters are not isodesmic. Because the number and character of the C–C bonds are not conserved in these reactions, significant electron correlation contributions to the energy differences are anticipated. In particular, isomer **1** with 30 bonds is expected to be stabilized

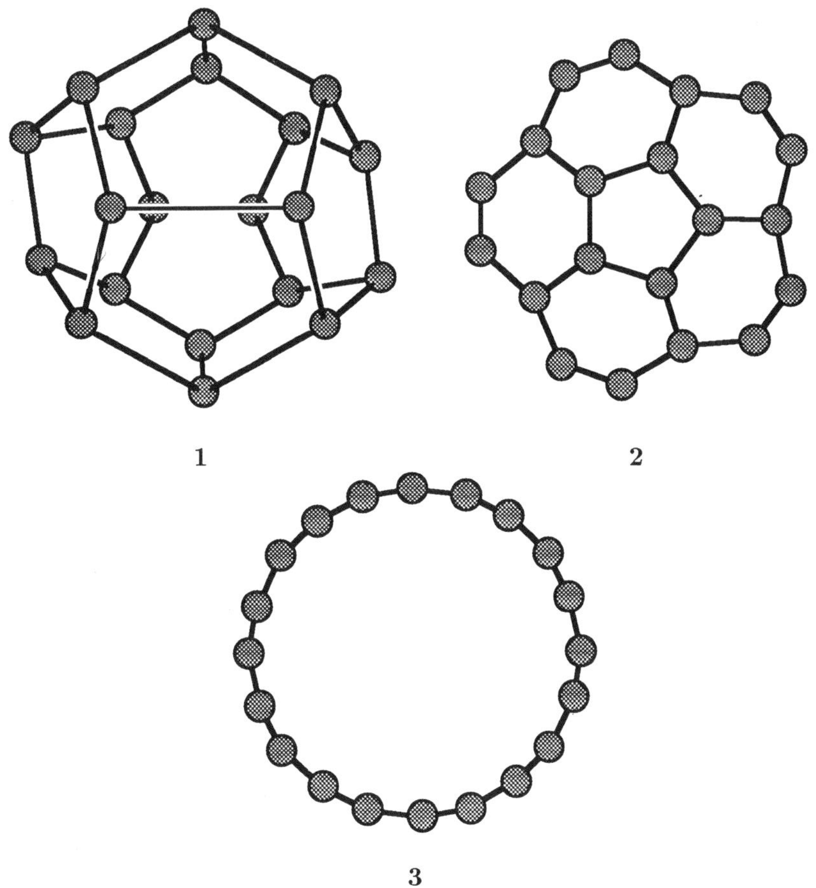

Fig. 7.1.  The fullerene **1**, bowl **2**, and cyclic **3**
isomers of the C$_{20}$ cluster.

by the electron correlation effects much more than either of isomers **2** and
**3**, which possess 25 and 20 bonds, respectively. This is reflected in the MP2
results. However, as clearly demonstrated by exhaustive calculations on the
cyclic structures of C$_{10}$ [25], the MP2 approximation often drastically overestimates the electron correlation effects. Application of higher-level approaches,
such as CCSD(T), brings the order of stabilities right back to that predicted
at the Hartree–Fock level of theory, although with reduced energy differences.
This observation, which casts some doubts on the validity of recent MP2 and
CISD calculations on carbon clusters [26, 27], appears to indicate that the stability predictions for the C$_{20}$ isomers obtained with the BLYP method are
reliable, as they not only are qualitatively similar to those resulting from

Hartree–Fock calculations, but also are in agreement with the available experimental data [10, 11, 28]. The fact that the LDA calculations favor the fullerene structure can be readily explained by the well-known tendency of the LDA formalism to overestimate bond energies.

Although LDA calculations [24] favor the fullerene structure **1** over its bowl counterpart **2** at 0 K, they predict a reversal of stabilities at high temperatures. This is so because vibrations in isomers **1** and **2** have very different contributions to their free energies. At very high temperatures, the cyclic structure **3** eventually becomes the most stable among the three [24].

## 7.2  The $C_{24}$ Clusters

The $C_{24}$ fullerene is not observed in the ion chromatography experiments [10, 11, 28]. Nevertheless, several electronic structure calculations on the $C_{24}$ cage **1** (Fig. 7.2) with 12 pentagons and two hexagons have been reported. Although such a structure has the topological $D_{6d}$ symmetry, its optimized MNDO geometry with six-membered rings located at the cage antipodes does not have any nontrivial symmetry elements [19]. The calculated standard enthalpy of formation is 850.2 kcal/mol, or 35.4 kcal/mol on a per-atom basis, indicating that $C_{24}$ is somewhat more stable than $C_{20}$. Despite the low symmetry, the predicted IR spectrum of **1** consists of only a few strong lines, the most prominent one being located at *ca.* 800 cm$^{-1}$ [20].

The HF/DZP calculations yield a $D_6$ structure for **1**, in which the six-membered rings are twisted by 32.9° with respect to each other (compared with the twist of 30.0° expected for $D_{6d}$ symmetry) [29]. The six-membered rings exhibit no bond alternation and the calculated bond length of 1.409 Å reflects their aromaticity. The $D_6$ structure is a minimum at the HF/STO-3G potential energy hypersurface. Even more importantly, $D_6$ symmetry persists in the MP2/DZ [30] optimized geometry. On the other hand, calculations involving the tight-binding Hamiltonian method predict a $D_{6d}$ structure [18].

Another cage structure **2** (Fig. 7.2), built from six four-membered and eight six-membered rings and possessing $O_h$ symmetry, has been the subject of numerous studies. At the MNDO level of theory, **2** is more stable than **1** by 34 kcal/mol [31]. This somewhat surprising order of stabilities is consistent with other MNDO calculations, which predict the process of replacing a pair of abutting pentagons by one four- and one six-membered ring to be exothermic for small carbon cages [32]. However, *ab initio* calculations contradict the MNDO results, placing **2** above **1** by 26.0 and 22.3 kcal/mol at the HF/DZP and MP2/DZP levels of theory, respectively [29]. Similarly, calculations carried out at the HF/6-31G* optimized geometries of **1** and **2** yield the corresponding energy differences of 25.4 (HF/6-311G*), 34.6 (LDA/6-311G*), and 23.3 kcal/mol (BLYP/6-311G*) [37]. Since both isomers have the same number of bonds and calculations based on the Harris functional also favor **1** over **2** [33], **2** is definitely less stable than **1**. The standard enthalpy of formation of **2** is most probably overestimated at 911 kcal/mol by the AM1 method, as the HF/6-31G* calculations afford the value of 769 kcal/mol [34].

Owing to its high symmetry, only two kinds of bonds are present in **2**. Each of the four-membered rings possesses four long bonds, whereas each six-membered ring has three short and three long bonds. The HF/DZP optimized bond lengths equal 1.355 and 1.487 Å [29], in agreement with the HF/6-31G* values of 1.352 and 1.485 Å [37], and the HF/3-21G values of 1.323 and 1.501 Å [34]. The AM1 optimized bond lengths of 1.385 and 1.464 Å [34] exhibit somewhat smaller alternation. The $O_h$ geometry of **2** corresponds to a local minimum, as attested by the HF/STO-3G and HF/3-21G vibrational

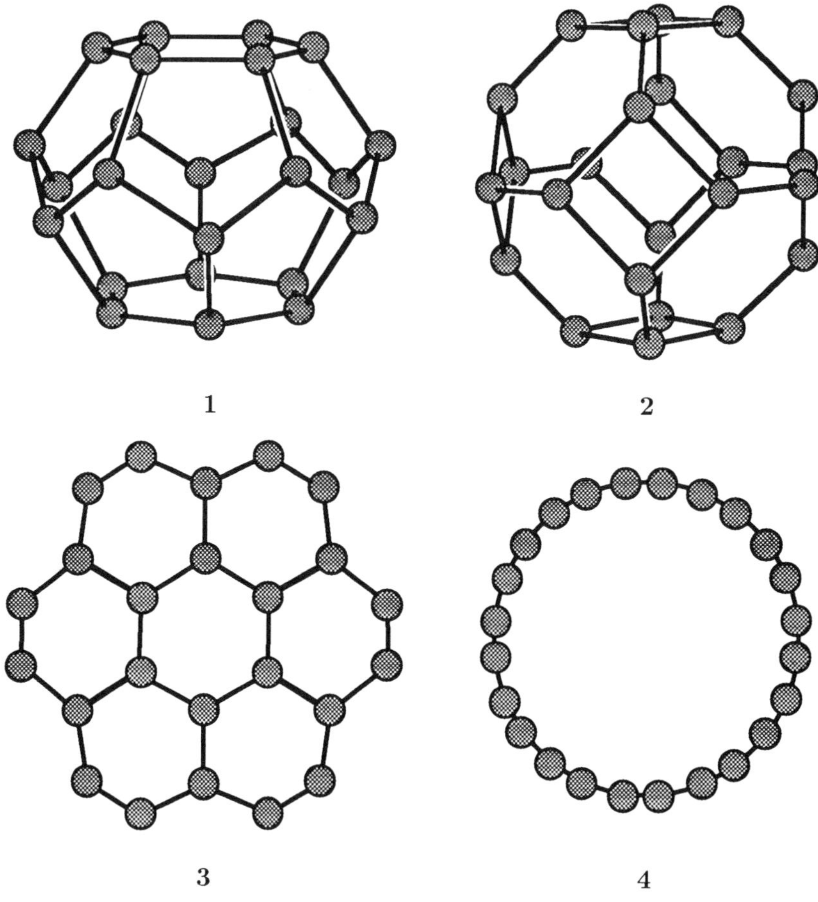

Fig. 7.2. *The fullerene* **1**, $O_h$ *cage* **2**, *sheet* **3**, *and cyclic* **4** *isomers of the* $C_{24}$ *cluster.*

frequencies that are all real [29, 34]. Other possible candidates for the structures of the $C_{24}$ cluster are a graphite sheet fragment **3** and a cyclic isomer **4** (Fig. 7.2). Structure **3**, which can be visualized as a coronene molecule stripped of all hydrogen atoms, has a $D_{6h}$ symmetry at the HF/DZP level of theory and is predicted to be more stable than **1** by 21.9 kcal/mol [29]. This energy difference is reduced to 7.5 kcal/mol at the MP2/DZP level of theory [29], whereas the MNDO estimate is 81 kcal/mol [35]. Structure **3** is also placed below **1** in energy by HF/6-311G* and BLYP/6-311G* calculations [37]. Except for those at the molecule's perimeter, bond lengths in the HF/DZP optimized geometry of **3** do not exhibit much alternation. Similar trends in bond lengths are observed in the HF/DZ [36] and HF/6-31G* [37] optimized geometries.

The cyclic structure **4** with an acetylenic geometry is predicted to be 6.9 kcal/mol more stable than **1** at the HF/DZP level of theory [29]. Although the inclusion of electron correlation effects at the MP2/DZP level of theory reverses that order of stabilities, in light of the discussion on $C_{20}$ (Section 7.1) this is most probably an artifact. This conclusion is supported by calculations involving the BLYP functional, in which **1** is predicted to be less stable than **4** by 46.4 kcal/mol [37].

## 7.3 The $C_{28}$ Clusters

A tetrahedral structure for $C_{28}$, composed of 12 five-membered and four six-membered rings (Fig. 7.3), has been proposed [30, 38] and has attracted considerable attention in recent literature. HF/dz calculations show that at $T_d$ symmetry the $C_{28}$ fullerene has the high-spin $^5A_2$ quintet electronic ground state with one electron occupying the $8a_1$ orbital and three electrons on the $14t_2$ orbital [39, 40]. The $^3A_2$, $^3B_1$, and $^3T_1$ triplet states are located at 1.6, 2.7, and 4.1 eV, respectively, above the ground state. The excitation energy to the $^1A_1$ singlet state equals 3.1 eV. Although not subject to a Jahn–Teller distortion, the quintet ground state is found to distort freely to $C_{2v}$ symmetry in the HF/dz calculations. However, both the energy lowering (2.8 kcal/mol) and the changes in bond lengths (0.006 Å) are small, and the distortion is not sustained at the BLYP/dz level of theory.

For the singlet state, which undergoes a Jahn–Teller distortion, the topological $T_d$ symmetry of $C_{28}$ is reduced all the way down to $C_1$ in MNDO [19] and tight-binding Hamiltonian [18] geometry optimizations. The heat of formation, calculated within the MNDO approximation, is 871.0 kcal/mol, or 31.1 kcal/mol on a per-atom basis. Therefore, $C_{28}$ fullerene appears to be more stable than either of its $C_{20}$ or $C_{24}$ counterparts. Another isomer of $C_{28}$, with $D_{7d}$ topological symmetry that is reduced to $C_1$ upon geometry optimization, lies 125 kcal/mol higher than the $T_d$ structure.

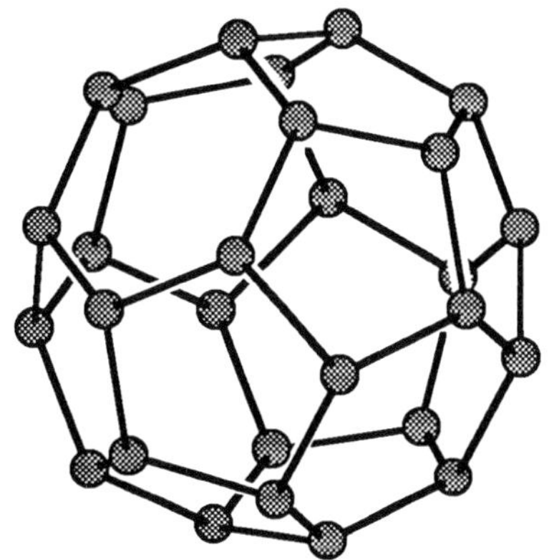

*Fig. 7.3. The $C_{28}$ fullerene.*

The $C_{28}$ fullerene is best visualized as a collection of four benzene rings that are connected through direct bonds as well as trivalent carbon atoms distributed in a tetrahedral fashion. The four trivalent atoms are located at the centers of triplets of fused pentagons. In the quintet electronic ground state of $C_{28}$, the four unpaired electrons are localized mostly on these carbon atoms, giving rise to four dangling bonds. Therefore, the $C_{28}$ fullerene can be viewed as an equivalent of a tetravalent atom with $sp^3$ directional bonding [39, 40–43]. The unsaturated valences can be directed either outward or inward with respect to the cage. In the first case, compounds such as $C_{28}H_4$ or $C_{28}F_4$ are formed. These molecules have singlet electronic ground states and $T_d$ symmetries [40]. At the HF/dz level of theory, the binding energies between the $C_{28}$ cage and the four substituents are estimated at 184 and 214 kcal/mol for the $C_{28}H_4$ and $C_{28}F_4$ species, respectively. The optimized geometries subscribe to the description of $C_{28}$ in terms of four interconnected benzene rings. In $C_{28}H_4$, the bonds within the benzene rings are all 1.391 Å long and the bonds connecting the rings have a length of 1.528 Å. The distance between each of the four carbon atoms that form the external C–H bonds and the respective benzene ring is 1.542 Å, indicating the presence of a single bond. The optimized C–H bond length is 1.077 Å. In $C_{28}F_4$, the corresponding bond lengths equal 1.388, 1.532, 1.541, and 1.382 Å.

Inward bonding is present in *endohedral compounds* that are formed when metal atoms are trapped inside the $C_{28}$ cage. In contrast to *endohedral complexes*, which are discussed in Chapter 8, bonding in endohedral compounds involves strong mixing of orbitals belonging to the trapped atom and the cage [44]. Formation of endohedral compounds is the probable reason for the presence of strong $C_{28}Ti^+$, $C_{28}Zr^+$, $C_{28}Hf^+$, and $C_{28}U^+$ peaks in the mass spectra of vapors obtained by laser vaporization of the respective graphite–metal oxide mixtures [39]. HF/dz calculations on endohedral compounds of $C_{28}$ with Si, Sn, Mg, Ca, Al, Sc, Ti, and Zr atoms [40] demonstrate that these species fall into three distinct categories. The $Sn@C_{28}$ and $Zr@C_{28}$ molecules have closed-shell electronic configurations and $T_d$ symmetry. The $Ti@C_{28}$, $Al@C_{28}$, and $Sc@C_{28}$ molecules have either closed- or open-shell electronic configurations that are not subject to a Jahn–Teller distortion, yet they spontaneously lower their symmetries to $C_{2v}$. Finally, the $Mg@C_{28}$, $Ca@C_{28}$, and $Si@C_{28}$ species have a degenerate triplet ground electronic state at $T_d$ symmetry, which they break because of the Jahn–Teller effect.

The HF/dz calculations predict that the $Sc@C_{28}$ and $Zr@C_{28}$ endohedral compounds are bound with respect to their constituents [40]. However, the $Ti@C_{28}$ molecule has a negative binding energy. At both the HF/dz [40] and LDA [45] levels of theory, the total energy of $Ti@C_{28}$ attains its minimum when the Ti atom is displaced from the cage center. The calculated displacement is close to 0.5 Å in both methods, with the Ti atom shifted toward one of the carbon atoms with dangling bonds. The stabilization due to this displacement is quite large, amounting to 41.5 and 47.3 kcal/mol at the HF/dz and LDA levels of theory, respectively. The LDA dipole moment is 2.82 D, indicating substantial charge transfer between the Ti atom and the $C_{28}$ cage. It should also be remarked that, in contrast to the HF/dz result, the LDA calculations predict a large, positive binding energy of 259 kcal/mol for $Ti@C_{28}$ [45]. However, taking into account the tendency of the LDA formalism to overestimate bond energies, this is most probably an artifact.

The $Ce@C_{28}$ species has also been studied with the LDA method [41, 42]. The host atom has been found to remain at the cage center. Again, the calculated binding energy of 316 kcal/mol seems rather unrealistic. The bonding in $Ce@C_{28}$ and other endohedral compounds of the $C_{28}$ fullerene has been rationalized with the help of a model in which the fullerene cage is approximated by a sphere [41, 42]. Finally, one should mention that the $Ti@C_{28}H_4$ molecule has been investigated at the HF/dz [40] and LDA [42] levels of theory, and a triplet open-shell configuration has been predicted in both cases.

## 7.4 Fullerenes with 50 or More Carbon Atoms

When irradiated with laser light of high fluence, the $C_{60}$ fullerene undergoes photofragmentation in which $C_2$ fragments are removed from the $C_{60}$ cage, leading to clusters $C_N$, where $34 \leq N \leq 58$ is an even number [46]. A mechanism for this process has been proposed [46]. This Rice mechanism (Fig. 7.4) involves an initial Stone–Wales transformation [47], which results in a $C_{60}$ cluster with an abutting pair of five-membered rings. This is followed by excision of the $C_2$ fragment, which turns the pentagon pair into a hexagon and the two adjacent hexagons into pentagons to yield the $C_{58}$ cage. Repeated application of these two steps leads to further loss of $C_2$ molecules and formation of even

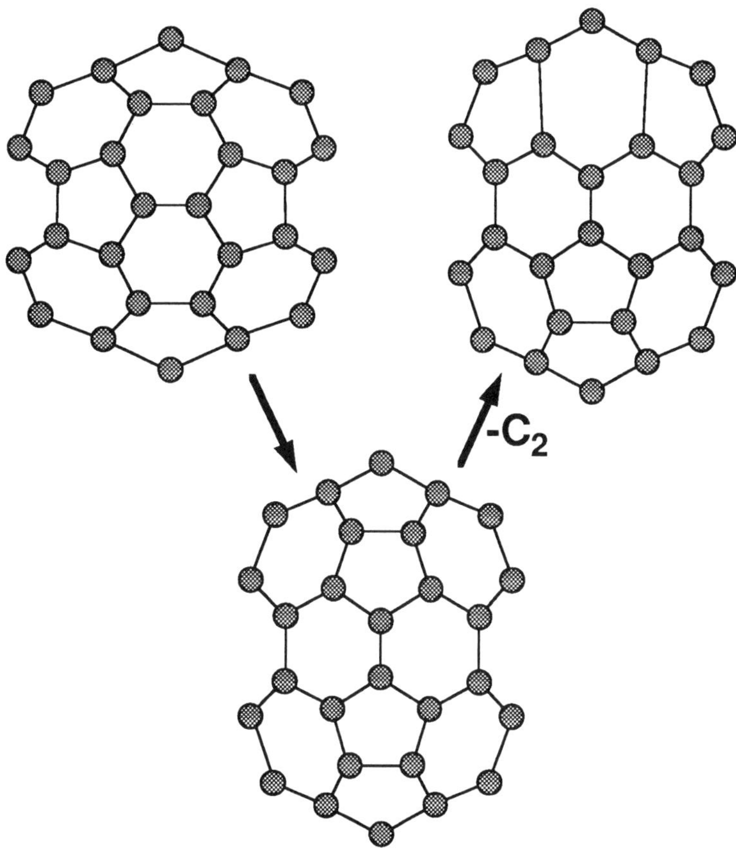

*Fig. 7.4. The Rice mechanism for elimination of
the $C_2$ fragment from fullerene cages.*

smaller fullerenes. It should be mentioned that in $C_{60}$ the first step of the
Rice mechanism has a rather large barrier (Chapter 3) and alternative routes
to the $C_2$ eliminations have been proposed [48, 49].

The first product of the $C_{60}$ photofragmentation is the $C_{58}$ cluster. There
is only one isomer of $C_{58}$ in which the number of abutting pentagons is minimized [50]. This low-energy structure (Fig. 7.5) possesses three pentagon–
pentagon edges and topological $C_{3v}$ symmetry that is subject to a Jahn–Teller
distortion. A $C_s$ minimum for $C_{58}$ is found in MNDO calculations, and the
calculated heat of formation equals 944 kcal/mol. The same symmetry is predicted by the tight-binding Hamiltonian approach [18].

The $C_{56}$ cluster has at least two isomers containing the minimal number
of abutting pentagons [50]. In each of these isomers four pentagon–pentagon

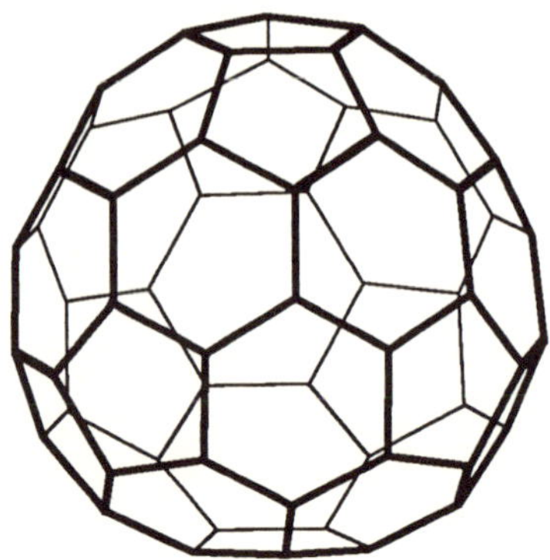

Fig. 7.5. The low-energy isomer of the $C_{58}$ cluster.

edges are present. The $C_2$ structure (**1**; Fig. 7.6), with the heat of formation estimated at 936 kcal/mol by the MNDO method, is more stable than the $C_{2v}$ isomer (**2**; Fig. 7.6), for which the predicted heat of formation is 963 kcal/mol. Structures incorporating either seven- ($C_1$ symmetry) [19] or four-membered rings ($D_{4h}$ symmetry) [32] are much higher in energy.

There are at least two low-energy isomers of the $C_{54}$ cluster, both of them with six pentagon–pentagon edges [50]. With the respective MNDO heats of formation equal to 970 and 965 kcal/mol, the $C_s$ (**1**; Fig. 7.7) and $C_{2v}$ (**2**; Fig. 7.7) structures are almost isoenergetic. The tight-binding Hamiltonian calculations [18] also predict a structure with $C_{2v}$ symmetry to have the lowest energy.

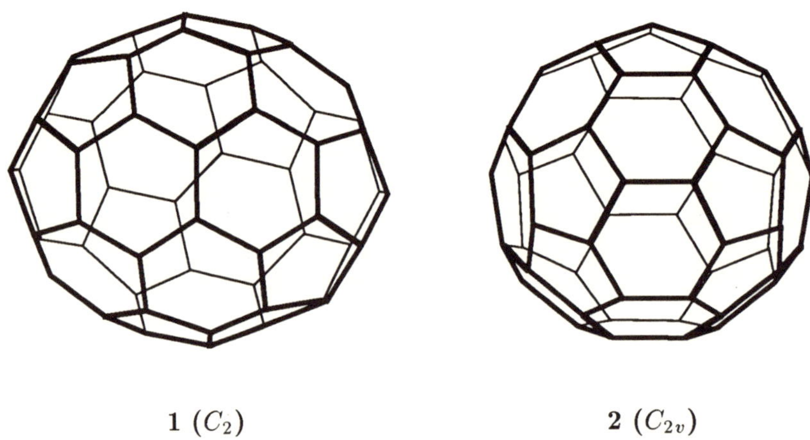

**1** ($C_2$)                                                    **2** ($C_{2v}$)

Fig. 7.6. The low-energy isomers of the $C_{56}$ cluster.

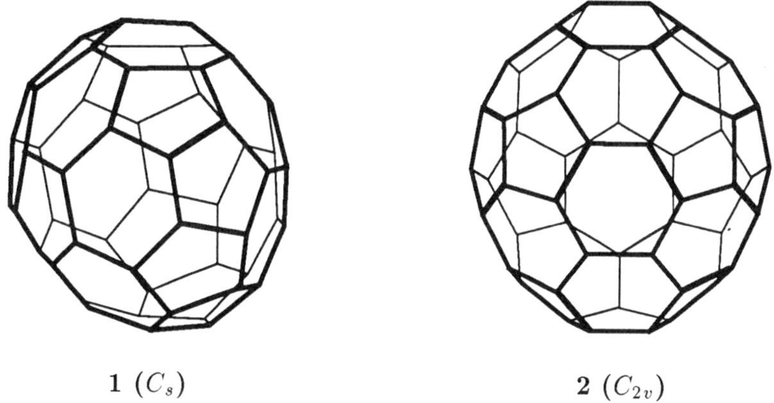

**1** ($C_s$)  **2** ($C_{2v}$)

Fig. 7.7. *The low-energy isomers of the $C_{54}$ cluster.*

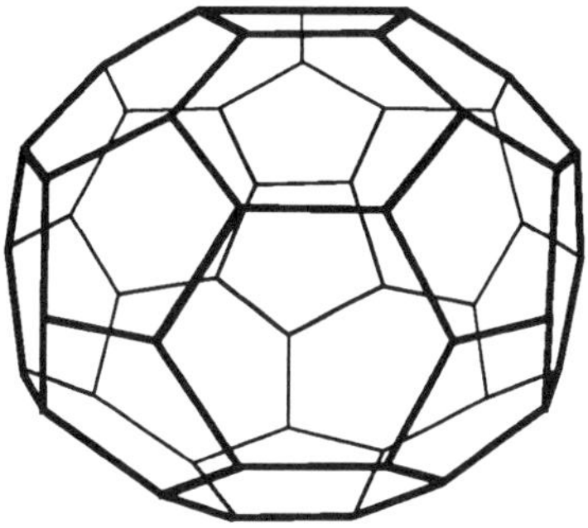

**1** ($D_{5h}$) 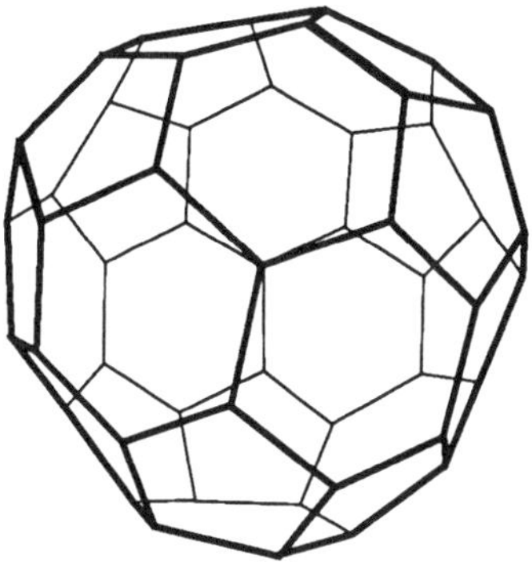  **2** ($D_3$)

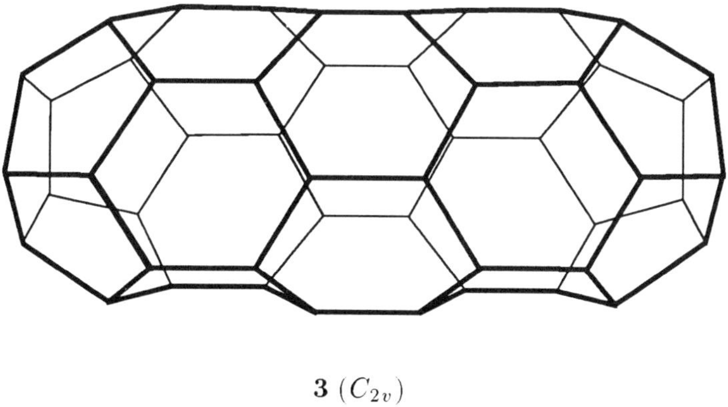

**3** ($C_{2v}$)

Fig. 7.8. *The low-energy isomers of the $C_{50}$ cluster.*

The $C_{50}$ cluster has been investigated with several theoretical approaches. Three isomers with $D_{5h}$, $D_3$, and $C_{2v}$ symmetries (**1**, **2**, and **3**; Fig. 7.8) have been studied with the MNDO method [19, 31]. The $D_{5h}$ structure **1** with five pairs of abutting pentagons [12], originally proposed for the $C_{50}$ cluster [38], has the lowest heat of formation (922.2 kcal/mol). The $D_3$ structure **2** with six pairs of abutting pentagons [12], suggested to be the most stable isomer on the basis of Hückel calculations [51], has a somewhat higher heat of formation (929.1 kcal/mol). The $C_{2v}$ isomer **3**, with the heat of formation amounting to 1300.1 kcal/mol, is very unfavorable energetically. Interestingly, because of substantial differences in the vibrational contributions to the free enthalpy, isomer **2** is predicted to be more stable than **1** at temperatures above *ca.* 1400 K [52]. Calculations involving the tight-binding Hamiltonian approximation favor the $D_{5h}$ isomer **1** of $C_{50}$ [18]. The geometry of this isomer has also been optimized at the MP2/DZ level of theory [30].

## 7.5  Other Small Fullerenes

Many small fullerenes with 30–50 carbon atoms have been studied with electronic structure methods. In particular, the important question of the stability of fullerenes versus that of cyclic structures with the same number of carbon atoms has been addressed with a series of MP2/DZ calculations [30]. On the basis of these calculations, it has been concluded that fullerenes become more stable as the cluster size reaches about 32 atoms. This conclusion is in good agreement with the experimental observations discussed at the beginning of this chapter.

Partial geometry optimizations at the MP2/DZ level of theory yield $D_{5h}$, $D_3$, and $D_{6h}$ geometries for the $C_{30}$, $C_{32}$, and $C_{36}$ fullerenes, respectively [30]. However, two different low-energy $C_{2v}$ structures that correspond to local energy minima are found for the $C_{30}$ species with the MNDO method [19, 20, 50]. The $C_{2v}$ geometry is favored as well by the HF/STO-3G, MM3 [53], and tight-binding Hamiltonian [18] methods. All three approaches also predict the $D_3$ structure to be the lowest-energy isomer of the $C_{32}$ fullerene, whereas MNDO calculations find a $C_1$ geometry [19, 20]. The $C_{36}$ fullerene is predicted to possess $D_{2d}$ symmetry by the tight-binding Hamiltonian calculations [18]. Gas-phase drift mobilities of the $D_{2d}$ and $C_{2v}$ isomers of the $C_{36}$ fullerene have been calculated from their MNDO, HF/STO-3G, and HF/DZ optimized geometries [54]. These mobilities have turned out to be virtually identical. Geometries and drift mobilities of the corresponding mono- and polycyclic, planar and "cup" $C_{36}$ clusters have also been predicted.

As mentioned in Section 5.2, the MNDO UHF method favors small carbon cages with four- and six-membered rings over those with abutting pentagons [32]. For the $C_{32}$ cluster, the $D_3$ structure with six hexagons and 12 pentagons is 10.9 kcal/mol higher than the $S_4$ isomer with eight six-, eight five-, and two four-membered rings. However, as noted earlier, the MNDO energy minimum for the $C_{32}$ fullerene is attained at $C_1$ rather than $D_3$ symmetry, the MNDO UHF wavefunctions for carbon clusters are usually heavily spin-contaminated, and the analogous energy ordering for the $C_{24}$ clusters is not confirmed with more accurate calculations. Hence, the preceding result is quite possibly an artifact.

Structures of somewhat larger $C_{40}$ and $C_{42}$ fullerenes have been studied with the MNDO [19, 20, 31] and tight-binding Hamiltonian [18] approaches. $D_{5d}$ and $D_2$ structures have been considered for the $C_{40}$ species together with the $C_s$ and $D_3$ structures for $C_{42}$. Full geometry optimizations have been carried out for all 87 isomers of the $C_{44}$ fullerene [55]. As expected, no structures with exceptionally high stabilities have been found in these molecular mechanics calculations. Finally, rather unusual $C_{44}$ and $C_{45}$ cages have been the subject of LDA calculations [56].

Despite some quantitative differences, results of electronic structure calculations on fullerenes, carried out with methods possessing various levels of sophistication [18, 19, 30, 33, 57, 58], show the same trend in calculated energies per carbon atom. The smaller fullerenes are invariably found to be less stable than their larger homologs, reflecting the diminution of strain with increasing cage size. The calculated heats of formation per carbon atom can be fitted for smaller fullerenes with a simple expression derived from elasticity theory [59], but large deviations are observed for clusters with fewer than 60 carbon atoms [58]. A more accurate correlation between the heats of formation per carbon atom calculated with the MNDO method and the average local curvatures of the optimized cage geometries has been reported [19]. The error in the heats of formation obtained from such a correlation amounts to only 1.3 kcal/mol for the computed values spanning the range of 4.8–41.3 kcal/mol.

In summary, calculations employing the contemporary methods of quantum chemistry successfully account for the observed trends in properties of small fullerenes. Whenever feasible, predictions based on *ab initio* approaches are preferable, especially for energy comparisons involving molecules with the same number of bonds. On the other hand, reliable (semi)quantitative estimates of energy differences for clusters with different numbers of bonds are not available at present, although methods involving approximate correlation energy functionals, such as BLYP, appear promising.

# References


1. W. Weltner, Jr., and R. J. Van Zee, *Carbon Molecules, Ions, and Clusters*, Chem. Rev. **89**, 1713 (1989).
2. K. S. Pitzer and E. Clementi, *Large Molecules in Carbon Vapor*, J. Am. Chem. Soc. **81**, 4477 (1959).

3. Z. Slanina and R. Zahradník, *MINDO/2 Study of Equilibrium Carbon Vapor*, J. Phys. Chem. **81**, 2252 (1977).

4. J. D. Watts, J. Gauss, J. F. Stanton, and R. J. Bartlett, *Linear and Cyclic Isomers of $C_4$. A Theoretical Study with Coupled-Cluster Methods and Large Basis Sets*, J. Chem. Phys. **97**, 8372 (1992).

5. K. Raghavachari and J. S. Binkley, *Structure, Stability, and Fragmentation of Small Carbon Clusters*, J. Chem. Phys. **87**, 2191 (1987).

6. R. A. Whiteside, R. Krishnan, M. J. Frisch, J. A. Pople, and P.v.R. Schleyer, *Cyclic $C_3$ Structures*, Chem. Phys. Lett. **80**, 547 (1981).

7. Z. Slanina, J. Kurtz, and L. Adamowicz, *A Computational Response to the $C_7$ Experimental Challenge: Low-Energy $C_{2v}$ Isomer*, Chem. Phys. Lett. **196**, 208 (1992).

8. W. Andreoni, D. Scharf, and P. Giannozzi, *Low-Temperature Structures of $C_4$ and $C_{10}$ from the Car-Parrinello Method: Singlet States*, Chem. Phys. Lett. **173**, 449 (1990).

9. V. Parasuk and J. Almlöf, *The Electronic and Molecular Structure of Carbon Clusters: $C_8$ and $C_{10}$*, Theor. Chim. Acta **83**, 227 (1992).

10. G. von Helden, M.-T. Hsu, P. R. Kemper, and M. T. Bowers, *Structures of Carbon Cluster Ions from 3 to 60 Atoms: Linears to Rings to Fullerenes*, J. Chem. Phys. **95**, 3835 (1991).

11. G. von Helden, M.-T. Hsu, N. Gotts, and M. T. Bowers, *Carbon Cluster Cations with up to 84 Atoms: Structures, Formation Mechanism, and Reactivity*, J. Phys. Chem. **97**, 8182 (1993).

12. T. G. Schmalz, W. A. Seitz, D. J. Klein, and G. E. Hite, *Elemental Carbon Cages*, J. Am. Chem. Soc. **110**, 1113 (1988).

13. P. W. Fowler, J. E. Cremona, and J. I. Steer, *Systematics of Bonding in Non-icosahedral Carbon Clusters*, Theor. Chim. Acta **73**, 1 (1988).

14. L. A. Paquette, D. W. Balogh, R. Usha, D. Kountz, and G. G. Christoph, *Crystal and Molecular Structure of a Pentagonal Dodecahedrane*, Science **211**, 575 (1981).

15. L. A. Paquette, R. J. Ternansky, D. W. Balogh, and G. Kentgen, *Total Synthesis of Dodecahedrane*, J. Am. Chem. Soc. **105**, 5446 (1983).

16. V. Parasuk and J. Almlöf, *$C_{20}$: The Smallest Fullerene?*, Chem. Phys. Lett. **184**, 187 (1991).

17. K. Raghavachari, D. L. Strout, G. K. Odom, G. E. Scuseria, J. A. Pople, B. G. Johnson, and P. M. W. Gill, *Isomers of $C_{20}$. Dramatic Effect of Gradient Corrections in Density Functional Theory*, Chem. Phys. Lett. **214**, 357 (1993).

18. B. L. Zhang, C. Z. Wang, K. M. Ho, C. H. Xu, and C. T. Chan, *The Geometry of Small Fullerene Cages: $C_{20}$ to $C_{70}$*, J. Chem. Phys. **97**, 5007 (1992).

19. D. Bakowies and W. Thiel, *MNDO Study of Large Carbon Clusters*, J. Am. Chem. Soc. **113**, 3704 (1991).

20. D. Bakowies and W. Thiel, *Theoretical Infrared Spectra of Large Carbon Clusters*, Chem. Phys. **151**, 309 (1991).

21. Z. Slanina and L. Adamowicz, *On the Relative Stabilities of Dodecahedron-Shaped and Bowl-Shaped Structures of $C_{20}$*, Thermochim. Acta **205**, 299 (1992).

22. Z. Slanina and L. Adamowicz, *One-, Two- and Three-Dimensional Structures of $C_{20}$*, Full. Sci. Tech. **1**, 1 (1993).

23. P. W. Fowler, P. Lazzeretti, and R. Zanasi, *Electric and Magnetic Properties of the Aromatic Sixty-Carbon Cage*, Chem. Phys. Lett. **165**, 79 (1990).

24. C. J. Brabec, E. B. Anderson, B. N. Davidson, S. A. Kajihara, Q.-M. Zhang, J. Bernholc, and D. Tománek, *Precursors to $C_{60}$ Fullerene Formation*, Phys. Rev. B **46**, 7326 (1992).

25. J. D. Watts and R. J. Bartlett, *The Nature of Monocyclic $C_{10}$. A Theoretical Investigation Using Coupled-Cluster Methods*, Chem. Phys. Lett. **190**, 19 (1992).

26. C. Liang and H. F. Schaefer III, *Carbon Clusters: The Structure of $C_{10}$ Studied with Configuration Interaction Methods*, J. Chem. Phys. **93**, 8844 (1990).

27. V. Parasuk, J. Almlöf, and M. W. Feyereisen, *The [18] All-Carbon Molecule: Cumulene or Polyacetylene?*, J. Am. Chem. Soc. **113**, 1049 (1991).

28. G. von Helden, M. T. Hsu, N. G. Gotts, P. R. Kemper, and M. T. Bowers, *Do Small Fullerenes Exist Only on the Computer? Experimental Results on $C_{20}^{+/-}$ and $C_{24}^{+/-}$*, Chem. Phys. Lett. **204**, 15 (1993).

29. F. Jensen and H. Toftlund, *Structure and Stability of $C_{24}$ and $B_{12}N_{12}$ Isomers*, Chem. Phys. Lett. **201**, 89 (1993).

30. M. Feyereisen, M. Gutowski, J. Simons, and J. Almlöf, *Relative Stabilities of Fullerene, Cumulene, and Polyacetylene Structures for $C_n$: $n = 18 - 60$*, J. Chem. Phys. **96**, 2926 (1992).

31. M. D. Newton and R. E. Stanton, *Stability of Buckminsterfullerene and Related Carbon Clusters*, J. Am. Chem. Soc. **108**, 2469 (1986).

32. Y.-D. Gao and W. C. Herndon, *Fullerenes with Four-Membered Rings*, J. Am. Chem. Soc. **115**, 8459 (1993).

33. N. Kurita, K. Kobayashi, H. Kumahora, K. Tago, and K. Ozawa, *Nonlocal Density Functional Calculations of Binding Energies of Carbon Fullerenes $C_n$, with $n = 10, 12, 16, 20, 24, 28, 32, 36, 50, 60, 70, 80, 90, 100, 110$ and $120$*, Chem. Phys. Lett. **188**, 181 (1992).

34. J. M. Schulman, R. L. Disch, M. A. Miller, and R. C. Peck, *Symmetrical Clusters of Carbon Atoms: The $C_{24}$ and $C_{60}$ Molecules*, Chem. Phys. Lett. **141**, 45 (1987).

35. M. L. McKee and W. C. Herndon, *Calculated Properties of $C_{60}$ Isomers and Fragments*, J. Mol. Struct. (Theochem) **153**, 75 (1987).

36. H. P. Lüthi and J. Almlöf, *Ab Initio Studies on the Thermodynamic Stability of the Icosahedral $C_{60}$ Molecule "Buckminsterfullerene"*, Chem. Phys. Lett. **135**, 357 (1987).

37. K. Raghavachari, B. Zhang, J. A. Pople, B. G. Johnson, and P. M. W. Gill, *Isomers of $C_{24}$. Density Functional Studies Including Gradient Corrections*, Chem. Phys. Lett. **220**, 385 (1994).

38. H. W. Kroto, *The Stability of the Fullerenes $C_n$, with $n = 24, 28, 32, 36, 50, 60,$ and 70*, Nature **329**, 529 (1987).

39. T. Guo, M. D. Diener, Y. Chai, M. J. Alford, R. E. Haufler, S. M. McClure, T. Ohno, J. H. Weaver, G. E. Scuseria, and R. E. Smalley, *Uranium Stabilization of $C_{28}$: A Tetravalent Fullerene*, Science **257**, 1661 (1992).

40. T. Guo, R. E. Smalley, and G. E. Scuseria, *Ab Initio Theoretical Predictions of $C_{28}$, $C_{28}H_4$, $C_{28}F_4$, $(Ti@C_{28})H_4$, and $M@C_{28}$ ($M = Mg, Al, Si, S, Ca, Sc, Ti, Ge, Zr,$ and $Sn$)*, J. Chem. Phys. **99**, 352 (1993).

41. N. Rösch, O. D. Häberlen, and B. I. Dunlap, *Bonding in Endohedral Metal-Fullerene Complexes: f-Orbital Covalency in $Ce@C_{28}$*, Angew. Chem. Int. Ed. Engl. **32**, 108 (1993).

42. O. D. Häberlen, N. Rösch, and B. I. Dunlap, *Are Endohedral Metal(IV) $C_{28}$ Compounds Hypervalent?*, Chem. Phys. Lett. **200**, 418 (1992).

43. G. E. Scuseria, *Ab Initio Theoretical Predictions of Fullerenes*, in *Buckminsterfullerenes* (W. E. Billups and M. A. Ciufolini, eds.), VCH Publishers, New York, 1993, Ch. 5, p. 103.

44. J. Cioslowski and K. Raghavachari, *Electrostatic Potential, Polarization, Shielding, and Charge Transfer in Endohedral Complexes of the $C_{60}$, $C_{70}$, $C_{76}$, $C_{78}$, $C_{82}$, and $C_{84}$ Clusters*, J. Chem. Phys. **98**, 8734 (1993).

45. B. I. Dunlap, O. D. Häberlen, and N. Rösch, *Asymmetric Localization of Titanium in $C_{28}$*, J. Phys. Chem. **96**, 9095 (1992).

46. S. C. O'Brien, J. R. Heath, R. F. Curl, and R. E. Smalley, *Photophysics of Buckminsterfullerene and Other Carbon Cluster Ions*, J. Chem. Phys. **88**, 220 (1988).

47. A. J. Stone and D. J. Wales, *Theoretical Studies of Icosahedral $C_{60}$ and Some Related Species*, Chem. Phys. Lett. **128**, 501 (1986).

48. R. Saito, G. Dresselhaus, and M. S. Dresselhaus, *Topological Defects in Large Fullerenes*, Chem. Phys. Lett. **195**, 537 (1992).

49. R. L. Murry, D. L. Strout, G. K. Odom, and G. E. Scuseria, *Role of $sp^3$ Carbon and 7-Membered Rings in Fullerene Annealing and Fragmentation*, Nature **366**, 665 (1993).

50. R. E. Stanton, *Fullerene Structures and Reactions: MNDO Calculations*, J. Phys. Chem. **96**, 111 (1992).

51. D. E. Manolopoulos, J. C. May, and S. E. Down, *Theoretical Studies of the Fullerenes: $C_{34}$ to $C_{70}$*, Chem. Phys. Lett. **181**, 105 (1991).

52. Z. Slanina, L. Adamowicz, D. Bakowies, and W. Thiel, *Fullerene $C_{50}$*

*Isomers: Temperature-Induced Interchange of Relative Stabilities*, Thermochim. Acta **202**, 249 (1992).

53. R. L. Murry, J. R. Colt, and G. E. Scuseria, *How Accurate Are Molecular Mechanics Predictions for Fullerenes? A Benchmark Comparison with Hartree–Fock Self-Consistent Field Results*, J. Phys. Chem. **97**, 4954 (1993).

54. L. D. Book, C. Xu, and G. E. Scuseria, *Carbon Cluster Ion Drift Mobilities. The Importance of Geometry and Vibrational Effects*, Chem. Phys. Lett. **222**, 281 (1994).

55. M. Lyons, B. I. Dunlap, D. W. Brenner, D. H. Robertson, R. C. Mowrey, J. W. Mintmire, and C. T. White, *Relative Energetics of $C_{44}$ Fullerene Isomers*, in *Physics and Chemistry of Finite Systems: From Clusters to Crystals*, Vol. II (P. Jena et al., eds.), Kluwer, Dordrecht, 1992, p. 1347.

56. D. W. Brenner, B. I. Dunlap, J. A. Harrison, J. W. Mintmire, R. C. Mowrey, D. H. Robertson, and C. T. White, *Group-IV Covalent Clusters: $Si_{45}$ and $C_{44}$ versus $Si_{44}$ and $C_{45}$*, Phys. Rev. B **44**, 3479 (1991).

57. B. L. Zhang, C. H. Xu, C. Z. Wang, C. T. Chan, and K. M. Ho, *Systematic Study of Structures and Stabilities of Fullerenes*, Phys. Rev. B **46**, 7333 (1992).

58. C. Z. Wang, B. L. Zhang, C. H. Xu, C. T. Chan, and K. M. Ho, *Structures and Stabilities of Carbon Fullerenes*, Int. J. Mod. Phys. B **6**, 3833 (1992).

59. J. Tersoff, *Energies of Fullerenes*, Phys. Rev. B **46**, 15546 (1992).

# Chapter 8

# Endohedral Complexes

One of the most interesting properties of fullerenes is their ability to encapsulate atoms, ions, and small molecules. The species resulting from such an encagement fall into one of two categories [1]. In *endohedral complexes* [2], mixing between molecular orbitals of the guest and those of the fullerene host cage is negligible. Therefore, bonding in endohedral complexes is dominated by electrostatic and polarization effects. On the other hand, in *endohedral compounds* [1], substantial orbital mixing is present, giving rise to direct chemical bonding. Endohedral complexes, such as $He@C_{60}$, are formed between small guests and fullerenes with closed-shell electronic configurations. Endohedral compounds, such as those between the $C_{28}$ fullerene and various metals (Chapter 7), ensue when the guests are large or fullerenes with dangling bonds are involved.

A wealth of experimental data on endohedral species has been accumulated in recent years. The first hints of endohedral encapsulation have been provided by the observation of the $C_{60}La^+$ ion in the mass spectrum of vapors produced by laser ablation of graphite impregnated with $LaCl_3$ [3]. Although, at first, the endohedral character of the observed ion has been only hypothesized (and subsequently disputed [4]), the "shrink-wrapping" experiments [5] have provided strong evidence for it. Further indirect evidence has been gathered by the gas-phase synthesis of ions in which metals, such as La, Rh, Cu, Fe, Co, and Ni [6–8], are exohedrally (externally) bound to the $C_{60}$ cage. Properties of such ions are very distinct from those of the original $C_{60}La^+$ species.

The first direct confirmation of the existence of endohedral complexes has been furnished by collisional insertion experiments [9]. In these experiments, high-energy (8 keV) $C_{60}^+$ ions are collided with a target gas, such as He, and the products of the collisions are analyzed by means of mass spectrometry. Since an exohedrally bound complex between $C_{60}^+$ and He is not expected to survive the drastic conditions of the experiment, the detection of a stable $C_{60}He^+$ ion among the collision products confirms its endohedral character [9–11]. The $C_{60}He^+$ ion can be neutralized to $C_{60}He$ [12]. Under similar conditions, Ne and $D_2$ have also been inserted into the $C_{60}$ cage [11, 13, 14], but attempts to incorporate the $N_2$, NO, or $O_2$ molecules have failed. Although the $C_{60}^{2+}$, $C_{60}^{3+}$, $C_{70}^{2+}$, and $C_{70}^{3+}$ ions can be used in place of the corresponding monocations [15–17], colliding neutral $C_{60}$ with $Ne^+$ does not produce intact $C_{60}Ne^+$ [18]. In contrast, the $C_{60} + Li^+$ and $C_{60} + Na^+$ collisions yield the corresponding endohedral complexes of $C_{60}$ [19]. Collisions between $C_{60}$ and $O^+$ have been speculated to produce the $CO@C_{58}^+$ endohedral complex [20].

Another route to insertion of the noble gas atoms into fullerene cages is offered by recent high-pressure experiments. Heating $C_{60}$ samples under an atmosphere of He or Ne to 600 °C results in the formation of the respective endohedral complexes, as evidenced by mass spectrometry [21]. However, the yields are extremely low (*ca.* $10^{-7}$–$10^{-6}$) at moderate pressures. On the other hand, increasing the pressure to over 200 MPa raises the yields to fractions of a percent for the complexes of He, Ne, Ar, and Kr with $C_{60}$ and $C_{70}$ [22]. The amount of the $He@C_{60}$ and $He@C_{70}$ complexes produced is large enough to make measurement of their $^3He$ NMR spectra possible [23]. The Xe atom can also be inserted into the $C_{60}$ and $C_{70}$ cages, but with lower yields.

ESR spectra of metallofullerenes formed between the $C_{60}$, $C_{80}$, or $C_{82}$ clusters and the transition or rare-earth metals [24–26] have been the subjects of numerous studies. The metal atoms in these metallofullerenes are believed to be located inside the fullerene cages, although an EXAFS study disputing this presumption has been published [27] and then rebutted [28]. The ESR spectra suggest that the $Sc@C_{82}$, $Y@C_{82}$, and $La@C_{82}$ species possess two interconvertible isomers [29, 30]. In addition, the spectra reveal that the dominant isomers of $Y@C_{82}$ and $La@C_{82}$ have the $Y^{3+}@C_{82}^{3-}$ [31] and $La^{3+}@C_{82}^{3-}$ [32] structures, respectively. ESR studies of the $Sc@C_{82}$, $Sc_2@C_{82}$, and $Sc_3@C_{82}$ metallofullerenes have also been published [33, 34]. Pure $La@C_{82}$ has been isolated [35] and its UV spectrum and electrochemical properties [36] have been measured. Pure samples of $Sc_2@C_{74}$, $Sc_2@C_{82}$, and $Sc_2@C_{84}$ have been prepared as well [37]. Properties of metallofullerenes have been recently reviewed [38].

Experimental evidence for endohedral species involving other guests, such as muonium [39], microcrystals of $LaC_2$ [40], and metal halides [41], is also available. In summary, it is fair to state that from the experimental standpoint the existence of endohedral compounds and complexes is now firmly established.

The host–guest interactions in endohedral compounds involve direct chemical bonding, the character of which changes from one species to another. On the other hand, a simple theory that affords quantitative predictions of electronic properties for a variety of endohedral complexes can be formulated.

## 8.1  Bonding in Endohedral Complexes

A simple, yet accurate picture of bonding in endohedral complexes is furnished by a mixed variational-perturbative approach [1, 2, 42, 43]. The perturbation theory can be successfully used to describe electronic structures of endohedral complexes because mixing between the orbitals of the host H and the guest G is usually negligible. On the other hand, the variational aspect of the theory makes it possible to construct effective Hamiltonians for the guests.

The aforementioned approach is based upon the expression for the wavefunction of the endohedral complex H·G in its electronic ground state, which reads [42]

$$\Psi_0(\text{H·G}) = \sum_i c_i \hat{A}[\Psi_i^0(\text{H})\Psi_0(\text{G})] \ . \tag{8.1}$$

In Eq. (8.1), $\hat{A}$ is an antisymmetrizer, $\Psi_i^0(\text{H})$ stands for the *unperturbed* wavefunction of the $i$th state of the host, and $\Psi_0(\text{G})$ is the variationally determined ground-state wavefunction of the guest. The linear combination coefficients $\{c_i\}$ and the corresponding ground-state energy are calculated with perturbation theory. The matrix elements of the electronic Hamiltonian that describes the endohedral complex and includes the nuclear–nuclear repulsion energy are given by

$$H_{ij} = \langle \hat{A}[\Psi_i^0(\text{H})\Psi_0(\text{G})]|\hat{H}|\hat{A}[\Psi_j^0(\text{H})\Psi_0(\text{G})]\rangle = H_{ij}^0 + V_{ij} \ . \tag{8.2}$$

The unperturbed Hamiltonian has matrix elements

$$H_{ij}^0 = \delta_{ij}\left[\langle \Psi_0(\text{G})|\hat{H}(\text{G})|\Psi_0(\text{G})\rangle + E_i^0(\text{H})\right] , \tag{8.3}$$

where $\hat{H}(\text{G})$ is the Hamiltonian of the guest *in vacuo* and $E_i^0(\text{H})$ stands for the energy of the $i$th electronic state of the unperturbed host. Provided the guest and the host are separated by a distance large enough to keep the exchange and nonorthogonality effects negligible, the matrix elements of the perturbation operator are equal to

$$V_{ij} = \int \rho_0(\vec{r})\Lambda_{ij}(\vec{r})\,d\vec{r} \ , \tag{8.4}$$

where $\rho_0(\vec{r})$ is the ground-state charge density of the guest (including the contribution from its nuclei) corresponding to $\Psi_0(\text{G})$ and

$$\Lambda_{ij}(\vec{r}) = \delta_{ij}\sum_{A\epsilon\text{H}} Z_A|\vec{r}-\vec{R}_A|^{-1} - \langle \Psi_i^0(\text{H})|\sum_{k\epsilon\text{H}}|\vec{r}-\vec{r}_k|^{-1}|\Psi_j^0(\text{H})\rangle \tag{8.5}$$

is related to the electrostatic potential generated by the host. In Eq. (8.5), the sums run over the nuclei $A$ and the electrons $k$ of the host.

With the help of perturbation theory, the ground-state energy of an endohedral complex can be expressed as

$$E_0(\text{H·G}) = H_{00}^0 + V_{00} + \sum_i{}' [E_0^0(\text{H}) - E_i^0(\text{H})]^{-1} V_{0i} V_{i0} + \cdots . \tag{8.6}$$

Eq. (8.6) can be rewritten in terms of many-body potentials,

$$E_0(\text{H·G}) = \langle \Psi_0(\text{G})|\hat{H}(\text{G})|\Psi_0(\text{G})\rangle + E_0^0(\text{H}) + \int \rho_0(\vec{r}_1) U^1(\vec{r}_1)\, d\vec{r}_1$$

$$+ \iint \rho_0(\vec{r}_1)\rho_0(\vec{r}_2) U^2(\vec{r}_1,\vec{r}_2)\, d\vec{r}_1 d\vec{r}_2 + \cdots , \tag{8.7}$$

with

$$U^1(\vec{r}_1) \equiv \Lambda_{00}(\vec{r}_1) , \tag{8.8}$$

$$U^2(\vec{r}_1,\vec{r}_2) \equiv \sum_i{}' [E_0^0(\text{H}) - E_i^0(\text{H})]^{-1} \Lambda_{0i}(\vec{r}_1)\Lambda_{i0}(\vec{r}_2) , \tag{8.9}$$

and so on, allowing one to calculate $\Psi_0(\text{G})$ as the ground-state eigenfunction of an effective nonlinear Hamiltonian,

$$\hat{H}_{eff}(\text{G}) = \hat{H}(\text{G}) + \sum_{k\in\text{G}} \hat{U}^1(\vec{r}_k) + 2 \sum_{k\in\text{G}} \int \rho_0(\vec{r}) \hat{U}^2(\vec{r}_k,\vec{r})\, d\vec{r} + \cdots , \tag{8.10}$$

where the sums run over the electrons $k$ of the guest.

The leading terms in the effective Hamiltonian, Eq. (8.10), can be deduced with the help of the multipole expansion. Provided the guest–host separation is large enough, the electrostatic potential generated by the charge density of the guest can be expanded with respect to the cage center, yielding

$$\int \rho_0(\vec{r}')|\vec{r} - \vec{r}'|^{-1} d\vec{r}' = r^{-1} Q(\text{G}) + r^{-3}\vec{\mu}_0(\text{G}) \cdot \vec{r} + \cdots , \tag{8.11}$$

where $Q(\text{G})$ is the electric charge of the guest and $\vec{\mu}_0(\text{G})$ is its ground-state dipole moment with respect to the cage center. By combining Eqs. (8.4), (8.5), and (8.11), one arrives at the following expression for the perturbation operator matrix elements,

$$V_{ij} = Q(\text{G}) \left[\delta_{ij} \sum_{A\in\text{H}} Z_A R_A^{-1} - \langle \Psi_i^0(\text{H})| \sum_{k\in\text{H}} \hat{r}_k^{-1} |\Psi_j^0(\text{H})\rangle\right]$$

$$+ \vec{\mu}_0(\text{G}) \cdot \left[\delta_{ij} \sum_{A\in\text{H}} Z_A R_A^{-3} \vec{R}_A - \langle \Psi_i^0(\text{H})| \sum_{k\in\text{H}} \hat{r}_k^{-3}\vec{r}_k |\Psi_j^0(\text{H})\rangle\right]$$

$$+ \cdots , \tag{8.12}$$

with all the position vectors relative to the cage center. The matrix elements that enter the first two terms in the expression for the ground-state energy of the complex, Eq. (8.6), are of particular interest,

$$V_{00} = Q(G) \left[ \sum_{A \epsilon H} Z_A R_A^{-1} - \left\langle \Psi_0^0(H) \Big| \sum_{k \epsilon H} \hat{r}_k^{-1} \Big| \Psi_0^0(H) \right\rangle \right]$$
$$+ \vec{\mu}_0(G) \cdot \left[ \sum_{A \epsilon H} Z_A R_A^{-3} \vec{R}_A - \left\langle \Psi_0^0(H) \Big| \sum_{k \epsilon H} \hat{r}_k^{-3} \vec{r}_k \Big| \Psi_0^0(H) \right\rangle \right]$$
$$+ \cdots , \tag{8.13}$$

and

$$V_{0i} = -Q(G) \left\langle \Psi_0^0(H) \Big| \sum_{k \epsilon H} \hat{r}_k^{-1} \Big| \Psi_i^0(H) \right\rangle$$
$$- \vec{\mu}_0(G) \cdot \left\langle \Psi_0^0(H) \Big| \sum_{k \epsilon H} \hat{r}_k^{-3} \vec{r}_k \Big| \Psi_i^0(H) \right\rangle - \cdots . \tag{8.14}$$

Extensive electronic structure calculations [1, 42–44] show that the electrostatic potential inside fullerenes is almost constant in the vicinities of their cage centers (Table 8.1). This endohedral potential $V_0(H)$ is always positive and decreases with the increasing size of the cage (Table 8.2) [1]. Since the regions around the centers of fullerene cages are almost completely devoid of electrons, the only possible explanation for the positive valuedness of $V_0(H)$ is that the probability of finding electrons inside fullerenes is noticeably smaller than that of finding them outside. This phenomenon, called the *endohedral effect*, can be explained with the help of the Thomas–Fermi theory [42].

### Table 8.1. The HF/DZP electrostatic potential in the $C_{60}$ cluster as a function of the distance from the cage center

| $r$ [Å] | Electrostatic potential $V(r)$ [V] | |
| --- | --- | --- |
| | Along the $C_5$ axis | In the direction of C atom |
| 0.00 | 0.767 | 0.767 |
| 0.25 | 0.767 | 0.767 |
| 0.50 | 0.767 | 0.767 |
| 0.75 | 0.770 | 0.770 |
| 1.00 | 0.784 | 0.778 |
| 1.25 | 0.824 | 0.808 |

## Table 8.2. The HF/DZP//MNDO endohedral potentials and monopole–monopole radial polarizabilities of selected fullerenes

| Cluster[a] | $V_0$ [V] | $V_0$ [kcal/mol] | $\alpha_{mm}$ [kcal/mol] |
|---|---|---|---|
| $C_{60}$ | 0.56 | 13.01 | 17.88 |
| $C_{70}$ | 0.45 | 10.39 | 16.37 |
| $C_{76}$ ($D_2$ isomer **2**) | 0.42 | 9.63 | 16.22 |
| $C_{78}$ ($C_{2v}$ isomer **4**) | 0.41 | 9.49 | 15.13 |
| $C_{78}$ ($C_{2v}$ isomer **5**) | 0.42 | 9.67 | 14.50 |
| $C_{82}$ ($C_2$ isomer **3**) | 0.40 | 9.30 | 14.21 |
| $C_{84}$ ($D_2$ isomer **22**) | 0.39 | 9.03 | 13.00 |
| $C_{84}$ ($D_{2d}$ isomer **23**) | 0.40 | 9.11 | 12.94 |

[a] See Chapter 5 for the description of fullerene isomers.

Taking the above considerations into account, one can neglect the higher-order multipole terms in Eq. (8.13) and write

$$V_{00} \approx Q(\mathrm{G})V_0(\mathrm{H}) \ . \tag{8.15}$$

By the same token,

$$U^1(\vec{r}) \approx V_0(\mathrm{H}) \ , \tag{8.16}$$

to a good degree of approximation. When combined with Eq. (8.14), the second-order term in Eq. (8.6) becomes

$$\sum_i{}' [E_0^0(\mathrm{H}) - E_i^0(\mathrm{H})]^{-1} V_{0i} V_{i0}$$

$$= -Q(\mathrm{G})^2 \alpha_{mm}(\mathrm{H}) - Q(\mathrm{G}) \ \vec{\alpha}_{md}(\mathrm{H}) \cdot \vec{\mu}_0(\mathrm{G})$$

$$- \vec{\mu}_0(\mathrm{G}) \cdot \overleftrightarrow{\alpha}_{dd}(\mathrm{H}) \cdot \vec{\mu}_0(\mathrm{G}) - \cdots \ , \tag{8.17}$$

where the monopole–monopole "radial" polarizability (Table 8.2),

$$\alpha_{mm}(\mathrm{H}) = \sum_i{}' [E_i^0(\mathrm{H}) - E_0^0(\mathrm{H})]^{-1}$$

$$\times \langle \Psi_0^0(\mathrm{H})|\hat{r}^{-1}|\Psi_i^0(\mathrm{H})\rangle \ \langle \Psi_i^0(\mathrm{H})|\hat{r}^{-1}|\Psi_0^0(\mathrm{H})\rangle \ , \tag{8.18}$$

and its monopole–dipole

### Table 8.3. The HF/DZP//MNDO dipole–dipole radial polarizabilities of selected fullerenes

| Cluster[a] | $\overleftrightarrow{\alpha}_{dd}$ [kcal/mol·D$^2$] | | |
|---|---|---|---|
| | xx | yy | zz |
| $C_{60}$ | 0.27 | 0.27 | 0.27 |
| $C_{70}$ | 0.24 | 0.24 | 0.18 |
| $C_{76}$ ($D_2$ isomer **2**) | 0.16 | 0.19 | 0.25 |
| $C_{84}$ ($D_2$ isomer **22**) | 0.14 | 0.15 | 0.15 |
| $C_{84}$ ($D_{2d}$ isomer **23**) | 0.15 | 0.15 | 0.14 |

[a] See Chapter 5 for the description of fullerene isomers.

$$\vec{\alpha}_{md}(\mathrm{H}) = 2 \sum_{i}{}' [E_i^0(\mathrm{H}) - E_0^0(\mathrm{H})]^{-1}$$

$$\times \langle \Psi_0^0(\mathrm{H})|\hat{r}^{-1}|\Psi_i^0(\mathrm{H})\rangle \, \langle \Psi_i^0(\mathrm{H})|\hat{r}^{-3}\vec{r}|\Psi_0^0(\mathrm{H})\rangle \, , \qquad (8.19)$$

and dipole–dipole counterparts (Table 8.3),

$$\overleftrightarrow{\alpha}_{dd}(\mathrm{H}) = \sum_{i}{}' [E_i^0(\mathrm{H}) - E_0^0(\mathrm{H})]^{-1}$$

$$\times \langle \Psi_0^0(\mathrm{H})|\hat{r}^{-3}\vec{r}|\Psi_i^0(\mathrm{H})\rangle \otimes \langle \Psi_i^0(\mathrm{H})|\hat{r}^{-3}\vec{r}|\Psi_0^0(\mathrm{H})\rangle \, , \qquad (8.20)$$

are all positive. One should note that the monopole–dipole term vanishes for many fullerene hosts because of their high symmetries.

In summary, the first-order correction to the Hamiltonian of the guest *in vacuo* is just a constant equal to the product of the endohedral potential of the host and the charge of the guest. This means that, within the first order of approximation, $\Psi_0(\mathrm{G})$ is the same as the ground-state wavefunction $\Psi_0^0(\mathrm{G})$ of the noninteracting guest. One can therefore approximate the complexation energy, defined as

$$E_{cmpl}(\mathrm{H\cdot G}) = E_0(\mathrm{H\cdot G}) - [E_0^0(\mathrm{G}) + E_0^0(\mathrm{H})] \, ,$$

$$E_0^0(\mathrm{G}) = \langle \Psi_0^0(\mathrm{G})|\hat{H}(\mathrm{G})|\Psi_0^0(\mathrm{G})\rangle \, , \qquad (8.21)$$

by [1, 2, 42, 43]

## Table 8.4.  The HF/DZP//MNDO
## dipole  screening  tensors

| Cluster[a] | $\overleftrightarrow{\sigma}$ | | |
|---|---|---|---|
| | xx | yy | zz |
| $C_{60}$ | 0.803 | 0.803 | 0.803 |
| $C_{70}$ | 0.852 | 0.852 | 0.763 |
| $C_{76}$ ($D_2$ isomer **2**) | 0.746 | 0.828 | 0.831 |
| $C_{84}$ ($D_2$ isomer **22**) | 0.815 | 0.811 | 0.780 |
| $C_{84}$ ($D_{2d}$ isomer **23**) | 0.808 | 0.808 | 0.781 |

[a] See Chapter 5 for the description of fullerene isomers.

$$E_{cmpl}(\text{H·G}) \approx Q(\text{G})V_0(\text{H}) - Q(\text{G})^2\alpha_{mm}(\text{H})$$
$$- Q(\text{G})\vec{\alpha}_{md}(\text{H}) \cdot \vec{\mu}_0(\text{G})$$
$$- \vec{\mu}_0(\text{G}) \cdot \overleftrightarrow{\alpha}_{dd}(\text{H}) \cdot \vec{\mu}_0(\text{G}) , \qquad (8.22)$$

provided the guest is small enough to maintain an adequate separation from the walls of the fullerene cage and its higher multipole moments are not too large. If necessary, the energy of steric repulsion between the host and the guest $E_{rep}(\text{H·G})$ can be added to the right-hand side of Eq. (8.22). The dispersion energy, neglected in Eq. (8.22), can be also included.

The above approximation is expected to break down for large values of $Q(\text{G})$ that necessitate inclusion of the higher-order terms in perturbation theory and for guests that either donate electrons to the cage or accept them from it. Eq. (8.22), whose resemblance to expressions for the solvation energy in the continuum medium theories [45, 46] is inescapable, gives rise to several interesting predictions. First of all, among guests with vanishing dipole moments, those with positive charges are expected to be stabilized less upon encagement than their negative counterparts. Second, as a consequence of the endohedral effect, the net electrostatic repulsion between the host cages and positively charged guests is anticipated to cause a slight increase in the cage sizes upon formation of endohedral complexes. Conversely, shrinkage of the cages is expected in endohedral complexes with negatively charged guests. Third, since displacing charged guests from the cage centers creates dipole moments, such guests are predicted to be located off-center in endohedral complexes at their equilibrium geometries. Fourth, formation of endohedral complexes with small guest molecules possessing permanent dipole moments is predicted to be an exothermic process. The guests in such complexes are expected to orient themselves along the shortest axes of fullerenes to maximize stabilization due to the radial dipole–dipole polarization effects. In contrast, because of the predominant steric interactions, nonpolar or weakly polar guests are predicted to be destabilized upon encapsulation, orienting themselves along the longest axes in order to minimize steric repulsion. All of these predictions are confirmed by rigorous electronic structure calculations (Sections 8.3, 8.4, and 8.6).

Approximate values of properties other than energies are also available from the aforedescribed theory. In particular, for fullerenes that do not possess permanent dipole moments, dipole moments of the corresponding endohedral complexes $\vec{\mu}(\mathrm{H \cdot G})$ are related to those of the guests *in vacuo* $\vec{\mu}_0(\mathrm{G})$ through the relation [1, 2, 43]

$$\vec{\mu}(\mathrm{H \cdot G}) \approx \vec{\mu}_0(\mathrm{G}) - \overleftrightarrow{\sigma}(\mathrm{H}) \cdot \vec{\mu}_0(\mathrm{G}) \, , \qquad (8.23)$$

where, like the dipole–dipole radial polarizability, the dipole–dipole screening tensor $\overleftrightarrow{\sigma}(\mathrm{H})$ (Table 8.4) is diagonal for highly symmetrical fullerenes.

## 8.2  Charge Transfer in Endohedral Complexes

The important issue of charge transfer between the hosts and the guests in endohedral complexes can also be addressed with the approximate theory of bonding. By extending the arguments employed in the derivation of Eqs. (8.1)–(8.23), one concludes that the endohedral potential not only affects the energetics of endohedral complexes, but also plays an important role in determining their propensity toward internal electron transfer [1, 42, 43]. The necessary and sufficient conditions for the ground state of an endohedral complex to be stable with respect to internal electron transfer are

$$E_{\mathrm{H \to G}} = E(\mathrm{H^+ \cdot G^-}) - E(\mathrm{H \cdot G}) > 0$$

and

$$E_{\mathrm{H \leftarrow G}} = E(\mathrm{H^- \cdot G^+}) - E(\mathrm{H \cdot G}) > 0 \, . \qquad (8.24)$$

In contrast to the case of molecules separated by a large distance, the charge transfer energies, $E_{\mathrm{H \to G}}$ and $E_{\mathrm{H \leftarrow G}}$, are not simply equal to the respective differences between the ionization potentials and the electron affinities. Instead, the suitably modified expressions for the complexation energy, Eqs. (8.21) and (8.22), have to be used. Let $IP(\mathrm{G})$ and $IP(\mathrm{H})$ be the ionization potentials of the guest and the host, respectively. Let $EA(\mathrm{G})$ and $EA(\mathrm{H})$ be the corresponding electron affinities. The conditions (8.24) become

$$E_{\text{H}\to\text{G}} \approx IP(\text{H}) - EA(\text{G}) - V_0(\text{H}^+) - \alpha_{mm}(\text{H}^+) + \vec{\alpha}_{md}(\text{H}^+) \cdot \vec{\mu}_0(\text{G}^-)$$
$$+ Q(\text{G})[V_0(\text{H}^+) - V_0(\text{H}) + 2\,\alpha_{mm}(\text{H}^+) + \vec{\alpha}_{md}(\text{H}) \cdot \vec{\mu}_0(\text{G})$$
$$- \vec{\alpha}_{md}(\text{H}^+) \cdot \vec{\mu}_0(\text{G}^-)] + Q(\text{G})^2\,[\alpha_{mm}(\text{H}) - \alpha_{mm}(\text{H}^+)]$$
$$+ \vec{\mu}_0(\text{G}) \cdot \overleftrightarrow{\alpha}_{dd}(\text{H}) \cdot \vec{\mu}_0(\text{G})$$
$$- \vec{\mu}_0(\text{G}^-) \cdot \overleftrightarrow{\alpha}_{dd}(\text{H}^+) \cdot \vec{\mu}_0(\text{G}^-) > 0\,, \tag{8.25}$$

and

$$E_{\text{G}\to\text{H}} \approx IP(\text{G}) - EA(\text{H}) + V_0(\text{H}^-) - \alpha_{mm}(\text{H}^-) - \vec{\alpha}_{md}(\text{H}^-) \cdot \vec{\mu}_0(\text{G}^+)$$
$$+ Q(\text{G})[V_0(\text{H}^-) - V_0(\text{H}) - 2\,\alpha_{mm}(\text{H}^-) + \vec{\alpha}_{md}(\text{H}) \cdot \vec{\mu}_0(\text{G})$$
$$- \vec{\alpha}_{md}(\text{H}^-) \cdot \vec{\mu}_0(\text{G}^+)] + Q(\text{G})^2\,[\alpha_{mm}(\text{H}) - \alpha_{mm}(\text{H}^-)]$$
$$+ \vec{\mu}_0(\text{G}) \cdot \overleftrightarrow{\alpha}_{dd}(\text{H}) \cdot \vec{\mu}_0(\text{G})$$
$$- \vec{\mu}_0(\text{G}^+) \cdot \overleftrightarrow{\alpha}_{dd}(\text{H}^-) \cdot \vec{\mu}_0(\text{G}^+) > 0\,. \tag{8.26}$$

Electronic structure calculations on selected fullerenes [1] indicate that the radial polarizabilities depend only very weakly on the host charge. Therefore, the inequalities (8.25) and (8.26) can be simplified to yield

$$E_{\text{H}\to\text{G}} \approx IP(\text{H}) - EA(\text{G}) - V_0(\text{H}^+) - \alpha_{mm}(\text{H}) + \vec{\alpha}_{md}(\text{H}) \cdot \vec{\mu}_0(\text{G}^-)$$
$$+ Q(\text{G})\,[V_0(\text{H}^+) - V_0(\text{H}) + 2\,\alpha_{mm}(\text{H}) + \vec{\alpha}_{md}(\text{H}) \cdot \vec{\mu}_0(\text{G})$$
$$- \vec{\alpha}_{md}(\text{H}) \cdot \vec{\mu}_0(\text{G}^-)] + \vec{\mu}_0(\text{G}) \cdot \overleftrightarrow{\alpha}_{dd}(\text{H}) \cdot \vec{\mu}_0(\text{G})$$
$$- \vec{\mu}_0(\text{G}^-) \cdot \overleftrightarrow{\alpha}_{dd}(\text{H}) \cdot \vec{\mu}_0(\text{G}^-) > 0\,, \tag{8.27}$$

and

$$E_{\text{G}\to\text{H}} \approx IP(\text{G}) - EA(\text{H}) + V_0(\text{H}^-) - \alpha_{mm}(\text{H}) - \vec{\alpha}_{md}(\text{H}) \cdot \vec{\mu}_0(\text{G}^+)$$
$$+ Q(\text{G})\,[V_0(\text{H}^-) - V_0(\text{H}) - 2\,\alpha_{mm}(\text{H}) + \vec{\alpha}_{md}(\text{H}) \cdot \vec{\mu}(\text{G})$$
$$- \vec{\alpha}_{md}(\text{H}) \cdot \vec{\mu}_0(\text{G}^+)] + \vec{\mu}_0(\text{G}) \cdot \overleftrightarrow{\alpha}_{dd}(\text{H}) \cdot \vec{\mu}_0(\text{G})$$
$$- \vec{\mu}_0(\text{G}^+) \cdot \overleftrightarrow{\alpha}_{dd}(\text{H}) \cdot \vec{\mu}_0(\text{G}^+) > 0\,. \tag{8.28}$$

Further simplification is possible for guests with vanishing dipole moments (which also implies no displacement from the cage center for charged guests), yielding

$$E_{\text{H}\to\text{G}} \approx IP(\text{H}) - EA(\text{G}) - V_0(\text{H}^+) - \alpha_{mm}(\text{H})$$
$$+ Q(\text{G})[V_0(\text{H}^+) - V_0(\text{H}) + 2\,\alpha_{mm}(\text{H})] > 0\,, \tag{8.29}$$

and

$$E_{\mathrm{G}\rightarrow\mathrm{H}} \approx IP(\mathrm{G}) - EA(\mathrm{H}) + V_0(\mathrm{H}^-) - \alpha_{mm}(\mathrm{H})$$
$$+ Q(\mathrm{G})\left[V_0(\mathrm{H}^-) - V_0(\mathrm{H}) - 2\,\alpha_{mm}(\mathrm{H})\right] > 0 \;. \qquad (8.30)$$

Expressions (8.29) and (8.30), used in conjunction with the values of $IP(\mathrm{H})$, $EA(\mathrm{H})$, $V_0(\mathrm{H})$, $V_0(\mathrm{H}^+)$, $V_0(\mathrm{H}^-)$, and $\alpha_{mm}(\mathrm{H})$ calculated at the HF/DZP level of theory [1], lead to the prediction that endohedral complexes of the $C_{60}$ fullerene do not experience internal charge transfer if

$$-EA(\mathrm{G}) + 2.71 + 4.95\,Q(\mathrm{G}) > 0 \quad\text{and}\quad IP(\mathrm{G}) - 6.14 - 4.93\,Q(\mathrm{G}) > 0 \;.$$
$$(8.31)$$

## Table 8.5. The stability conditions (8.31) [eV] for selected endohedral complexes of the $C_{60}$ fullerene

| Guest | $IP(\mathrm{G})$ | $EA(\mathrm{G})$ | $-EA(\mathrm{G})+2.7+5.0Q(\mathrm{G})$ | $IP(\mathrm{G})-6.1-4.9Q(\mathrm{G})$ |
|---|---|---|---|---|
| F | 17.4 | 3.4 | −0.7 | 11.3 |
| O | 13.6 | 1.5 | 1.2 | 7.5 |
| K | 4.4 | 0.5 | 2.2 | −1.7 |
| Ca | 6.2 | −1.8 | 4.5 | 0.1 |
| Mn | 7.4 | 0.0 | 2.7 | 1.3 |
| La | 5.7 | 0.5 | 2.2 | −0.4 |
| Cs | 3.9 | 0.5 | 2.2 | −2.2 |
| Ba | 5.3 | −0.5 | 3.2 | −0.8 |
| $H_2$ | 15.4 | −2.0 | 4.7 | 7.3 |
| HF | 16.0 | −6.0 | 8.7 | 9.9 |
| CO | 14.0 | −1.8 | 4.5 | 7.9 |
| $N_2$ | 15.6 | −2.2 | 4.9 | 9.5 |
| $H^+$ | n/a | 13.6 | −5.9 | n/a |
| $Li^+$ | 75.6 | 5.4 | 2.3 | 64.6 |
| $Na^+$ | 47.3 | 5.1 | 2.6 | 36.3 |
| $Mg^{2+}$ | 80.1 | 15.0 | −2.3 | 64.2 |
| $Al^{3+}$ | 120.0 | 28.5 | −10.8 | 99.2 |

In Table 8.5, the conditions (8.31) are tested for several guest atoms, ions, and molecules (note that less accurate values of the constants entering these conditions have been used previously [43]). Endohedral complexes with K, La, Cs, and Ba are predicted to be composed of the metal cations and a negatively charged $C_{60}$ cage. Electron transfer is also expected to occur in the endohedrally protonated fullerene, as well as in the complexes with $Mg^{2+}$ and $Al^{3+}$. The Ca atom represents a borderline case. The $Li^+$, $Na^+$, Mn, O, $H_2$, HF, CO, and $N_2$ guests are expected to neither lose nor gain electrons. Results of accurate electronic structure calculations (Sections 8.4, 8.5, and 8.6) are in line with a majority of these predictions.

One interesting consequence of conditions (8.27) and (8.28) is the prediction that internal charge transfer can be triggered by displacing guests from the cage center. In other words, for a given host–guest pair, the curves describing the dependence of the total energy on the guest displacement in the complexes with and without charge transfer can cross under certain conditions. This phenomenon is caused by the fact that the internal charge transfer produces a charged guest. Upon displacement, this leads to a dipole moment with respect to the cage center and stabilization due to the monopole–dipole and dipole–dipole polarization effects, which can be substantial. The displacement-induced charge transfer has been observed in the electronic structure calculations on the $La@C_{82}$ species [47, 48] (Section 8.5).

A property intimately related to the issue of electron transfer is the ionization potential of a guest encapsulated inside the host cage. The orbital energies of the guest are obtained by solving the modified Hartree–Fock equations arising from the effective Hamiltonian given by Eq. (8.10). Neglecting the monopole–dipole and dipole–dipole terms, the difference between the orbital energies of the encapsulated and bare guest can be approximated by [42, 43]

$$\Delta\epsilon_i = \epsilon_i(H{\cdot}G) - \epsilon_i(G) \approx -V_0(H) + 2\,\alpha_{mm}(H)\,Q(G)\,. \tag{8.32}$$

However, application of Koopmans' theorem to the orbital energies of the guest is expected to result in serious errors in the respective ionization potentials. The source of these errors lies in the neglect of the relaxation of the host orbitals. This fact can be demonstrated by examining the formula for the $i$th ionization potential of the encapsulated guest, derived in a manner analogous to Eq. (8.25),

$$\begin{aligned}
IP_i(H{\cdot}G) &= E_i(H{\cdot}G^+) - E(H{\cdot}G) \\
&\approx E_i^0(G^+) + Q(G^+)V_0(H) - Q(G^+)^2\alpha_{mm}(H) \\
&\quad - E^0(G) - Q(G)V_0(H) + Q(G)^2\alpha_{mm}(H) \\
&= IP_i(G) + V_0(H) - \alpha_{mm}(H) - 2\,\alpha_{mm}(H)Q(G)\,, \tag{8.33}
\end{aligned}$$

or

$$\begin{aligned}
\Delta IP_i(H{\cdot}G) &= IP_i(H{\cdot}G) - IP_i(G) \\
&\approx V_0(H) - \alpha_{mm}(H) - 2\,\alpha_{mm}(H)Q(G) \neq -\Delta\epsilon_i\,. \tag{8.34}
\end{aligned}$$

Finally, it should be mentioned that other approximate descriptions of bonding in endohedral complexes have been published. Among those, the most sophisticated approach [49] employs the Born–Haber cycle to calculate the complexation energies. However, since the endohedral potential is completely neglected in this approach, quantitatively correct results are not expected. The other models of bonding [50, 51] are based on even more drastic approximations, putting their validity in serious doubt.

## 8.3  Endohedral Complexes with Noble Gases and Hydrogen

Endohedral complexes with noble gases as the guests provide an opportunity to verify the predictions of the aforedescribed theory. The results of electronic structure calculations on the $He@C_{60}$, $Ne@C_{60}$, and $Ar@C_{60}$ complexes [43], carried out at the HF/DZP level of theory, are presented in Table 8.6. The complexation energies of the first two species are small and positive. Since the noble gas atoms possess neither a net electric charge nor a permanent dipole moment, Eq. (8.22) predicts vanishing complexation energies. The calculated slight destabilization is the result of steric repulsion between the guests and their host cages. Because of this repulsion, the energy minima correspond to the guests located at the cage center. However, this does not imply confinement of the guests to the center, since the calculated positive force constants are quite small. For example, the RMS amplitude of the Ne atom "rattling" is estimated at 0.11 Å in the vibrational ground state of $Ne@C_{60}$ [2]. The smallness of the steric repulsion between the guests and the hosts in the $He@C_{60}$ and $Ne@C_{60}$ species is also evident in the optimized bond lengths, $r_1$ and $r_2$, and the radius $R$ of the host, which are virtually identical to those of the pristine $C_{60}$. In contrast, the guest in the $Ar@C_{60}$ complex is large enough to affect the cage radius, which increases by *ca.* 0.001 Å. The underlying steric repulsion is substantial, amounting to destabilization by 5.21 kcal/mol.

The complexation energies listed in Table 8.6 are in good agreement with those of 0.6, 4.3, 10.1, and 31.0 kcal/mol calculated at the HF/STO-3G level of theory for the complexes with Ne, Ar, Kr, and Xe, respectively [52]. For the $He@C_{60}$ species, the MNDO approach yields the heat of complexation equal to 5.04 kcal/mol [53]. This figure, as well as those of 5.02 and 4.97 kcal/mol obtained for the complexes with the $C_{60}^+$ and $C_{60}^{2+}$ cations, are believed to be too high due to the well-known exaggeration of the core–core repulsion in the MNDO method [53]. Calculations employing the 6-12 potential predict complexation energies of $-7.61$, $-17.66$, $-33.26$, $-29.92$, and $-13.10$ kcal/mol for the respective endohedral complexes of $C_{60}$ with He, Ne, Ar, Kr, and Xe [54]. Although these energies include contributions from attractive dispersion forces, which are completely neglected in the Hartree–Fock approximation, their values are strongly influenced by the parameterization of the potential.

## Table 8.6. The HF/DZP properties of endohedral complexes of $C_{60}$ with noble gases

| Guest | $E_{HF}$ [au] | $E_{cmpl}$ [kcal/mol] | $R$ [Å] | $r_1$ [Å] | $r_2$ [Å] | $IP^a$ [eV] |
|---|---|---|---|---|---|---|
| None | −2271.671352 | n/a | 3.5229 | 1.3735 | 1.4488 | 8.03 |
| He | −2274.526091 | 0.26 | 3.5230 | 1.3735 | 1.4488 | 8.03 |
| Ne | −2400.192874 | 0.52 | 3.5230 | 1.3735 | 1.4488 | 8.03 |
| Ar | −2798.415204 | 5.21 | 3.5241 | 1.3739 | 1.4493 | 8.04 |

$^a$ Ionization potentials from Koopmans' theorem.

Vibrations of the noble gas atoms inside the $C_{60}$ cage have been the subject of several studies. The harmonic vibrational frequency of 182 cm$^{-1}$ has been calculated for the He atom with the MNDO method [53]. In conjunction with much more sophisticated HF/6-311G calculations, the dependence of the total energy $E(r)$ on the guest displacement $r$ has been accurately described by the formula

$$E(r) \approx E(0) + A\left[(ar)^{-1}\sinh(ar) - 1\right], \tag{8.35}$$

with $A = 296$ cm$^{-1}$ and $a = 3.59$ Å$^{-1}$ [55]. Analysis of the corresponding rovibrational spectrum has yielded 116.3 cm$^{-1}$ for the frequency of the lowest-energy transition.

Using the 6-12 potential, vibrational frequencies of 70, 65, and 120 cm$^{-1}$ have been predicted for the He, Ne, and Ar guests, respectively, [56]. The estimate for the vibrational frequency of the Ne guest is in good agreement with that of 68 cm$^{-1}$ obtained at the HF/4-31G&DZP level of theory [2]. Molecular dynamics simulations (with the 6-12 potential), in which coupling between the Ne libration and the vibrations of the host cage is found to result in irregular "chaotic" motion of the guest, yield the vibrational frequency of $ca.$ 90 cm$^{-1}$ [57]. Analogous calculations for the Xe@$C_{60}$ complex afford the frequencies of 97 and 143 cm$^{-1}$, depending on the model potential used [52].

Prompted by the experimentally observed collisional and thermal insertions of the noble gas atoms into $C_{60}$ and $C_{70}$, several calculations have aimed at estimating the energy necessary to penetrate fullerene cages. Molecular dynamics calculations produce an estimate of 216 kcal/mol for the energy required to push a helium atom through the center of a six-membered ring in $C_{60}$ [58]. The energy barrier of 302 kcal/mol is predicted for the penetration through a five-membered ring. Corresponding barriers of 265 and 359 kcal/mol are found in MNDO calculations [53], with somewhat smaller barriers of 262

## Table 8.7. The HF/DZP orbital energies of noble gases *in vacuo* and in endohedral complexes with the $C_{60}$ fullerene

| Guest | Orbital | Negative orbital energy $-\epsilon$ [eV] | | $\Delta\epsilon$ [eV] |
|---|---|---|---|---|
| | | *In vacuo* | In endohedral complex | |
| He | $1s$ | 24.87 | 25.70 | $-0.83$ |
| Ne | $1s$ | 891.76 | 892.66 | $-0.90$ |
| | $2s$ | 52.28 | 53.17 | $-0.88$ |
| | $2p$ | 22.96 | 23.86 | $-0.90$ |
| Ar | $1s$ | 3227.23 | 3228.13 | $-0.91$ |
| | $2s$ | 335.20 | 336.12 | $-0.92$ |
| | $2p$ | 260.33 | 261.25 | $-0.92$ |
| | $3s$ | 34.59 | (35.37, 35.86) | $(-0.78, -1.27)$ |
| | $3p$ | 15.84 | (16.12, 17.83) | $(-0.27, -1.99)$ |

and 358 kcal/mol predicted for the $C_{60}^{+}$ cation. These barriers are much higher than those of 171 and 238 kcal/mol calculated for the insertion of $Li^{+}$ into $C_{60}$ [59]. Using the $C_6H_6 + He$ system as a model, the energy required for the penetration through a six-membered ring has been estimated at 247 kcal/mol [60] at the MP2/6-31G** level of theory. The high reaction barriers predicted by these calculations are in variance with the experimental estimate of the activation energy for the release of helium from the He@$C_{60}$ complex, which equals only *ca.* 80 kcal/mol [21]. This discrepancy rules out the ring-penetration mechanism and suggests that an alternative route, involving breaking of a single bond and creation of a diradical with a "window", is operative. Indeed, recent calculations [61] indicate that the barrier for such a process amounts to only *ca.* 120 kcal/mol.

As one may conclude from inspection of Table 8.6, HF/DZP calculations [43] predict that the first ionization potential of the $C_{60}$ cluster remains unchanged upon the formation of endohedral complexes with noble gases. This is exactly what should be expected taking into account the vanishing net electric charge of the guests. In the complexes of He and Ne, the energy levels of the orbitals originating from the host are almost the same as those of the empty $C_{60}$ cage (Fig. 8.1). Analysis of the data listed in Table 8.7 reveals that the orbital energies of the guests are shifted downward due to the endohedral potential [compare Eq. (8.32)]. The shift is almost constant and slightly larger in magnitude than $V_0(\mathrm{H})$. The difference is easily explained by the fact that the orbitals of the guests extend over the entire Cartesian space, which brings parts of them to the regions where the potential is larger than $V_0$ (compare Table 8.1).

On the other hand, in the $\mathrm{Ar@C_{60}}$ complex only the $1s$, $2s$, and $2p$ core orbitals conform to the uniform shift. The valence $3s$ and $3p$ orbitals are diffuse enough to mix with the orbitals of the $\mathrm{C_{60}}$ cage and lose their clear-cut identities. For this reason, the $\mathrm{Ar@C_{60}}$ species constitutes a borderline case between

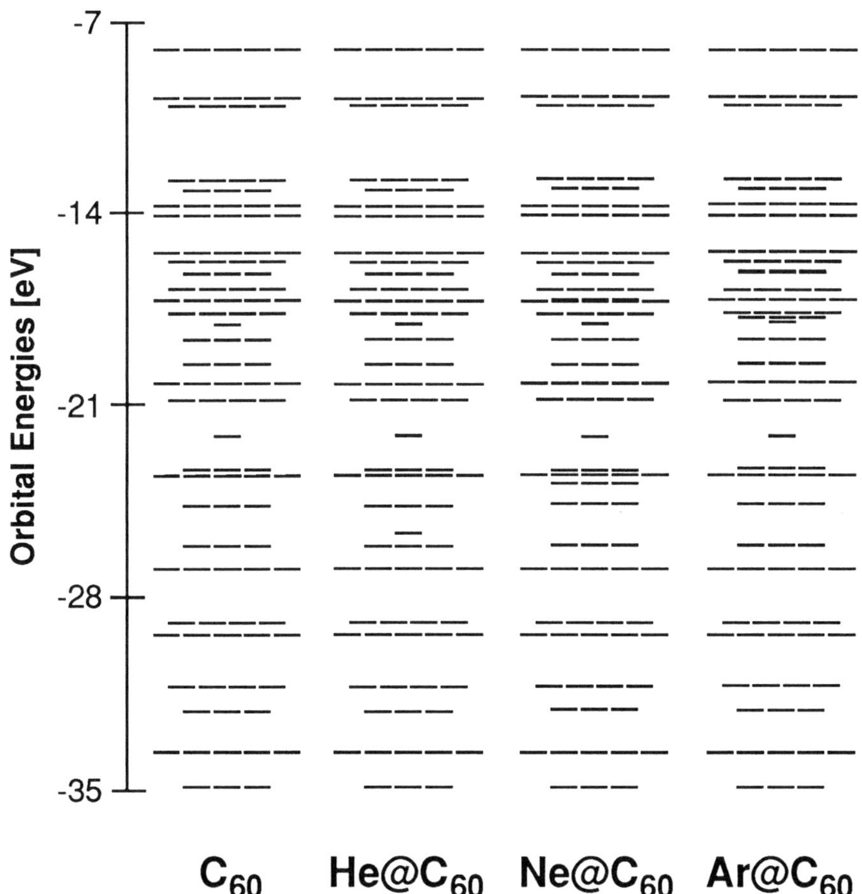


*Fig. 8.1. The HF/DZP occupied orbital energy levels in the $C_{60}$ fullerene and its endohedral complexes with noble gases. Only the orbitals with energies greater than $-35$ eV are shown.*

an endohedral complex and an endohedral compound. The orbital energies and shifts listed in parentheses in Table 8.7 refer to the molecular orbitals that have the largest overlap with the $3s$ and $3p$ orbitals of the host. One should note that the mixing is not severe for the HOMO of the $C_{60}$ cluster, leaving the first ionization potential almost unchanged upon formation of the $Ar@C_{60}$ endohedral complex.

Despite the negligible mixing between orbitals of their constituting hosts and guests, electron densities of endohedral complexes possess interesting topological features. In the $Ne@C_{60}$ complex, there are 30 bond points lying on the attractor interaction lines that connect the central Ne atom with the bond points of 30 C–C bonds linking five-membered rings in the $C_{60}$ host [2, 62]. The host–guest interactions in $Ne@C_{60}$ also give rise to 60 ring points and 32 cage points, all in addition to 90 bond points, 32 ring points, and one cage point of the host itself. Thanks to the underlying $I_h$ symmetry, the entire manifold of critical points forms an intricate and aesthetically pleasing pattern.

The existence of the $H@C_{60}$ species has been investigated with semiempirical quantum-mechanical methods. Both the MNDO [63] and the PRDDO [64] unrestricted Hartree–Fock calculations find the exohedral $C_{60}H$ molecule with the hydrogen atom bound to a $sp^3$-hybridized carbon to be the lowest energy isomer. The endohedral complex with the H guest located at the cage center is predicted to be 72 (MNDO) and 67 kcal/mol (PRDDO) higher in energy. The endohedral compound with the hydrogen atom bound to one of the carbons from the inside is even less favorable and lies 78 (MNDO) and 104 kcal/mol (PRDDO) above the exohedral isomer. The barrier to interconversion between the two endohedral species amounts to 18 kcal/mol at the MNDO level of theory. The PRDDO calculations yield the estimates of 88 and 249 kcal/mol for the barriers to insertion of the hydrogen atom through the six- and five-membered rings, respectively, of the $C_{60}$ cage.

## 8.4 Endohedral Complexes with Alkali and Alkaline Earth Metals and Their Cations

Endohedral complexes with the cations of alkali and alkaline earth metals as guests have been the subject of several theoretical studies. Early MNDO calculations [59, 65] have predicted the formation of the $Li^+@C_{60}$ complex to be an endothermic process, as reflected by the large positive complexation energy of 49.7 kcal/mol for the guest located at the cage center. Displacing the $Li^+$ cation toward the centers of the five- and six-membered rings has been found to reduce the complexation energy to 28.3 and 27.8 kcal/mol, respectively. However, the formation of the corresponding exohedral complexes has been predicted to be much more favorable, with the calculated complexation energies of $-53.0$ and $-54.5$ kcal/mol.

## Table 8.8. The HF/DZP properties of the complexes of $C_{60}$ with the $Li^+$ and $Na^+$ cations

| Guest | Complex | $E_{HF}$ [au] | $E_{cmpl}$ [kcal/mol] | $r_1$ [Å] | $r_2$ [Å] | $IP^a$ [eV] |
|---|---|---|---|---|---|---|
| $Li^+$ | (endo, $I_h$) | $-2278.914477$ | $-4.42$ | 1.3749 | 1.4504 | 11.39 |
| | (endo, $C_{5v}$) | $-2278.925978$ | $-11.63$ | see Fig. 8.2 | | 11.38, 11.35, 11.32 |
| | (exo, $C_{5v}$) | $-2278.947039$ | $-24.85$ | see Fig. 8.2 | | 10.75, 10.63, 10.61 |
| $Na^+$ | (endo, $I_h$) | $-2433.314901$ | $-3.77$ | 1.3749 | 1.4505 | 11.39 |
| | (endo, $C_{5v}$) | $-2433.315631$ | $-4.23$ | see Fig. 8.3 | | 11.39, 11.38, 11.37 |
| | (exo, $C_{5v}$) | $-2433.334717$ | $-16.20$ | see Fig. 8.3 | | 10.52, 10.44, 10.41 |

$^a$ Ionization potentials from Koopmans' theorem listed for complexes with $C_{5v}$ symmetry in the order $E_1$, $E_2$, $A_2$.

Although results of *ab initio* electronic structure calculations on the $Li^+@C_{60}$ complex show trends similar to those observed in the MNDO data, they are quantitatively very different. At the HF/DZP level of theory, the complexes with $Li^+$ and $Na^+$ located at the center of the $C_{60}$ cage are found to be maxima on the corresponding potential energy hypersurfaces [43]. However, these maxima, which are characterized by almost identical bond lengths and ionization potentials, have very similar *negative* complexation energies (Table 8.8). The complexes with $Li^+$ and $Na^+$ differ in the magnitudes of guest displacement at the equilibrium geometry and the corresponding energy lowerings. The $Li^+$ cation is displaced by 1.297 Å from the cage center, whereas the corresponding figure for the $Na^+$ cation is 0.574 Å, in reasonable agreement with the previous HF/4-31G&DZP estimate of 0.66 Å [2]. The $Li^+@C_{60}$ complex gains as much as 7.21 kcal/mol in stabilization by moving its guest from the center, whereas the $Na^+@C_{60}$ complex has its energy lowered by a mere 0.46 kcal/mol.

Similar results for the $Li^+@C_{60}$ and $Na^+@C_{60}$ species are obtained with the LDA formalism [66, 67]. The $Li^+$ guest is found to move by 1.2–1.4 Å, lowering the energy of the complex by 11.8 kcal/mol. The corresponding values for the $Na^+$ guest are 0.6–0.7 Å and 2.8 kcal/mol. The $K^+$ guest is predicted to remain at the cage center. The complexation energy for all three species is estimated at 27 kcal/mol for the guests located at the cage center. This figure is most probably too large because of the well-known tendency of the LDA method to overbind. Both the guest displacement and the exothermicity of the complex formation are also qualitatively reproduced by calculations carried out at the HF/STO-3G level of theory [44]. Displacements of 1.47 and 1.29 Å, corresponding to the energy lowerings of 17.2 and 15.1 kcal/mol, are computed for the $Li^+$ and $Na^+$ guests, respectively. A displacement of 1.53 Å is predicted for the guest in the $Mg^{2+}@C_{60}$ complex.

The trends in guest displacement are easily rationalized with the approximate theory of bonding discussed in Section 8.1. Displacing a charged guest from the center of the $C_{60}$ cage creates a dipole moment that, according to Eq. (8.22), results in extra stabilization of the endohedral complex. The observed displacements represent compromises between the stabilization originating from the dipole–dipole radial polarizability and the steric repulsion from the cage walls. Hence, the magnitude of the displacement is expected to increase with the increasing electric charge of the guest and to decrease with its increasing size.

As in the cases of noble gases (Section 8.3) and small molecules (Section 8.6), the computed energies of endohedral complexes with alkali metal cations exhibit only very weak angular dependence. For this reason, librational motions of guests in these complexes can be readily described by radial potentials fitted to a few calculated energy points. The "reflected Morse" potential of the form [68]

$$E(r) \approx E(0) + D\left\{1 - \exp\left[a(r - r_0)\right]\right\}^2 , \qquad (8.36)$$

has been found particularly suitable for this purpose. The constants $D$, $a$, and $r_0$ that enter Eq. (8.36) have been estimated at 0.666 eV, 1.54 Å$^{-1}$, and 1.36 Å for the $Li^+@C_{60}$ complex [69, 70]; 0.254 eV, 1.68 Å$^{-1}$, and 0.71 Å for the $Na^+@C_{60}$ complex [68]; and 0.018 eV, 1.95 Å$^{-1}$, and 0.0 Å for the $K^+@C_{60}$ complex [71]. The calculated rovibrational spectra of the three complexes are very complicated, exhibiting several intense transitions in the far-infrared region.

HF/DZP calculations [43] show that the effect of the $Li^+$ and $Na^+$ guests on the geometry of the $C_{60}$ cage is small (Figs. 8.2 and 8.3, note that the guests are displaced upward along the $C_5$ symmetry axes that are perpendicular to the plane of the paper). The cage relaxation is more pronounced in the $Li^+@C_{60}$ complex, in which the guest nucleus is positioned at 2.012 Å from the top pentagon, than in the $Na^+@C_{60}$ species, in which the respective distance is 2.731 Å. In both cases, bond lengths of the top hemisphere are longer than those of the bottom one. The HF/DZP calculations also reveal that exohedral complexes (the species with the $Li^+$ and $Na^+$ cations located outside the $C_{60}$ cage) have energies significantly lower than their endohedral counterparts (Table 8.8). The enhanced stability of the exohedral complexes is easily accounted for by noting that the endohedral potential is larger than the electrostatic potential at a comparable distance outside of the host cage. However, these energy differences do not imply instability of the endohedral complexes, as

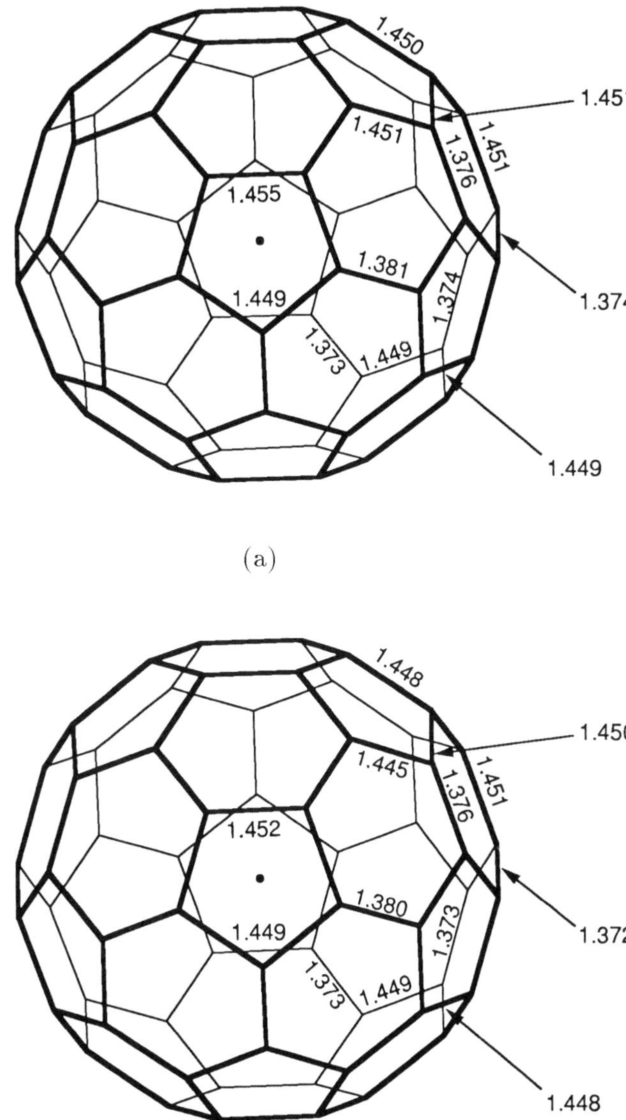



Fig. 8.2. *The HF/DZP $C_{5v}$ geometries of the complexes of $C_{60}$ with $Li^+$: (a) endohedral, (b) exohedral.*

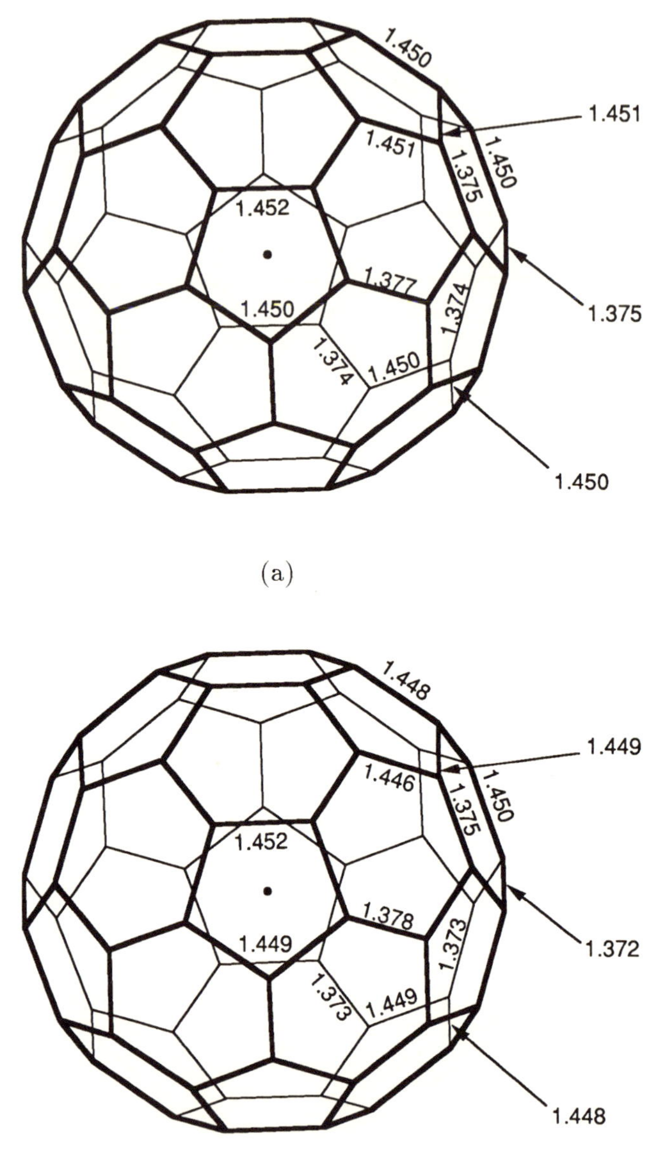

*Fig. 8.3. The HF/DZP $C_{5v}$ geometries of the complexes of $C_{60}$ with Na$^+$: (a) endohedral, (b) exohedral.*

the barriers between the endohedral and exohedral worlds are expected to be rather high, rendering escape of the guests trapped inside the $C_{60}$ cage unlikely at ambient temperatures. When compared with the results of previously published semiempirical calculations [59, 65], the data listed in Table 8.8 demonstrate that the MNDO method does not predict the energetics of formation for endohedral complexes in even a semiquantitatively correct manner.

The HF/DZP optimized geometries of the exohedral species (Figs. 8.2 and 8.3) exhibit bond length patterns quite different from those of their endohedral counterparts. In the exohedral complex of $Na^+$, the guest nucleus hovers 2.571 Å over the five-membered ring. This distance is reduced to 2.088 Å in the complex of the smaller $Li^+$ cation. The exohedral species can be easily told apart from their endohedral counterparts by measuring the ionization potentials. Taking into account that the ionization potential of a pristine $C_{60}$ is overestimated by 0.42 eV at the HF/DZP level [43], one predicts the ionization potentials of 10.90 eV for $Li^+@C_{60}$ and 10.19 eV for the exohedral $C_{60} \cdot Li^+$. The difference is even larger for the complexes of $Na^+$, for which the expected ionization potentials are 10.95 and 9.99 eV, respectively.

The HF/DZP orbital energy levels of the guests (Table 8.9) also show distinct patterns for the two types of complexes. Shifts in orbital energies are larger for the exohedral species thanks to the absence of the endohedral potential. The shifts in endohedral complexes increase when the guests are displaced from the cage center. For the complexes with guests at the cage center, the orbital energy shifts are in good agreement with the predictions of Section 8.2 [compare Eq. (8.32) and Table 8.2]. A similarly good agreement between predicted and calculated shifts is observed in the results of HF/4-31G&DZP calculations on the $F^-@C_{60}$, $Ne@C_{60}$, $Na^+@C_{60}$, and $Mg^{2+}@C_{60}$ series [2, 42].

Endohedral complexes with alkali and alkaline earth metals as guests have been extensively studied. Hartree–Fock calculations in which the guests are located at the cage center predict that the K, Cs, and Ca guests donate one electron to the $C_{60}$ host, whereas Ba donates two [72]. The charge transfer results in energy lowering of 1.71 eV for both the K and Cs guests, in good agreement (Table 8.5) with the approximate predictions presented in Section 8.2. Transfer of one electron to the $C_{60}$ host is also predicted for endohedral complexes with Li, Na, K, and Cs by LDA calculations [73, 74].

In reality, the guests in the $Li^+@C_{60}^-$ and $Na^+@C_{60}^-$ complexes are displaced from the cage center. As one would expect from the arguments outlined in Section 8.1, the displacements and the resulting energy lowerings are almost the same as for the corresponding complexes of neutral $C_{60}$ with metal cations. Thus, the displacement of the guest in the $Li^+@C_{60}^-$ species amounts to 1.2–1.4 Å and the energy lowering equals 13.0 kcal/mol according to LDA calculations [66], indicating that the dipole–dipole radial polarizability of the $C_{60}$ cage does not change significantly upon attachment of one electron.

A combined experimental and theoretical study of the $Ca@C_{60}$ species has been published [75]. The earlier prediction of single-electron charge trans-

## Table 8.9. The HF/DZP orbital energies of the $Li^+$ and $Na^+$ cations *in vacuo* and in complexes with the $C_{60}$ fullerene

|  |  | Negative orbital energy $-\epsilon$ [eV] | | $\Delta\epsilon^a$ |
|---|---|---|---|---|
| Guest | Orbital | *In vacuo* | In endohedral complex[a] | [eV] |
| $Li^+$ (endo, $I_h$) | $1s$ | 75.97 | 74.76 | 1.21 |
| (endo, $C_{5v}$) | $1s$ | 75.97 | 73.40 | 2.57 |
| (exo, $C_{5v}$) | $1s$ | 75.97 | 72.87 | 3.10 |
| $Na^+$ (endo, $I_h$) | $1s$ | 1109.40 | 1108.25 | 1.14 |
| | $2s$ | 83.41 | 82.26 | 1.15 |
| | $2p$ | 48.94 | 47.80 | 1.15 |
| (endo, $C_{5v}$) | $1s$ | 1109.40 | 1108.02 | 1.38 |
| | $2s$ | 83.41 | 82.05 | 1.36 |
| | $2p$ | 48.94 | 47.59, 47.58 | 1.35, 1.36 |
| (exo, $C_{5v}$) | $1s$ | 1109.40 | 1107.43 | 1.97 |
| | $2s$ | 83.41 | 81.49 | 1.92 |
| | $2p$ | 48.94 | 47.03, 47.02 | 1.91, 1.92 |

[a] The $2p$ orbitals in $C_{5v}$ complexes listed in the order $A_1$, $E_1$.

fer [72] has been found inconsistent with the experimental data. For the complex with the guest at the cage center, the $t_{1u}^2$ electronic configuration gives rise at the HF/dz level of theory to the $^3T_{1g}$ ground state in which two electrons are transferred from Ca to $C_{60}$. The previously considered [72] $a_g^1 t_{1u}^1$ state is found to lie 1.31 eV above the ground state. When the guest atom is allowed to displace from the cage center along the $C_5$ axis, the ground state remains a triplet, with the lowest-energy singlet state located 0.12, 0.24, and 0.38 eV above the ground state at the HF/dz, HF/dzP, and BLYP/dzP levels of theory, respectively. Comparing with endohedral complexes involving transition metals (Section 8.5), both the guest displacement (0.7 Å) and the resulting energy lowering (11.3 kcal/mol) are moderate.

Hartree–Fock and correlation energy functional calculations on endohedral complexes of $C_{60}$ with Li, Na, K, $Li^+$, $Na^+$, $K^+$, and Ca have also been published [76]. However, for some unknown reasons, the results of these calculations are in variance with both the approximate theory of bonding in endohedral complexes and the other theoretical and computational studies discussed earlier. Finally, EHT calculations on endohedral complexes of graphitic microtubules incorporating Li and K atoms as guests [77] should also be mentioned here.

## 8.5 Endohedral Complexes with Transition Metals

The experimental observations of metallofullerenes with transition metals have prompted several theoretical studies of the corresponding endohedral complexes. Hartree–Fock calculations employing relativistic pseudopotentials in conjunction with basis sets of a double-zeta quality for the valence electrons [72] predict the La@$C_{60}$ endohedral complex to have the $t_{1u}^2 h_g^1$ ground-state electronic configuration, corresponding to two electrons transferred from the 6s orbital of La to the LUMO of $C_{60}$. However, another $t_{1u}^1 h_g^2$ state, arising from the guest-to-host transfer of a single electron, is found to be almost degenerate with the ground state. In both cases, three unpaired electrons are present. The computed electronic configuration of the La$^+$@$C_{60}$ species is $h_g^2$, consistent with the La$^+$ guest located inside a neutral $C_{60}$ cage. Both of these predictions are in variance with the results of earlier LDA calculations [78, 79], which find the $t_{1u}^3$ and $t_{1u}^2$ configurations for the ground states of La@$C_{60}$ and La$^+$@$C_{60}$, respectively.

In analogy with La@$C_{60}$, the Eu@$C_{60}$ complex is predicted to have the ground-state electronic structure of Eu$^{2+}$@$C_{60}^{2-}$. However, in this case the excited state corresponding to the Eu$^+$@$C_{60}^-$ structure lies 0.62 eV above the ground state. The complex with the uranium atom is best formulated as U$^+$@$C_{60}^-$, whereas the host–guest charge transfer is absent in the Mn@$C_{60}$ species.

In all of the aforementioned calculations, the guest atoms have been placed at the cage center. However, both the approximate theory outlined in Section 8.1 and the results of calculations on endohedral complexes with charged guests (Section 8.4) demonstrate that such geometries correspond to energy maxima. Indeed, studies of the La@$C_{82}$ species show that at the equilibrium geometry the La guest is displaced from the cage center. Hartree–Fock calculations on the endohedral complex formed between the La atom and the $C_2$ isomer **3** of the $C_{82}$ fullerene (Chapter 5) have been carried out with the 3-21G basis set for the carbon atoms, whereas the DZ basis set with a relativistic pseudopotential has been employed for the guest [48]. Displacing the La atom by 1.35 Å along the $C_2$ axis toward the center of one of the six-membered rings results in an energy lowering of *ca.* 40 kcal/mol. At this energy minimum, as well as at the maximum corresponding to the guest located at the cage center, two electrons are transferred from the guest to the host. Further displacement involves an energy barrier of *ca.* 8 kcal/mol, finally leading to a second minimum at the displacement of 1.75 Å. At the second minimum, the energy lowering amounts to *ca.* 54 kcal/mol and the electronic structure of the complex can best be described as $La^{3+}@C_{82}^{3-}$. In other words, the displacement-induced charge transfer mentioned in Section 8.2 is observed in the $La@C_{82}$ species. This charge transfer is also triggered by displacing the La atom along the $C_2$ axis in the opposite direction, although the corresponding minimum at *ca.* 1.3 Å is shallow, the energy lowering amounting to only *ca.* 8 kcal/mol.

Two minima have also been found in analogous calculations on the $Sc@C_{82}$, $Y@C_{82}$ [80], and $Ce@C_{82}$ [81] species. In all three cases, the metal atoms donate approximately two electrons to the host cage, resulting in the $Sc^{2+}@C_{82}^{2-}$, $Y^{2+}@C_{82}^{2-}$, and $Ce^{2+}@C_{82}^{2-}$ electronic configurations. Displacement of the guests along the $C_2$ axis toward one of the hexagons of $C_{82}$ is strongly favored as it lowers the total energy by as much as 59.2, 37.6, and 40.3 kcal/mol in the complexes with Sc, Y, and Ce, respectively. At the energy minima, which correspond to the guest displacements of 1.55, 1.50, and 1.30 Å (relative to the cage center), strong mixing of the guest and host orbitals occurs, reflecting direct bonding of the metal atoms to a six-membered ring of the fullerene cage. In this sense, these metallofullerenes are endohedral compounds rather than endohedral complexes. The secondary minima are located along the $C_2$ axis in the direction opposite to one of the hexagons of $C_{82}$. They are relatively shallow, as indicated by the energy lowerings that equal 27.3, 12.4, and 10.4 kcal/mol, and the corresponding displacements of 1.25, 1.10, and 0.90 Å. The predicted presence of two minima may explain the experimentally observed existence of two isomers of the $Sc@C_{82}$, $Y@C_{82}$, and $Ce@C_{82}$ species [29, 30].

Displacement-induced charge transfer is also observed in the complex of La with the $C_{3v}$ isomer **7** of the $C_{82}$ fullerene (Chapter 5), for which LDA calculations have been performed [47]. Two electrons are transferred from the La atom to the host cage when the guest is located at the cage center. This electronic configuration persists at the energy minimum that occurs when the guest is displaced along the $C_3$ axis by *ca.* 1.6 Å from the cage center. At this minimum, the energy lowering due to guest displacement amounts to *ca.* 58 kcal/mol. When the guest is displaced by 2.5 Å from the cage center on the $C_s$ symmetry plane, transfer of an additional electron ensues and the energy lowering equals *ca.* 81 kcal/mol. As in the previously described complex of the $C_2$ isomer **3** of $C_{82}$, the guest atom is fairly close to the cage walls in the second minimum, causing a strong host–guest orbital mixing.

Finally, it should be mentioned that the electronic structure of $Sc_3@C_{82}$ has been the subject of EHT calculations [82]. The position of the guest inside the host cage has not been optimized. The low level of theory employed in these calculations precludes any quantitative conclusions concerning the magnitude of the guest–host charge transfer.

## 8.6 Endohedral Complexes with Molecules

The fullerene cages are large enough to accommodate many small molecules. Significant steric host–guest interactions are expected to be present in the ensuing complexes. However, as discussed in Section 8.1, for polar guests the destabilization due to steric repulsion can be at least partially compensated by the stabilization arising from the polarization effects. Electronic structure calculations on endohedral complexes of $C_{60}$ with selected diatomic molecules confirm these predictions.

Properties of the $H_2@C_{60}$, $N_2@C_{60}$, $CO@C_{60}$, $HF@C_{60}$, $LiH@C_{60}$, and $LiF@C_{60}$ endohedral complexes, computed within the Hartree–Fock approximation (with the 4-31G basis set for the host, the DZP basis set for the guests, and the host geometry frozen) [83], are presented in Table 8.10. The endohedral complexes fall into two broad categories. Encagement of nonpolar or slightly polar species, such as the $H_2$, $N_2$, and CO molecules, is an endothermic process. This is so because in these complexes stabilization due to the polarization effects is either nonexistent or very small. Hence, the bulk of the complexation energy comes from the host–guest repulsion. The magnitude of this repulsion increases with the increasing size of the guest. On the other hand, as a result of the host polarization, strongly polar molecules, such as HF, LiH, and LiF, are stabilized upon encapsulation in the $C_{60}$ cluster.

As one can deduce from the comparisons between two geometries of the complexes with the $H_2$ and HF molecules, the guests exhibit only very weak orientational preferences. Moreover, there is a marked tendency for heteronuclear guests to move from the cage center. The displacement is a consequence of several phenomena, including the slight increase of the electrostatic potential toward the cage walls (Table 8.1), which pulls the negatively charged ends of the guests from the cage center, steric repulsion from the walls, which favors displacement of the "leaner" ends, and the cage polarizability, which pulls both ends of the guests. Still, it does not take a lot of energy to displace the guests from their equilibrium positions — a fact reflected in the small values of the calculated frequencies of librational motion. It is fair to say that the diatomic molecules listed in Table 8.10 are free to tumble and rattle without much hindrance inside endohedral complexes of the $C_{60}$ fullerene.

The reduced symmetry of the endohedral complexes lifts the fivefold degeneracy of the HOMO of the $C_{60}$ cage. However, as reflected by the computed ionization potentials, the splitting is very small. Moreover, the ionization potentials are very close to that of the pristine $C_{60}$ cluster and span a narrow (0.05 eV) range.

Properties such as the equilibrium bond length $r$, the vibrational frequency of bond stretching $\nu$ and the dipole moment $\mu$ are all affected by encagement (Table 8.11). The trends in bond lengths and vibrational frequencies follow closely those in complexation energies. The steric repulsion from the cage walls brings about a decrease in bond lengths and an increase in vi-

### Table 8.10. The HF/4-31G&DZP properties of endohedral complexes of $C_{60}$ with selected diatomic molecules

| Guest[a] | $E_{HF}$ [au] | $E_{cmpl}$ [kcal/mol] | $\nu_{libr}$ [cm$^{-1}$] | $z_1/z_2^{\,b}$ [Å] | $IP^c$ [eV] |
|---|---|---|---|---|---|
| $H_2$ | $-2269.646755$ | $1.22$ | $222.2$ | $-0.3717/0.3717$ | $7.97, 7.97, 7.96$ |
| $H_2^{\,d}$ | $-2269.646756$ | $1.22$ | $220.6$ | $-0.3717/0.3717$ | $7.96, 7.96, 7.97$ |
| $N_2$ | $-2377.440808$ | $9.60$ | $182.4$ | $-0.5347/0.5347$ | $7.97, 7.98, 8.00$ |
| $CO$ | $-2381.229873$ | $11.20$ | $194.9$ | $-0.4558/0.6486$ | $7.97, 7.98, 8.00$ |
| $HF$ | $-2368.520646$ | $-1.94$ | $84.5$ | $-0.8481/0.0546$ | $7.97, 7.96, 7.94$ |
| $HF^d$ | $-2268.520684$ | $-1.96$ | $81.1$ | $-0.8614/0.0413$ | $7.97, 7.97, 7.96$ |
| $LiH$ | $-2276.502500$ | $-2.39$ | $293.2$ | $-1.1488/0.4450$ | $7.95, 7.93, 7.90$ |
| $LiF$ | $-2375.445815$ | $-14.94$ | $154.2$ | $-1.0626/0.5269$ | $7.93, 7.92, 7.92$ |

[a] Guests aligned along the fivefold symmetry axis unless indicated otherwise.

[b] Positions of guests' nuclei relative to the cage center.

[c] Ionization potentials from Koopmans' theorem listed in the order $E_1$, $E_2$, $A_2$ (or $E$, $E$, $A$).

[d] The guest aligned along the threefold symmetry axis.

brational frequencies of bond stretching in nonpolar and slightly polar guests. Effects to the opposite (such as the lengthening of the Li–F bond by 0.042 Å upon encagement) that are observed for polar molecules are readily explained with Eq. (8.22), which predicts a quadratic dependence of the stabilization energy on the dipole moment of the guest. Dipole moments usually increase with increasing bond lengths in molecules that have ionic or strongly polar covalent bonds. Hence the bond weakening is a direct consequence of the high dipole–dipole radial polarizability of the $C_{60}$ cage.

The computed dipole moments of endohedral complexes with polar guests are in line with the predictions based on Eq. (8.23). The screening constants $\sigma$ calculated from the dipole moments compiled in Table 8.11 equal 0.779/0.777, 0.753, and 0.760 for the complexes with the HF, LiH, and LiF molecules, respectively (compare Table 8.4). Thus, only between one-fifth and one-fourth of the guest's dipole moment remains detectable from the outside of the host cage.

It should be mentioned that the endohedral complexes discussed above

## Table 8.11. The HF/4-31G&DZP equilibrium bond lengths, stretching vibrational frequencies, and dipole moments of selected diatomic molecules *in vacuo* and in endohedral complexes with the $C_{60}$ fullerene

| | In vacuo | | | In endohedral complex[a] | | |
|---|---|---|---|---|---|---|
| Molecule | $r$ [Å] | $\nu$ [cm$^{-1}$] | $\mu$ [D] | $r$ [Å] | $\nu$ [cm$^{-1}$] | $\mu$ [D] |
| $H_2$ | 0.7466 | 4591.8 | 0.000 | 0.7434 | 4681.3 | 0.000 |
| $H_2^b$ | 0.7466 | 4591.8 | 0.000 | 0.7434 | 4687.4 | 0.000 |
| $N_2$ | 1.0723 | 2776.9 | 0.000 | 1.0694 | 2811.3 | 0.000 |
| CO | 1.1074 | 2458.2 | −0.183 | 1.1045 | 2488.3 | −0.086 |
| HF | 0.9020 | 4452.9 | −1.934 | 0.9027 | 4448.0 | −0.427 |
| HF$^b$ | 0.9020 | 4452.9 | −1.934 | 0.9028 | 4461.4 | −0.432 |
| LiH | 1.6002 | 1464.0 | −5.854 | 1.5938 | 1526.1 | −1.444 |
| LiF | 1.5474 | 1007.1 | −6.479 | 1.5896 | 909.7 | −1.553 |

[a] Guests aligned along the fivefold symmetry axis unless indicated otherwise.

[b] The guest aligned along the threefold symmetry axis.

have also been studied with empirical atom–atom potentials [84, 85]. The latter investigation has demonstrated that results of *ab initio* electronic structure calculations on endohedral complexes can be accurately reproduced with a formalism employing simple Lennard–Jones potential energy expressions, provided the aforedescribed polarization effects are taken into account for guests with permanent dipole moments.

One exciting application of endohedral complexes is their potential use as a new class of ferroelectric materials [86]. The known ferroelectric materials are traditionally divided into two classes that reflect the mechanisms responsible for the transitions to the low-temperature ordered phases. The first class encompasses many ionic crystals in which the individual ions move in potentials with several closely spaced minima. In these displacive-type ferroelectrics, of which barium titanate $BaTiO_3$ is a well-known example, the high-temperature phase is associated with the vibrationally averaged positions of ions, whereas the low-temperature (anti)ferroelectric phase corresponds to ions at the individual minima of the potential energy hypersurface. On the other hand, in the order–disorder-type ferroelectrics, such as potassium dihydrogen phosphate $KH_2PO_4$, the appearance of the low-temperature phase is a direct consequence of an order–disorder transition that often involves infinite networks of hydrogen bonds. Materials composed of endohedral complexes of the $C_{60}$ cluster with polar molecules (endohedral fullerites) are also expected to be (anti)ferroelectric, but due to a completely different mechanism. Since, as discussed above, the barriers to rotation for sufficiently small guest molecules inside the $C_{60}$ cage are small and the $C_{60}$ cage screens the dipole moments of the guest molecules only partially, endohedral fullerites can be regarded as the first practical realization of ideal electric dipolar lattices. It should be noted that such lattices, in which the low-temperature ordering stems from dipole–dipole interactions, cannot be realized with bare guests, as their close packing in molecular crystals hinders free rotation of molecular dipoles.

The temperature below which the transition to a ferroelectric phase occurs in endohedral fullerites is given by [86]

$$T_0 = c\,\mu^2 \ , \tag{8.37}$$

where $\mu$ is the dipole moment of the guest and $c$ is a constant estimated at $0.63\ K/D^2$. With the dipole moments of the LiF, LiCl, NaF, and NaCl molecules equal to 6.3, 7.6, 8.1, and 9.8 D, respectively, this gives the values of 25, 36, 41, and 60 K for the transition temperatures of the respective endohedral fullerites. One should note that all of these molecules are expected to fit easily inside the $C_{60}$ cage, with the NaCl molecule being perhaps slightly hindered in its rotation.

The endohedral fullerites are predicted to possess properties significantly different from those of both the displacive and the order–disorder ferroelectrics. First of all, as shown above, the transition temperatures of the endohedral fullerites are expected to be much lower than those of common ferroelectric materials. Second, since the onset of the ordered phase is not related to any vibrational mechanism, the isotopic effect should be absent in endohedral fullerites (except for solids with very low $T_0$, in which quantum effects may play an important role). On the other hand, since endohedral fullerites (like the $C_{60}$ crystals) are expected to be very compressible at low pressures and the parameter $c$ in Eq. (8.37) for $T_0$ is inversely proportional to the third power of the lattice parameter, there is going to be a pronounced positive pressure effect on $T_0$.

## 8.7  NMR Spectra of Endohedral Complexes

The recent availability [22] of macroscopic quantities of the $He@C_{60}$ and $He@C_{70}$ endohedral complexes has made it possible to measure the NMR spectra of their $^3He$ guests for the first time [23]. The nucleus of $^3He$ trapped inside the $C_{60}$ cage has been found shielded $6.3 \pm 0.2$ ppm more than in a free atom. Much higher relative shielding of $28.8 \pm 0.2$ ppm has been measured for the $^3He$ nucleus in the $He@C_{70}$ complex.

The key to understanding the origin of the observed shifts in the NMR spectra lies in the expression for the isotropically averaged shielding tensor of a guest nucleus located at the center of the cage host, which reads [87]

$$\sigma = \frac{\alpha^2}{2} \langle \Psi_0(\text{H·G})|\hat{r}^{-1}|\Psi_0(\text{H·G})\rangle$$

$$- \alpha^2 \sum_j{}' [E_j(\text{H·G}) - E_0(\text{H·G})]^{-1}$$

$$\times |\langle \Psi_0(\text{H·G})|\hat{L}|\Psi_j(\text{H·G})\rangle \cdot \langle \Psi_j(\text{H·G})|\hat{L}\hat{r}^{-3}|\Psi_0(\text{H·G})\rangle| \,, \qquad (8.38)$$

where $\alpha$ is the fine structure constant and $\hat{L}$ is the angular momentum operator. The wavefunctions $\Psi_0(\text{H·G})$ and $\Psi_j(\text{H·G})$ of the ground and the $j$th excited states of the complex, as well as the corresponding energies $E_0(\text{H·G})$ and $E_j(\text{H·G})$, can be accurately estimated with the approximate theory outlined in Section 8.1. Neglecting the orbital mixing between the guest and the host cage, one obtains [87]

$$\Delta\delta = -\frac{\alpha^2}{2} \langle \Psi_0^\circ(\text{H})|\hat{r}^{-1}|\Psi_0^\circ(\text{H})\rangle$$

$$+ \alpha^2 \sum_j{}' [E_j^\circ(\text{H}) - E_0^\circ(\text{H})]^{-1}$$

$$\times |\langle \Psi_0^\circ(\text{H})|\hat{L}|\Psi_j^\circ(\text{H})\rangle \cdot \langle \Psi_j^\circ(\text{H})|\hat{L}\hat{r}^{-3}|\Psi_0^\circ(\text{H})\rangle| \qquad (8.39)$$

for the change in the chemical shift upon encapsulation. Thus, the value of $\Delta\delta$ is a guest-independent property characteristic of a given host.

Eq. (8.39) is somewhat similar to the expression for the isotropically averaged magnetic susceptibility of the host, which reads

$$\chi = -\frac{\alpha^2}{4} \langle \Psi_0^\circ(\text{H})|\hat{r}^2|\Psi_0^\circ(\text{H})\rangle$$

$$+ \frac{\alpha^2}{2} \sum_j{}' [E_j^\circ(\text{H}) - E_0^\circ(\text{H})]^{-1}$$

$$\times |\langle \Psi_0^\circ(\text{H})|\hat{L}|\Psi_j^\circ(\text{H})\rangle \cdot \langle \Psi_j^\circ(\text{H})|\hat{L}|\Psi_0^\circ(\text{H})\rangle| \qquad (8.40)$$

In the case of the $\text{C}_{60}$ host, one might be tempted to assume that the electrons are largely confined to the vicinity of the cage surface and write

$$\Delta\delta \approx 2\,\chi\,R^{-3} \,, \qquad (8.41)$$

where $R$ is the cage radius. Unfortunately, Eq. (8.41), which can also be derived with arguments of classical electrodynamics, yields very poor results.

With the experimentally determined $\chi$ of $-2.9 \times 10^{-3}$ au [88, 89] and the cage radius of $C_{60}$ equal to *ca.* 6.66 au, Eq. (8.41) predicts the endohedral shielding of 19.6 ppm, which is more than three times larger than the experimental value. Predictions based on the London theory do not fare any better, estimating $\Delta\delta$ at less than 1 ppm [90].

The reasons for the failure of Eq. (8.41) are easy to understand [87]. In $C_{60}$, the diamagnetic component of shielding, which is the first term in Eq. (8.39), amounts to about $-1440$ ppm and is almost exactly canceled by the paramagnetic component (the second term). Therefore even a small error in either of these two components, introduced by the assumption that the electron cloud in $C_{60}$ is localized near the cage surface, results in a large deviation of the estimated shift from its exact value.

GIAO CPHF calculations are successful in predicting the observed $^3$He chemical shift in the He@$C_{60}$ endohedral complex [87]. The choice of the 6-31G basis set for the host molecule and the DZP basis set for the guest atom keeps the computational cost of these calculations within reasonable limits. At that level of theory, the shielding of 59.8 ppm is calculated for the nucleus of a free $^3$He atom. For the endohedral helium atom located at the center of the $C_{60}$ cage with the MNDO-optimized geometry, the shielding of 66.5 ppm is found. The difference between these two numbers amounts to an upfield shift of 6.7 ppm. The agreement between the computed shift and the experimental value of 6.3 ppm is very good, taking into account that electron correlation effects and librational motion of the guest are completely neglected in the calculations.

Several other GIAO CPHF calculations have been carried out for endohedral complexes of both $C_{60}$ and higher fullerenes [91, 92]. In accordance with Eq. (8.39), the computed values of $\Delta\delta$ are almost independent of the guest. However, the shifts are very sensitive to geometries of the host cages, depending on the degree of bond alternation in an approximately linear manner. The shieldings in endohedral complexes of the $D_2$ isomer **2** of $C_{76}$ (Chapter 5) are predicted to fall between those in analogous complexes of the $C_{60}$ and $C_{70}$ fullerenes. According to the GIAO CPHF calculations, the diamagnetic $C_{60}^{6-}$ anion (Chapter 3) induces a very large upfield shift of *ca.* 56 ppm in the encaged guests. The computed endohedral shieldings in complexes of the $C_{60}H_{36}$ and $C_{60}H_{60}$ molecules (Chapter 9) are expected to guide experimentalists in their efforts directed toward identification of these fullerene derivatives.

# References

1.   J. Cioslowski and K. Raghavachari, *Electrostatic Potential, Polarization, Shielding, and Charge Transfer in Endohedral Complexes of the $C_{60}$, $C_{70}$, $C_{76}$, $C_{78}$, $C_{82}$, and $C_{84}$ Clusters*, J. Chem. Phys. **98**, 8734 (1993).

2. J. Cioslowski and E. D. Fleischmann, *Endohedral Complexes: Atoms and Ions Inside the $C_{60}$ Cage*, J. Chem. Phys. **94**, 3730 (1991).

3. J. R. Heath, S. C. O'Brien, Q. Zhang, Y. Liu, R. F. Curl, H. W. Kroto, F. K. Tittel, and R. E. Smalley, *Lanthanum Complexes of Spheroidal Carbon Shells*, J. Am. Chem. Soc. **107**, 7779 (1985).

4. D. M. Cox, D. J. Trevor, K. C. Reichmann, and A. Kaldor, *$C_{60}La$: A Deflated Soccerball?*, J. Am. Chem. Soc. **108**, 2457 (1986).

5. F. D. Weiss, J. L. Elkind, S. C. O'Brien, R. F. Curl, and R. E. Smalley, *Photophysics of Metal Complexes of Spheroidal Carbon Shells*, J. Am. Chem. Soc. **110**, 4464 (1988).

6. Y. Huang and B. S. Freiser, *Externally Bound Metal Ion Complexes of Buckminsterfullerene, $MC_{60}^+$, in the Gas Phase*, J. Am. Chem. Soc. **113**, 9418 (1991).

7. Y. Huang and B. S. Freiser, *Synthesis of Bis(buckminsterfullerene)nickel Cation, $Ni(C_{60})_2^+$, in the Gas Phase*, J. Am. Chem. Soc. **113**, 8186 (1991).

8. L. M. Roth, Y. Huang, J. T. Schwedler, C. J. Cassady, D. Ben-Amotz, B. Kahr, and B. S. Freiser, *Evidence for an Externally Bound $Fe^+$–Buckminsterfullerene Complex, $FeC_{60}^+$, in the Gas Phase*, J. Am. Chem. Soc. **113**, 6298 (1991).

9. T. Weiske, D. K. Böhme, J. Hrušák, W. Krätschmer, and H. Schwarz, *Endohedral Cluster Compounds: Inclusion of Helium within $C_{60}^{\bullet\oplus}$ and $C_{70}^{\bullet\oplus}$ through Collision Experiments*, Angew. Chem. Int. Ed. Engl. **30**, 884 (1991).

10. M. M. Ross and J. H. Callahan, *Formation and Characterization of $C_{60}He^+$*, J. Phys. Chem. **95**, 5720 (1991).

11. K. A. Caldwell, D. E. Giblin, C. S. Hsu, D. Cox, and M. L. Gross, *Endohedral Complexes of Fullerene Radical Cations*, J. Am. Chem. Soc. **113**, 8519 (1991).

12. T. Weiske, T. Wong, W. Krätschmer, J. K. Terlouw, and H. Schwarz, *The Neutralization of $HeC_{60}^{\bullet+}$ in the Gas Phase: Compelling Evidence for the Existence of an Endohedral Structure for $He@C_{60}$*, Angew. Chem. Int. Ed. Engl. **31**, 183 (1992).

13. K. A. Caldwell, D. E. Giblin, and M. L. Gross, *High-Energy Collisions of Fullerene Radical Cations with Target Gases: Capture of the Target Gas and Charge Stripping of $C_{60}^{\bullet+}$, $C_{70}^{\bullet+}$, and $C_{84}^{\bullet+}$*, J. Am. Chem. Soc. **114**, 3743 (1992).

14. E. E. B. Campbell, R. Ehlich, A. Hielscher, J. M. A. Frazao, and I. V. Hertel, *Collision Energy Dependence of He and Ne Capture by $C_{60}^+$*, Z. Phys. D **23**, 1 (1992).

15. T. Weiske, J. Hrusak, D. K. Böhme, and H. Schwarz, *Formation of Endohedral Carbon-Cluster Noble-Gas Compounds with High-Energy Bimolecular Reactions: $C_{60}He^{n+}$ (n = 1, 2)*, Chem. Phys. Lett. **186**, 459 (1991).

16. T. Weiske, D. K. Böhme, and H. Schwarz, *Injection of Helium Atoms into Doubly and Triply Charged $C_{60}$ Cations*, J. Phys. Chem. **95**, 8451 (1991).

17. T. Weiske, J. Hrušák, D. K. Bohme, and H. Schwarz, 5. *Endohedral Fullerene-Noble Gas Clusters Formed with High-Energy Bimolecular Reactions of $C_x^{n+}$ (x = 60, 70; n = 1, 2, 3)*, Helv. Chim. Acta **75**, 79 (1992).

18. Z. Wan, J. F. Christian, and S. L. Anderson, $Ne^+ + C_{60}$: *Collision Energy and Impact Parameter Dependence for Endohedral Complex Formation, Fragmentation, and Charge Transfer*, J. Chem. Phys. **96**, 3344 (1992).

19. Z. Wan, J. F. Christian, and S. L. Anderson, *Collision of $Li^+$ and $Na^+$ with $C_{60}$: Insertion, Fragmentation, and Thermionic Emission*, Phys. Rev. Lett. **69**, 1352 (1992).

20. J. F. Christian, Z. Wan, and S. L. Anderson, $O^+ + C_{60}$. $C_{60}O^+$ *Production and Decomposition, Charge Transfer, and Formation of $C_{59}O^+$. Dopeyball or $[CO@C_{58}]^+$*, Chem. Phys. Lett. **199**, 373 (1992).

21. M. Saunders, H. A. Jiménez-Vázquez, R. J. Cross, and R. J. Poreda, *Stable Compounds of Helium and Neon: $He@C_{60}$ and $Ne@C_{60}$*, Science **259**, 1428 (1993).

22. M. Saunders, H. A. Jiménez-Vázquez, R. J. Cross, S. Mroczkowski, M. L. Gross, D. E. Giblin, and R. J. Poreda, *Incorporation of Helium, Neon, Argon, Krypton and Xenon into Fullerenes Using High Pressure*, J. Am. Chem. Soc. **116**, 2193 (1994).

23. M. Saunders, H. A. Jiménez-Vázquez, R. J. Cross, S. Mroczkowski, D. I. Freedberg, and F. A. L. Anet, *Probing the Interior of Fullerenes by $^3He$ NMR Spectroscopy of Endohedral $^3He@C_{60}$ and $^3He@C_{70}$*, Nature **367**, 256 (1994).

24. E. G. Gillan, C. Yeretzian, K. S. Min, M. M. Alvarez, R. L. Whetten, and R. B. Kaner, *Endohedral Rare-Earth Fullerene Complexes*, J. Phys. Chem. **96**, 6869 (1992).

25. M. M. Ross, H. H. Nelson, J. H. Callahan, and S. W. McElvany, *Production and Characterization of Metallofullerenes*, J. Phys. Chem. **96**, 5231 (1992).

26. C. Yeretzian, K. Hansen, M. M. Alvarez, K. S. Min, E. G. Gillan, K. Holczer, R. B. Kaner, and R. L. Whetten, *Collisional Probes and Possible Structures of $La_2C_{80}$*, Chem. Phys. Lett. **196**, 337 (1992).

27. L. Soderholm, P. Wurz, K. R. Lykke, D. H. Parker, and F. W. Lytle, *An EXAFS Study of the Metallofullerene $YC_{82}$: Is the Yttrium Inside the Cage?*, J. Phys. Chem. **96**, 7153 (1992).

28. C. -H. Park, B. O. Wells, J. DiCarlo, Z. -X. Shen, J. R. Salem, D. S. Bethune, C. S. Yannoni, R. D. Johnson, M. S. de Vries, C. Booth, F. Bridges, and P. Pianetta, *Structural Information on Y ions in $C_{82}$ from EXAFS Experiments*, Chem. Phys. Lett. **213**, 196 (1993).

29. S. Suzuki, S. Kawata, H. Shiromaru, K. Yamauchi, K. Kikuchi, T. Kato, and Y. Achiba, *Isomers and $^{13}C$ Hyperfine Structures of Metal-Encapsulated Fullerenes $M@C_{82}$ (M = Sc, Y, and La)*, J. Phys. Chem. **96**, 7159 (1992).

30. M. Hoinkis, C. S. Yannoni, D. S. Bethune, J. R. Salem, R. D. Johnson, M. S. Crowder, and M. S. de Vries, *Multiple Species of $La@C_{82}$ and $Y@C_{82}$: Mass Spectroscopic and Solution EPR Studies*, Chem. Phys. Lett. **198**, 461 (1992).

31. H. Shinohara, H. Sato, Y. Saito, M. Ohkohchi, and Y. Ando, *Mass Spectroscopic and ESR Characterization of Soluble Yttrium-Containing Metallofullerenes $YC_{82}$ and $Y_2C_{82}$*, J. Phys. Chem. **96**, 3571 (1992).

32. R. D. Johnson, M. S. de Vries, J. Salem, D. S. Bethune, and C. S. Yannoni, *Electron Paramagnetic Resonance Studies of Lanthanum-Containing $C_{82}$*, Nature **355**, 239 (1992).

33. C. S. Yannoni, M. Hoinkis, M. S. de Vries, D. S. Bethune, J. R. Salem, M. S. Crowder, and R. D. Johnson, *Scandium Clusters in Fullerene Cages*, Science **256**, 1191 (1992).

34. H. Shinohara, H. Sato, M. Ohkohchi, Y. Ando, T. Kodama, T. Shida, T. Kato, and Y. Saito, *Encapsulation of a Scandium Trimer in $C_{82}$*, Nature **357**, 52 (1992).

35. K. Kikuchi, S. Suzuki, Y. Nakao, N. Nakahara, T. Wakabayashi, H. Shiromaru, K. Saito, I. Ikemoto, and Y. Achiba, *Isolation and Characterization of the Metallofullerene $LaC_{82}$*, Chem. Phys. Lett. **216**, 67 (1993).

36. T. Suzuki, Y. Maruyama, T. Kato, K. Kikuchi, and Y. Achiba, *Electrochemical Properties of $La@C_{82}$*, J. Am. Chem. Soc. **115**, 11006 (1993).

37. H. Shinohara, H. Yamaguchi, N. Hayashi, H. Sato, M. Ohkohchi, Y. Ando, and Y. Saito, *Isolation and Spectroscopic Properties of $Sc_2@C_{74}$, $Sc_2@C_{82}$, and $Sc_2@C_{84}$*, J. Phys. Chem. **97**, 4259 (1993).

38. D. S. Bethune, R. D. Johnson, J. R. Salem, M. S. de Vries, and C. S. Yannoni, *Atoms in Carbon Cages: The Structure and Properties of Endohedral Fullerenes*, Nature **366**, 123 (1993).

39. K. Prassides, T. J. S. Dennis, C. Christides, E. Roduner, H. W. Kroto, R. Taylor, and D. R. M. Walton, *$Mu@C_{70}$: Monitoring the Dynamics of Fullerenes from Inside the Cage*, J. Phys. Chem. **96**, 10600 (1992).

40. R. S. Ruoff, D. C. Lorents, B. Chan, R. Malhotra, and S. Subramoney, *Single Crystal Metals Encapsulated in Carbon Nanoparticles*, Science **259**, 346 (1993).

41. T. P. Martin, M. Heinebrodt, U. Näher, H. Göhlich, T. Lange, and H. Schaber, *Fullerenes Doped with Metal Halides*, Int. J. Mod. Phys. B **6**, 3871 (1992).

42. J. Cioslowski and A. Nanayakkara, *Endohedral Effect in Inclusion Complexes of the $C_{60}$ Cluster*, J. Chem. Phys. **96**, 8354 (1992).

43. J. Cioslowski, *Ab Initio Electronic Structure Calculations on Endohedral Complexes of the $C_{60}$ Cluster*, in *Spectroscopic and Computational Studies of Supramolecular Systems* (J. E. D. Davies, ed.), Kluwer Academic Publishers, Dordrecht, 1992, Ch. 10, p. 269.

44. D. B. Whitehouse and A. D. Buckingham, *The Vibrational Contribution to the Polarisability of Endohedral $[C_{60}M]^{n+}$ Complexes (where $M = Metal Atom$)*, Chem. Phys. Lett. **207**, 332 (1993).

45. L. Onsager, *Electric Moments of Molecules in Liquids*, J. Am. Chem. Soc. **58**, 1486 (1936).

46. J. Hylton, R. E. Christoffersen, and G. G. Hall, *A Model for the ab Initio Calculation of Some Solvent Effects*, Chem. Phys. Lett. **24**, 501 (1974).

47. K. Laasonen, W. Andreoni, and M. Parrinello, *Structural and Electronic Properties of $La@C_{82}$*, Science **258**, 1916 (1992).

48. S. Nagase, K. Kobayashi, T. Kato, and Y. Achiba, *A Theoretical Approach to $C_{82}$ and $LaC_{82}$*, Chem. Phys. Lett. **201**, 475 (1993).

49. Y. Wang, D. Tománek, and R. S. Ruoff, *Stability of $M@C_{60}$ Endohedral Complexes*, Chem. Phys. Lett. **208**, 79 (1993).

50. R. Rabilizirov, *The Role of Cavities and Mantles in the Ultraviolet Extinction Peak of Graphite Spheres with Particular Preference to a Possibly Discovered $C_{60}$ Structure*, Astrophys. Space Sci. **125**, 331 (1986).

51. J. L. Ballester, P. R. Antoniewicz, and R. Smoluchowski, *Atoms in Carbon Cages as a Source of Interstellar Diffuse Lines*, Astrophys. J. **356**, 507 (1990).

52. G. Cardini, P. Procacci, P. R. Salvi, and V. Schettino, *Vibrational Properties of Xe–Fullerene Adducts. A Molecular Dynamics Approach*, Chem. Phys. Lett. **200**, 39 (1992).

53. M. Kolb and W. Thiel, *MNDO Parameters for Helium: Optimization, Tests, and Application to Endohedral Fullerene-Helium Complexes*, J. Comp. Chem. **14**, 37 (1993).

54. L. Pang and F. Brisse, *Endohedral Energies and Translation of Fullerene-Noble Gas Clusters $G@C_n$ ($G = He$, $Ne$, $Ar$, $Kr$, and $Xe$; $n = 60$ and $70$)*, J. Phys. Chem. **97**, 8562 (1993).

55. C. G. Joslin, C. G. Gray, J. D. Goddard, S. Goldman, J. Yang, and J. D. Poll, *Infrared Rotation and Vibration-Rotation Bands of Endohedral Fullerene Complexes. $He@C_{60}$*, Chem. Phys. Lett. **213**, 377 (1993).

56. J. Breton, J. Gonzalez-Platas, and C. Girardet, *Endohedral and Exohedral Adsorption in $C_{60}$: An Analytical Model*, J. Chem. Phys. **99**, 4036 (1993).

57. A. L. R. Bug, A. Wilson, and G. A. Voth, *Nonlinear Vibrational Dynamics of a Neon Atom in $C_{60}$*, J. Phys. Chem. **96**, 7864 (1992).

58. R. C. Mowrey, M. M. Ross, and J. H. Callahan, *Molecular Dynamics Simulations and Experimental Studies of the Formation of Endohedral Complexes of Buckminsterfullerene*, J. Phys. Chem. **96**, 4755 (1992).

59. D. Bakowies and W. Thiel, *MNDO Study of Large Carbon Clusters*, J. Am. Chem. Soc. **113**, 3704 (1991).

60. J. Hrušák, D. K. Böhme, T. Weiske, and H. Schwarz, *Ab Initio MO Calculation on the Energy Barrier for the Penetration of a Benzene Ring by a Helium Atom. Model Studies for the Formation of Endohedral $He@C_{60}^{+\bullet}$ Complexes by High-Energy Bimolecular Reactions*, Chem. Phys. Lett. **193**, 97 (1992).

61. R. L. Murry and G. E. Scuseria, *Theoretical Evidence for a $C_{60}$ "Window" Mechanism*, Science **263**, 791 (1994).

62. J. Cioslowski, S. T. Mixon, and W. D. Edwards, *Weak Bonds in the Topological Theory of Atoms in Molecules*, J. Am. Chem. Soc. **113**, 1083 (1991).

63. P. W. Percival and S. Wlodek, *The Structure of $C_{60}Mu$ and Other Fullerenyl Radicals*, Chem. Phys. Lett. **196**, 317 (1992).

64. S. K. Estreicher, C. D. Latham, M. I. Heggie, R. Jones, and S. Öberg, *Stable and Metastable States of $C_{60}H$: Buckminsterfullerene Monohydride*, Chem. Phys. Lett. **196**, 311 (1992).

65. M. L. McKee and W. C. Herndon, *Calculated Properties of $C_{60}$ Isomers and Fragments*, J. Mol. Struct. (Theochem) **153**, 75 (1987).

66. B. I. Dunlap, J. L. Ballester, and P. P. Schmidt, *Interactions Between $C_{60}$ and Endohedral Alkali Atoms*, J. Phys. Chem. **96**, 9781 (1992).

67. P. P. Schmidt, B. I. Dunlap, and C. T. White, *Endohedral Vibrations of $Na^+$ in $C_{60}$*, J. Phys. Chem. **95**, 10537 (1991).

68. J. L. Ballester and B. I. Dunlap, *Radial Vibrations of a Sodium Ion Inside Icosahedral $C_{60}$*, Phys. Rev. A **45**, 7985 (1992).

69. C. G. Joslin, J. Yang, C. G. Gray, S. Goldman, and J. D. Poll, *Infrared Rotation and Vibration-Rotation Bands of Endohedral Fullerene Complexes. Absorption Spectrum of $Li^+@C_{60}$ in the Range $1 - 1000$ $cm^{-1}$*, Chem. Phys. Lett. **208**, 86 (1993).

70. C. G. Joslin, C. G. Gray, S. Goldman, J. Yang, and J. D. Poll, *Raman Spectra of Endohedral Fullerenes. $Li^+@C_{60}$*, Chem. Phys. Lett. **215**, 144 (1993).

71. C. G. Joslin, J. Yang, C. G. Gray, S. Goldman, and J. D. Poll, *Infrared Rotation and Vibration-Rotation Bands of Endohedral Fullerene Complexes. $K^+@C_{60}$*, Chem. Phys. Lett. **211**, 587 (1993).

72. A. H. H. Chang, W. C. Ermler, and R. M. Pitzer, *The Ground and Excited States of $C_{60}M$ and $C_{60}M^+$ (M = O, F, K, Ca, Mn, Cs, Ba, La, Eu, U)*, J. Chem. Phys. **94**, 5004 (1991).

73. B. Wästberg and A. Rosén, *Electronic Structure of $C_{60}$ and the Alkali Atom Containing Compounds $MC_{60}$, M = Li, Na, K, Rb, Cs and $K_2C_{60}$*, Physica Scripta **44**, 276 (1991).

74. J. Guo, D. E. Ellis, and D. J. Lam, *Electronic Structure of Pure and K-Doped $C_{60}$ Clusters*, Chem. Phys. Lett. **184**, 418 (1991).

75.  L. S. Wang, J. M. Alford, Y. Chai, M. Diener, J. Zhang, S. M. McClure, T. Guo, G. E. Scuseria, and R. E. Smalley, *The Electronic Structure of* $Ca@C_{60}$, Chem. Phys. Lett. **207**, 354 (1993).

76.  G. Corongiu and E. Clementi, *Comments on the Carbon Cluster* $C_{60}$ *and on Its Complexes with Alkaline Elements*, Int. J. Quant. Chem. **42**, 1185 (1992).

77.  E. G. Gal'pern, I. V. Stankevich, A. L. Chistykov, and L. A. Chernozatonskii, *Carbon Nanotubes with Metal Inside: Electron Structure of Tubelenes* $[Li@C_{24}]_n$ *and* $[K@C_{36}]_n$, Chem. Phys. Lett. **214**, 345 (1993).

78.  A. Rosén and B. Wästberg, *First-principle Calculations of the Ionization Potentials and Electron Affinities of the Spheroidal Molecules* $C_{60}$ *and* $LaC_{60}$, J. Am. Chem. Soc. **110**, 8701 (1988).

79.  A. Rosén and B. Wästberg, *Electronic Structure of Spheroidal Metal Containing Carbon Shells: Study of the* $LaC_{60}$ *and* $C_{60}$ *Clusters and Their Ions within the Local Density Approximation*, Z. Phys. D **12**, 387 (1989).

80.  S. Nagase and K. Kobayashi, *Metallofullerenes* $MC_{82}$ *(M = Sc, Y, and La). A Theoretical Study of the Electronic and Structural Aspects*, Chem. Phys. Lett. **214**, 57 (1993).

81.  S. Nagase and K. Kobayashi, *Theoretical Study of the Lantanide Fullerene* $CeC_{82}$. *Comparison with* $ScC_{82}$, $YC_{82}$ *and* $LaC_{82}$, Chem. Phys. Lett. **228**, 106 (1994).

82.  J. R. Ungerer and T. Hughbanks, *The Electronic Structure of* $Sc_3@C_{82}$, J. Am. Chem. Soc. **115**, 2054 (1993).

83.  J. Cioslowski, *Endohedral Chemistry: Electronic Structures of Molecules Trapped Inside the* $C_{60}$ *Cage*, J. Am. Chem. Soc. **113**, 4139 (1991).

84.  C. I. Williams, M. A. Whitehead, and L. Pang, *Interaction and Dynamics of Endohedral Gas Molecules in* $C_{60}$ *Isomers and* $C_{70}$, J. Phys. Chem. **97**, 11652 (1993).

85.  J. Hernández-Rojas, J. Bretón, and J. M. Gomez Llorente, *A Semi-Empirical Analytical Potential for Diatomic Molecules at Spherical Fullerenes*, Chem. Phys. Lett. **222**, 88 (1994).

86.  J. Cioslowski and A. Nanayakkara, *Endohedral Fullerites: A New Class of Ferroelectric Materials*, Phys. Rev. Lett. **69**, 2871 (1992).

87.  J. Cioslowski, *Endohedral Magnetic Shielding in the* $C_{60}$ *Cluster*, J. Am. Chem. Soc. **116**, 3619 (1994).

88.  R. C. Haddon, L. F. Schneemeyer, J. V. Waszczak, S. H. Glarum, R. Tycko, G. Dabbagh, A. R. Kortan, A. J. Muller, A. M. Mujsce, M. J. Rosseinsky, S. M. Zahurak, A. V. Makhija, F. A. Thiel, K. Raghavachari, E. Cockayne, and V. Elser, *Experimental and Theoretical Determination of the Magnetic Susceptibility of* $C_{60}$ *and* $C_{70}$, Nature **350**, 46 (1991).

89.  R. S. Ruoff, D. Beach, J. Cuomo, T. McGuire, R. L. Whetten, and F. Diederich, *Confirmation of a Vanishingly Small Ring-Current Magnetic Susceptibility of Icosahedral* $C_{60}$, J. Phys. Chem. **95**, 3457 (1991).

90. V. Elser and R. C. Haddon, *Icosahedral $C_{60}$: An Aromatic Molecule with a Vanishingly Small Ring Current Magnetic Susceptibility*, Nature **325**, 792 (1987).

91. M. Bühl, W. Thiel, H. Jiao, P.v.R. Schleyer, M. Saunders, and F. A. L. Auet, *Helium and Lithium NMR Chemical Shifts of Endohedral Fullerene Compaunds: An ab Initio Study*, J. Am. Chem. Soc. **116**, 6005 (1994).

92. J. Cioslowski, *Endohedral Magnetic Shielding in Fullerenes. A GIAO CPHF Study*, Chem Phys. Lett. **227**, 361 (1994).

# Chapter 9

# Heterofullerenes
# and Fullerene Derivatives

Fullerenes undergo diverse reactions that produce a plethora of substances with interesting chemical and physical properties [1]. The large number of derivatives of the $C_{60}$ cluster synthesized thus far testifies to the rich chemistry of fullerenes. To name a few, reactions of $C_{60}$ with azides [2], benzyne [3], diamines [4], diazoacetates [5], *N,N*-diethylpropynylamine [6], the nitronium cation [7], and various dienes [8, 9], including *o*-quinodimethane [10], have been reported. The $C_{60}$ cluster has also been linked with sugars [11]. In addition, the reactivity of the $C_{60}^{2+}$ cation toward a variety of molecules has been studied [12].

Electronic structure calculations have been found particularly useful in investigating products of hydrogenation, halogenation, and oxidation of fullerenes. A few other addition reactions, such as the formation of fulleroids and complexes with transition metals, have been studied. Properties of *heterofullerenes*, that is, clusters in which some of the carbons are replaced by heteroatoms, have also been calculated.

## 9.1 Heterofullerenes Derived from $C_{60}$

Heterocyclic analogs of $C_{60}$ have attracted considerable attention in the chemical literature. The theoretical interest in these species has been prompted by mass-spectroscopic detection of the $C_{60-n}B_n$ ($n = 1-6$) clusters in the products of covaporization of graphite and BN [13] (on the other hand, the evidence for aza-analogs of fullerenes has been so far less clear-cut [14]).

Replacement of a single carbon in a fullerene by either a boron or a nitrogen atom furnishes a cluster with at least one unpaired electron. Results of LDA calculations [15] show that in the open-shell $C_{59}B$ and $C_{59}N$ "dopey balls" the perturbation due to the heteroatom is largely localized. The $C_{59}B$ molecule is predicted to be a Lewis acid. The singly occupied HOMO of $C_{59}B$ lies within the HOMO–LUMO gap of the unperturbed $C_{60}$ and is confined mostly to the B–C bonds, which are 7.5–10% longer than the C–C bonds in $C_{60}$. The C–C bond lengths in $C_{59}B$ are within 2% of those in $C_{60}$. When compared with the positions of the carbons, the boron atom is found to protrude slightly from the cage. The computed dipole moment of $C_{59}B$ equals 1.4 D and is directed toward the cage center. Substituting one of the carbons in $C_{60}$ by a nitrogen atom brings about distortions in geometry that are less severe than those in $C_{59}B$. The lengths of the C–N and C–C bonds in $C_{59}N$ are within 2% of the corresponding C–C bond lengths in $C_{60}$. The computed dipole moment, which is directed outward from the cage, amounts to 2.2 D. Non-self-consistent LDA calculations involving the Harris functional yield similar results, although the predicted difference between the B–C and C–C bond lengths in $C_{59}B$ is smaller [16].

Replacement of two carbons in $C_{60}$ may lead to molecules with either open- or closed-shell electronic configurations. MNDO calculations indicate that closed-shell $C_{58}B_2$, $C_{58}N_2$, and $C_{58}BN$ molecules result from substituting two adjacent carbons in $C_{60}$ with the respective heteroatoms [17]. The same conclusion is reached within the Harris functional approximation [18].

The $C_{60}$ cluster is a nonalternant hydrocarbon (Chapter 3). Therefore, all isomers of the $B_{30}N_{30}$ cluster, obtained by replacing half of the carbons in $C_{60}$ by borons and the other half by nitrogen atoms, must possess both N–N and B–B bonds. Such bonds are believed to be destabilizing [13, 17]. Two isomers of $B_{30}N_{30}$, which is isoelectronic with $C_{60}$, have been considered [17]. The first isomer, in which six B–B and six N–N bonds are present, is predicted by the MNDO method to be 193 kcal/mol more stable than the second isomer with nine B–B and nine N–N bonds that is constructed by placing 30 nitrogen atoms within the upper hemisphere and 30 boron atoms within the lower hemisphere of the fullerene cage. It should also be noted that several B–B bonds are present in the $B_{36}N_{24}$ cluster, the existence of which has recently been speculated upon [19].

Adjacency of the same heteroatoms is avoided in the $C_{12}B_{24}N_{24}$ structure [20] (Fig. 9.1), which has been called a CBN ball [17]. The $C_{12}B_{24}N_{24}$ heterofullerene possesses six C–C, 12 C–B, 12 C–N, and 60 B–N bonds, arranged in a cage with $S_6$ symmetry. Interestingly, heterofullerenes with the composition $C_{60-2n}B_nN_n$ and $n > 12$ cannot be constructed unless one allows the B–B and N–N bonds to be present. Both MNDO [17] and Harris functional [21] calculations predict the CBN heterofullerene to be a stable molecule. The latter method finds the HOMO–LUMO gap in $C_{12}B_{24}N_{24}$ to be wider than in $C_{60}$. This widening can be attributed to the strong polarities of the C–B, C–N, and B–N bonds that impede electron delocalization. For this reason, $C_{12}B_{24}N_{24}$ is expected to be less aromatic than its unsubstituted prototype.

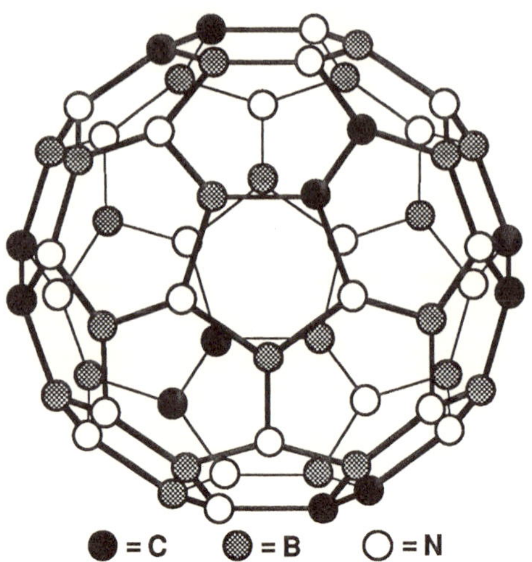



Fig. 9.1. The $S_6$ isomer of the $C_{12}B_{24}N_{24}$ heterofullerene.

Several experimental studies on silicon clusters have been published [22–26]. In order to explain the observed periodicity in reactivities and fragmentation patterns, structures composed of stacked six-membered rings, complemented by smaller caps if necessary, have been proposed for these clusters [27]. Limited tight-binding Hamiltonian calculations [28] appear to support the existence of such structures for silicon clusters. The $Si_{60}$ cluster has also been hypothesized to have a structure of six naphthalene-like rings stacked one upon another [29, 30]. Recent AM1 studies [31] find the benzene-stack isomer of the $Si_{60}$ cluster to be less stable by 70 kcal/mol than the naphthalene-stack isomer. However, the latter structure is predicted to lie 158 kcal/mol higher than the silicon analog of the $C_{60}$ fullerene. Results of HF/DZ calculations with effective core potentials (at the AM1 geometries) confirm this order of stabilities, the energy difference between the naphthalene-stack and fullerene isomers amounting to as much as 606 kcal/mol [31]. The fullerene structure for the $Si_{60}$ cluster is not inconsistent with its measured mobility [25, 26]. However, since the benzene-stack, naphthalene-stack, and fullerene structures possess 114, 116, and 90 Si–Si bonds, respectively, both quantitative and qualitative changes in the relative stabilities of the $Si_{60}$ isomers upon the inclusion of electron correlation cannot be ruled out.

The $Si_{60}$ fullerene has several properties in common with its $C_{60}$ prototype [32]. Like $C_{60}$, $Si_{60}$ possesses a fivefold degenerate HOMO and a threefold degenerate LUMO. Its bonds have two distinct lengths [2.189 and 2.266 Å at the HF/DZ(ECP) level of theory]. Moreover, the fullerene isomer of $Si_{60}$ is predicted to form endohedral complexes. However, there are also some important differences. The isomer of $Si_{60}$ with two abutting five-membered rings lies only 16.3 (AM1) and 21.1 kcal/mol [HF/DZ(ECP)//AM1] above the IPR structure, from which it is obtained by a single Stone–Wales transformation [31]. This value is less than half of the corresponding energy difference calculated for the $C_{60}$ cluster (Chapter 3). There is also some evidence provided by tight-binding Hamiltonian calculations that the $I_h$ $Si_{60}$ cage might be unstable to distortions and the energy minimum could be actually attained at $C_{2h}$ geometry [33, 34]. Finally, the $Si_{60}H_{60}$ molecule is predicted [31] to be much less strained than $C_{60}H_{60}$ (Section 9.3).

## 9.2  Metallocarbohedrenes

Heteroanalogs of small fullerene cages, such as $N_{20}$ [35] and $B_{12}N_{12}$ [36], have occasionally been studied with electronic structure methods. Recently, however, the interests of many experimentalists and theoreticians have been directed toward clusters called *metallocarbohedrenes*, or *"metcars"*, composed of 12 carbons and eight atoms of transition metals. This attention has been triggered by the observation of remarkably stable $Ti_8C_{12}^+$ species, abundant in the products of gas-phase reactions between titanium atoms and simple hydrocarbons [37]. In order to account for its stability, a dodecahedral structure (**1**; Fig. 9.2) with six C–C and 24 Ti–C bonds has been proposed for $Ti_8C_{12}$ [37, 38]. Other metallocarbohedrenes, such as $V_8C_{12}$, $Zr_8C_{12}$, and $Hf_8C_{12}$ [38, 39], as well as multicage $Zr_{13}C_{22}$, $Zr_{14}C_{21}$, $Zr_{14}C_{23}$, $Zr_{18}C_{29}$, and $Zr_{22}C_{35}$ [40] have also been detected in the gas phase.

Since metallocarbohedrenes still await isolation in macroscopic quantities, their structures have been subjects of controversy. A $D_{3d}$ structure with 12 C–C, 12 Ti–C, and six Ti–Ti bonds (**2**; Fig. 9.2) has been put forward for $Ti_8C_{12}$ on the basis of simple Hückel calculations [41]. Resonance theory has been invoked to argue that the $T_h$ structure **1** in fact possesses $O_h$ geometry (**3**; Fig. 9.2) [42]. A $T_d$ structure with the titanium atoms located at the vertices of a tetracapped tetrahedron and the $C_2$ units lying on the resulting six $Ti_4$ faces (**4**; Fig. 9.2) is another viable possibility for $Ti_8C_{12}$ [43].

With 80 valence electrons, the originally proposed structure **1** of the $Ti_8C_{12}$ cluster is formally isoelectronic with the $C_{20}$ fullerene (Chapter 7). Electronic structure calculations, carried out at the Hartree–Fock level of theory with the (13s8p5d)/[5s3p3d] basis set for Ti and the (9s5p)/[3s2p] basis set for C, identify the $(4a_g)^2(1a_u)^2(3e_g)^4(1e_u)^4(3t_g)^6(6t_u)^6$ electronic configuration as the $^1A_g(1)$ state that has the lowest energy (within $T_h$ symmetry) among all singlet states [44]. However, $^3T_g(1)$ and $^3T_g(2)$ triplet states with lower energies ensue when either two electrons from the $4a_g$ orbital or four electrons from the $4a_g$ and $1a_u$ orbitals are moved to the $4t_g$ orbital. The $^3T_g(1)$ and $^3T_g(2)$ states lie 107 and 146 kcal/mol, respectively, below the $^1A_g(1)$ singlet state. Both the $^3T_g(1)$ and the $^3T_g(2)$ states are subject to a Jahn–Teller distortion. Upon reducing the molecular symmetry of $Ti_8C_{12}$ to $D_{2h}$, a new

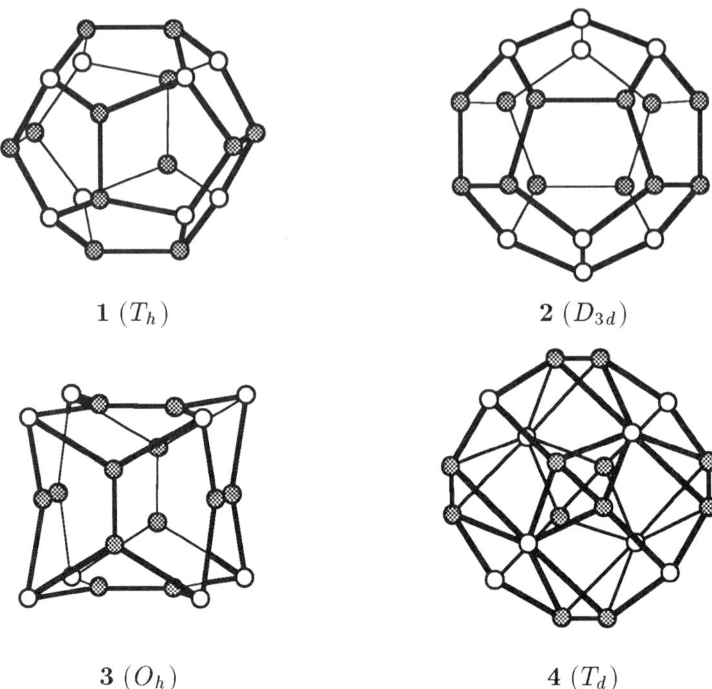

1 $(T_h)$                                           2 $(D_{3d})$

3 $(O_h)$                                           4 $(T_d)$

*Fig. 9.2. Four possible structures of $Ti_8C_{12}$.*
*Filled circles – carbon atoms;*
*empty circles – titanium atoms.*

$^1A_g(2)$ singlet state emerges, corresponding to the $(9a_g)^2(4b_{1g})^2(4b_{2g})^2(3b_{3g})^2$ $(2a_u)^2(6b_{1u})^2(6b_{2u})^2(6b_{3u})^2$ electronic configuration. This state is 175 kcal/mol lower in energy than its $^1A_g(1)$ counterpart. The corresponding optimized geometry exhibits a wide variation in bond lengths, with three sets of C–C (1.299, 1.392, 1.478 Å) and Ti–C (1.952, 2.067, 2.145 Å) bonds. With the pertinent force constants presently unavailable, it is unclear whether this geometry corresponds to a local minimum. The $6b_{2u}$ HOMO and $3a_u$ LUMO of the $^1A_g(2)$ state have the orbital energies equal to $-4.35$ and $0.79$ eV, respectively.

The C–C and Ti–C bond lengths of 1.40 and 1.98 Å in the $^3T_g(2)$ state of $Ti_8C_{12}$ are obtained within the LDA formalism [45]. These values are in good agreement with their Hartree–Fock counterparts (1.402/2.013 Å) [44], and are reproduced by other published LDA calculations, performed for unspecified electronic states, that afford the bond lengths of 1.40/1.99 [43], 1.39/1.99 [46], and 1.51/1.96 Å [47]. Kohn–Sham calculations employing the Hedin–Lundqvist exchange-correlation functional yield the bond lengths of 1.40 and 1.97 Å for the $^3T_g(1)$ electronic state [48].

As demonstrated by the above data, the optimized geometry of the $T_h$ structure **1** does not change much from one electronic state to another. However, the issue of the lowest-energy electronic configuration of $Ti_8C_{12}$ is very important for understanding the reactivity and stability of this cluster. Although the aforementioned Hartree–Fock calculations [44] would indicate the $^1A_g(2)$ singlet to be the ground state (corresponding to $D_{2h}$ geometry), a $^9A_g$ state with eight unpaired electrons and the $(4a_g)^1(2a_u)^1(3e_g)^4(1e_u)^4(4t_g)^3(6t_u)^3$ electronic configuration has been found to lie 297 kcal/mol below the $^3T_g(1)$ state at the HF/DZ(ECP) level of theory [49]. This high-spin state is not subject to a Jahn–Teller distortion. Thus, the questions about the ground-state symmetry and multiplicity of $Ti_8C_{12}$ remain unanswered at present.

The issue of the lowest-energy structures for $Ti_8C_{12}$ and other metallocarbohedrenes is further complicated by the fact that, according to several electronic structure calculations, the alternative isomers **2** and **4** have lower energies than the conventional isomer **1**. In particular, within the LDA formalism the $T_d$ structure **4** of $Ti_8C_{12}$ is found to lie 351 kcal/mol lower than **1** [43]. The same preference is observed in Hartree–Fock calculations that predict the $^5A_2$ electronic ground state for **4** [50]. Isomer **4** has the lowest energy within the family of seven topologically distinct structures in which the titanium atoms are located at the vertices of strongly distorted cubes with the $C_2$ units aligned along diagonals of the resulting six $Ti_4$ faces. Inclusion of electron correlation effects through a limited CI does not alter this energy ordering, although the ground state of **4** turns out to be the totally symmetric $^1A_1$ singlet rather then the $^5A_2$ quintet. However, the predicted singlet–quintet splitting in **4** is small (6.5 kcal/mol) and the singlet–triplet energy difference is even smaller (2.1 kcal/mol), meaning that the definitive determination of the ground-state multiplicity of $Ti_8C_{12}$ will require extensive CI calculations carried out with large basis sets [51].

Density-functional calculations establish that **3** cannot be considered a reasonable structure for $Ti_8C_{12}$ [52]. Hartree–Fock studies on the $Y_8C_{12}$ and $Nb_8C_{12}$ species indicate that the $D_{3d}$ structure **2** is either substantially lower (by 20.1 kcal/mol for the former cluster) or only slightly higher (by 3.6 kcal/mol for the latter cluster) in energy than the conventional structure **1** [53]. However, for metallocarbohedrenes with the Y, Zr, and Mo atoms, the $T_d$ isomers **4** are more stable than their $T_h$ counterparts **1** by 176–256 kcal/mol.

Electronic structures of other metallocarbohedrenes and their analogs have been studied as well, although the calculations have been limited to the conventional isomers **1** that, in light of the preceding discussion, may or may not be the lowest-energy structures. Bond lengths of 1.38 (C–C) and 1.88 Å (Si–C) in the quintet electronic state of the $Si_8C_{12}$ species have been predicted by LDA calculations [45]. The analogous bond lengths of 1.36 and 2.38 Å in $Y_8C_{12}$ have been calculated within the Hartree–Fock approximation [53]. At $T_h$ symmetry, the $^3T_g$ ground states have been found for the $Ca_8C_{12}$

and $Zn_8C_{12}$ species by Kohn–Sham calculations [48], whereas the $Sc_8B_{12}$ and $Al_8N_{12}$ clusters have been predicted to possess closed-shell electronic configurations.

## 9.3 The $C_{60}H_{60}$ and $C_{60}F_{60}$ Molecules

Since fullerenes are composed of $sp^2$-hybridized carbon atoms, exhaustive hydrogenation or fluorination, leading to the $C_nH_n$ or $C_nF_n$ species, is possible in principle. In the case of $C_{60}$, the ensuing $C_{60}H_{60}$ and $C_{60}F_{60}$ molecules are of particular interest, as their high symmetries make them easily amenable to electronic structure calculations. However, comprehensive experiments show that the maximum intake of fluorine is limited to 46 [54], 48 [55], and 52 [56, 57] atoms per molecule of $C_{60}$. Similarly, the $C_{70}$ fullerene can add only 46 [56], 48 [54], and 56 [55] fluorine atoms. The initial claim [58] of synthesis of $C_{60}F_{60}$ has never been confirmed, although traces ($< 10^{-3}\%$) of the perfluoro derivative have been reportedly observed [57]. Attempts to perhydrogenate $C_{60}$ and $C_{70}$ have resulted in the addition of only 36 hydrogen atoms in each case [59].

Geometries of the $C_{60}H_{60}$ and $C_{60}F_{60}$ molecules have been optimized at $I_h$ symmetry with a variety of quantum-chemical methods (Table 9.1). The results of HF/6-31G**(5d) calculations, corrected for the basis set deficiencies and electron correlation effects [60], are representative of the optimized geometries. In both $C_{60}H_{60}$ and $C_{60}F_{60}$, the C–C bonds within the five-membered rings are slightly longer than those connecting the pentagons. The C–C bonds are substantially longer than those in ordinary saturated hydrocarbons and their perfluorinated derivatives, the lengthening being caused by strong steric H–H and F–F repulsions rather than rehybridization effects [60]. This conclusion is corroborated by the presence of F–F attractor interaction lines in $C_{60}F_{60}$ that are indicative of strong nonbonded interactions between the fluorine atoms. On the other hand, the rehybridization effects slightly enhance the polarity of the C–H bonds in $C_{60}H_{60}$, as evidenced by the HF/6-31G**(5d) atomic charges on hydrogens equal to $-0.120$ [66].

The large steric repulsion between the fluorine atoms is $C_{60}F_{60}$ is well reflected in the energy of the following isodesmic reaction,

$$C_{60}H_{60} + 10\ C_2F_6 \rightarrow C_{60}F_{60} + 10\ C_2H_6\ . \tag{9.1}$$

At the HF/6-31G**(5d) [60], HF/6-311G** [61], MNDO, AM1, and PM3 [64] levels of theory, the calculated energies of reaction (9.1) amount to 1010, 1024, 1710, 1212, and 960 kcal/mol, respectively, meaning that each of the C–F bonds in $C_{60}F_{60}$ is at least 16 kcal/mol weaker than the analogous bond in $C_2F_6$. Similar weakening of the C–F bonds is predicted by LDA calculations [63].

## Table 9.1. Bond lengths [Å] in exohedral $I_h$ isomers of the $C_{60}H_{60}$ and $C_{60}F_{60}$ molecules

| Method | $C_{60}H_{60}$ | | | $C_{60}F_{60}$ | | |
|---|---|---|---|---|---|---|
| | $C-C^a$ | $C-C^b$ | $C-H$ | $C-C^a$ | $C-C^b$ | $C-F$ |
| MNDO [64] | 1.554 | 1.537 | 1.132 | 1.645 | 1.633 | 1.362 |
| AM1 [64, 65] | 1.530 | 1.504 | 1.146 | 1.584 | 1.565 | 1.412 |
| PM3 [64] | 1.527 | 1.508 | 1.122 | 1.585 | 1.569 | 1.381 |
| HF/STO-3G [62] | 1.568 | 1.561 | 1.084 | 1.600 | 1.594 | 1.394 |
| HF/dz [62] | 1.567 | 1.551 | 1.078 | 1.596 | 1.589 | 1.376 |
| HF/dzP [62] | 1.565 | 1.553 | 1.082 | 1.597 | 1.591 | 1.339 |
| HF/4-31G [60] | 1.563 | 1.550 | 1.081 | 1.593 | 1.589 | 1.381 |
| HF/6-31G**(5d) [60] | 1.561 | 1.552 | 1.080 | 1.602 | 1.601 | 1.336 |
| HF/6-31G**(5d)(corr.) [60] | 1.568 | 1.560 | 1.088 | 1.627 | 1.627 | 1.339 |
| HF/6-311G** [61] | 1.561 | 1.553 | 1.080 | 1.603 | 1.604 | 1.330 |
| LDA [63] | 1.55 | 1.54 | 1.10 | 1.61 | 1.60 | 1.35 |

[a] Bonds within five-membered rings.

[b] Bonds between five-membered rings.

One potential route to relieving the steric overcrowding in $C_{60}F_{60}$ without affecting the sixtyfold equivalence of carbon and fluorine atoms is to distort its geometry to $I$ symmetry [67]. However, HF/dz calculations show that such a distortion does not result in energy lowering [68]. The same conclusion is reached (for both $C_{60}H_{60}$ and $C_{60}F_{60}$) with the MNDO, AM1, and PM3 methods [64].

Another possible energy-lowering alteration of the $C_{60}H_{60}$ molecule is provided by placing some or all of its hydrogen atoms inside the fullerene cage. Such an alteration eliminates some of the H–H steric repulsions and reduces the strain due to the 120° C–C–C bond angles [69]. MM3 calculations indicate that placing just one hydrogen of $C_{60}H_{60}$ inside the cage reduces the total energy by as much as 53 kcal/mol. One of the 23 possible isomers [70] with two endohedral hydrogens is 106 kcal/mol lower in energy than its fully exohedral counterpart. The placement of further hydrogens inside the cage results in even more pronounced energy lowerings until an energy minimum is reached for a structure with 10 endohedral hydrogen atoms. This structure lies a remarkable 402 kcal/mol below the $I_h$ isomer of $C_{60}H_{60}$.

Electronic structure calculations confirm these predictions, although the computed energy differences are much smaller. Within the MNDO approximation, the isomer with 10 endohedral hydrogens is found to be 120 kcal/mol more stable than the $I_h$ structure, whereas the AM1 and PM3 calculations yield the energy differences of 264 and 258 kcal/mol, respectively [64]. The relative energy of 272 kcal/mol is calculated at the HF/DZ level of theory for the $T_h$ isomer possessing 12 hydrogens inside the cage [71].

Similar trends are observed for the $C_{60}F_{60}$ molecule. According to MNDO calculations [64], placing 10 fluorines inside the cage results in an energy decrease of 557 kcal/mol. The corresponding AM1 and PM3 estimates of 360 and 365 kcal/mol are substantially lower. At the HF/dz level of theory, the $T_h$ isomer of $C_{60}F_{60}$ with 12 fluorine atoms placed inside the cage is more stable than the fully exohedral $I_h$ structure by 372 kcal/mol [68].

The exohedral $I_h$ isomers of $C_{60}H_{60}$ and $C_{60}F_{60}$ possess substantial HOMO–LUMO gaps. The first ionization potential of $C_{60}H_{60}$ has been estimated at 9.60 (HF/dzP) [62], 5.45 (LDA) [63], 6.75 (LDA) [72], and 9.73 eV (AM1) [65]. For $C_{60}F_{60}$, the respective values are 13.88 (HF/dzP) [62] and 9.50 eV (LDA) [63].

Finally, it should be mentioned that vibrational frequencies have been calculated for the fully exohedral $I_h$ isomers of both $C_{60}H_{60}$ and $C_{60}F_{60}$. Because of high symmetry, the IR spectra of these species are predicted to consist of only nine peaks [64, 73].

## 9.4  Hydrofullerenes

Hydrogenation of fullerenes results in a plethora of products. The $C_{60}H^{\bullet}$ radical, which is the simplest hydrofullerene, has been already discussed in the previous chapter (Section 8.3). Several isomers are possible for polyhydrogenated fullerenes. In particular, there are 23 isomers of the $C_{60}H_2$ hydrofullerene with both hydrogens outside the cage [74, 75], whereas the corresponding numbers for the $C_{70}H$, $C_{70}H_2$, and $C_{70}H_3$ derivatives are five, 143, and 2792, respectively [75, 76]. In order to facilitate systematic nomenclature of these isomers, a general numbering scheme for carbon atoms in $C_{60}$ (Fig. 9.3), $C_{70}$ (Fig. 9.4), and other fullerenes has been developed [77].

Reaction of $C_{60}$ with $BH_3$ followed by hydrolysis produces exclusively the 1,2-isomer of $C_{60}H_2$, in addition to adducts with a higher number of hydrogens [78, 79]. This derivative, which corresponds to a 1,2-addition across the "double" C–C bond in $C_{60}$ (Chapter 3), is the lowest-energy isomer according to MNDO [64], AM1 [64, 65], and PM3 calculations [74, 75] (Table 9.2). The product of a 1,2-addition across the "single" bond (the 1,6- isomer) is energetically much less favorable and lies above both the 1,4-$C_{60}H_2$ and the 1,16-$C_{60}H_2$ molecules that ensue from 1,4- and 2,6-additions across the benzene

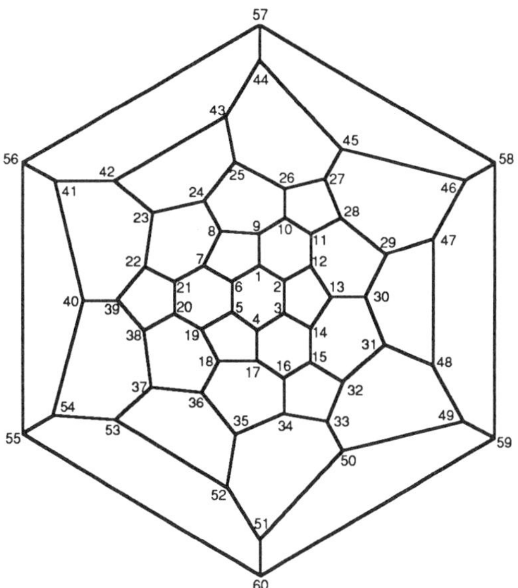

Fig. 9.3. The Schlegel diagram of the $C_{60}$ fullerene with atoms numbered according to [77].

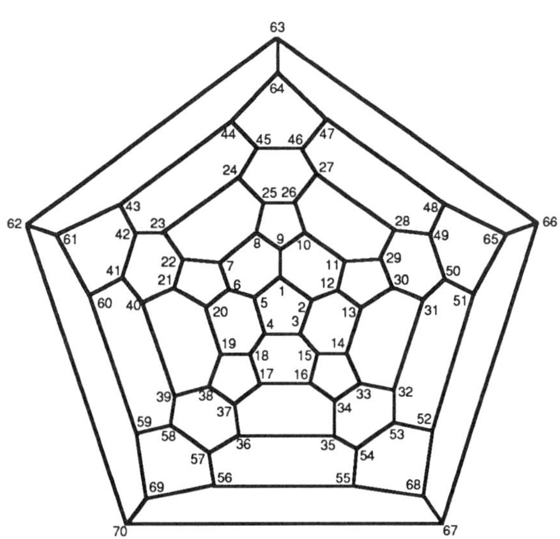

Fig. 9.4. The Schlegel diagram of the $C_{70}$ fullerene with atoms numbered according to [77].

## Table 9.2.  Relative energies [kcal/mol] of the $C_{60}H_2$ isomers

|  | Isomer[a] | | | |
| Method | 1,2- | 1,4- | 1,16- | 1,6- |
| --- | --- | --- | --- | --- |
| MNDO [64] | 0.0 | n/a | n/a | 16.1 |
| AM1 [64, 65, 83, 84] | 0.0 | 4.4 | n/a | 18.8 |
| PM3 [64, 74, 75] | 0.0 | 3.8 | 15.5 | 18.4 |
| HF/3-21G [75] | 0.0 | 7.8 | 23.1 | 26.4 |
| HF/6-31G*(5d) [75] | 0.0 | 7.6 | 20.9 | 24.0 |
| LDA [83] | 0.0 | 6.8 | n/a | n/a |

[a] See Fig. 9.3.

and naphthalene moieties, respectively. *Ab initio* electronic structure calculations yield the same order of stabilities, although the energy differences are substantially larger [75]. The fact that the 1,4- isomer is less stable than its 1,2- counterpart means that it is energetically less costly to put two hydrogens in close proximity than to introduce a double bond into a five-membered ring.

An analogous reaction between $C_{70}$ and $BH_3$ produces only two isomers, namely 1,9-$C_{70}H_2$ and 7,8-$C_{70}H_2$ [80]. These two isomers are also present in the products of $C_{70}$ reduction with diimide [81]. The 7,8- isomer is thermodynamically unstable and isomerizes to the 1,9- species in the presence of a platinum-over-silica catalyst [80]. The experimental value of $\Delta G_{295}$ for the

## Table 9.3.  Relative energies [kcal/mol] of the $C_{70}H_2$ isomers

|  | Isomer[a] | | | |
| Method | 1,9- | 7,8- | 21,42- | 1,7- |
| --- | --- | --- | --- | --- |
| AM1 [82] | 0.0 | −1.1 | 0.7 | 2.1 |
| PM3 [75, 80] | 0.0 | −1.1 | 0.3 | 1.4 |
| HF/3-21G [75, 80] | 0.0 | 0.2 | 2.1 | 5.8 |
| HF/6-31G*(5d) [75, 80] | 0.0 | 1.3 | 4.5 | 6.4 |

[a] See Fig. 9.4.

isomerization reaction equals $1.4\pm0.2$ kcal/mol, in excellent agreement with the HF/6-31G*(5d) prediction of energy difference [75] (Table 9.3). The agreement at the HF/3-21G level of theory is much worse, indicating the necessity of including polarization functions in calculations on relative stabilities of fullerene derivatives. The orders of stability obtained with the AM1 [82] and PM3 [75] methods are in variance with the experimental data.

A similar disagreement between the semiempirical predictions on one side, and the experimental data together with the results of *ab initio* electronic structure calculations on the other, is observed in the case of the $C_{60}H_4$ adducts. The 1,2,3,4- isomer of $C_{60}H_4$ is predominantly formed by hydroboration of either $C_{60}$ or 1,2-$C_{60}H_2$, and is believed to be thermodynamically stable with respect to isomerization [79]. The same isomer is also the main product of the reaction of $C_{60}$ with diimide [81]. The experimental data are in agreement with the relative stabilities obtained at the HF/3-21G and HF/6-31G* levels of theory that place the observed isomer, respectively, 3.2 and 4.0 kcal/mol below the next most stable 1,2,31,32- isomer [79]. This order of stabilities agrees with a qualitative argument that the 1,2-addition fixes the bond alternation pattern within the six-membered ring of $C_{60}$, facilitating the subsequent 3,4-addition [85]. In contrast, AM1 calculations predict the 1,2,3,4- isomer to be 3.3 kcal/mol less stable than the 1,2,31,32- one [65]. Even worse, five other structures are calculated to be lower in energy than 1,2,3,4-$C_{60}H_4$. Such qualitative discrepancies put the usefulness of the published predictions for the hydrogenation patterns [65, 86] in serious doubt.

The $C_{60}$ fullerene can be reduced with a variety of other reagents. In particular, reactions with $Cr(CH_3COO)_2$ [81], $(\eta^5\text{-}C_5H_5)_2Zr(H)Cl$ [87], and high-pressure hydrogen gas (without a catalyst) [88] have been reported. Exhaustive hydrogenation of $C_{60}$ furnishes the $C_{60}H_{36}$ species. The hydrogenation can be accomplished by the Birch reduction (Li in liquid $NH_3$) [89], heating with hydrogen under pressure (in the presence of $C_2H_5I$ [59] or Ru, Pd, Ni, or Pt catalysts [90]), or hydrogen transfer involving 9,10-dihydroanthracene [91]. The last method has been reported to produce pure $C_{60}H_{18}$ and $C_{60}H_{36}$. However, structures of these reaction products have yet to be determined. Solid-state $^{13}C$ NMR and X-ray diffraction studies on $C_{60}H_{36}$ samples obtained by free-radical hydrogenation of $C_{60}$ have yielded results consistent with $D_{3d}$ molecular symmetry, ruling out a more symmetrical $T_h$ structure [59, 92].

The experimental observations notwithstanding, several electronic structure calculations have been carried out for the $T_h$ isomer of the $C_{60}H_{36}$ hydrofullerene. The approaches of the MNDO family favor this isomer by 6–29 kcal/mol over three less symmetrical structures [64]. Since the tetrahedral $C_{60}H_{36}$ has a substantial HOMO–LUMO gap, it is not subject to a Jahn–Teller distortion. The computed values of the first ionization potential of the $T_h$ isomer of $C_{60}H_{36}$ range from 5.5 (LDA) [93] to 6.36 (LDA) [72], 7.65 (AM1) [65], and 8.14 eV (HF/DZ) [71]. A less symmetrical $T$ structure of $C_{60}H_{36}$ is more stable than its $T_h$ counterpart by as much as 97.4 kcal/mol at the BLYP/DZ//HF/STO-3G level of theory [94]. However, AM1 calculations predict another $D_{3d}$ isomer to be even lower in energy [95]. Analogous calculations on other $C_{60}H_n$ ($n = 12$–$36$) species have been published [95, 96].

## 9.5　Halogenofullerenes

As already mentioned in Section 9.3, fluorination of $C_{60}$ furnishes a complex mixture of thermally stable adducts. Under special conditions the $C_{60}F_{48}$ species, believed to be a single isomer, is formed [97]. On the other hand, the chloro- and bromo- derivatives of $C_{60}$ revert to the parent hydrocarbon upon heating [98–100]. Although chlorination of $C_{60}$ yields a poorly characterized mixture of products containing as many as 26 [98] chlorine atoms, bromination results in isolable $C_{60}Br_6$, $C_{60}Br_8$ [99], and $C_{60}Br_{24}$ [99, 100] adducts. The X-ray diffraction studies identify the $C_{60}Br_6$ and $C_{60}Br_8$ species as the 1,2,4,11,15,30-hexabromo and 1,4,7,10,16,19,24,36-octabromo derivatives, respectively. The $C_{60}Br_{24}$ adduct is a 1,4,7,10,12,14,16,19,22,24,27,29,31,33,36,38,41,43,46,49,52,54,57,60-tetracosa derivative [99, .100]. Its remarkable structure possesses $T_h$ symmetry, meaning that all of the bromine atoms are equivalent. The arrangement of bonds around the carbon atoms attached to the bromines is approximately tetrahedral, with the C–Br bond lengths measured at 1.993 Å [100]. In contrast with the lower halogens, iodine does not react with $C_{60}$, forming $C_{60} \cdot I_2$ and $C_{60} \cdot (I_2)_2$ intercalation compounds instead [101–103].

As already noted in the previous section, the 1,2-addition to $C_{60}$ produces molecules with closely positioned addents. On the other hand, the 1,4-addition introduces an energetically disadvantageous double bond in one of the five-membered rings. Therefore, one may expect the 1,2-addition to be preferred for small addents and the 1,4-addition to be favorable for the bulky ones. Results of various electronic structure calculations, compiled in Table 9.4, confirm these predictions. Within the LDA formalism, 1,2-additions are preferred for fluorine and chlorine [83]. The marginal preference for the 1,4-addition of bromine is reflected in the structure of the $C_{60}Br_6$ species, which possesses the 1,2-addition patterns, and those of $C_{60}Br_8$ and $C_{60}Br_{24}$ that do not. The 1,4-addition is favored for iodine, but the reaction is predicted to be endothermic by 26.8 kcal/mol, explaining the lack of reactivity toward $C_{60}$. The addition of bromine is endothermic by 10.1 kcal/mol and that of chlorine barely exothermic by 3.1 kcal/mol, accounting for the experimentally observed thermal instability of the respective adducts. As expected, addition of two fluorine atoms releases a substantial amount of energy (84.5 kcal/mol). As in the case of hydrogen, the 1,6-addition of fluorine across the "single" C–C bond is highly unfavorable [64].

Photolysis of a $C_{70}$–$SF_5Cl$ mixture yields the $C_{70}F^{\bullet}$ species [104]. The

## Table 9.4. Energetics of addition patterns in $C_{60}$ [kcal/mol] relative to the 1,2-addition

| Method | 1,4-addition | | | | 1,6-addition | | | |
|---|---|---|---|---|---|---|---|---|
| | F | Cl | Br | I | F | Cl | Br | I |
| MNDO [64] | n/a | n/a | n/a | n/a | 14.5 | n/a | n/a | n/a |
| AM1 [64, 83] | 4.2 | −0.9 | −1.2 | 0.2 | 18.2 | n/a | n/a | n/a |
| PM3 [64] | n/a | n/a | n/a | n/a | 17.8 | n/a | n/a | n/a |
| LDA [83] | 6.6 | 2.5 | −0.5 | −9.0 | n/a | n/a | n/a | n/a |

EPR spectrum of the reaction products is consistent with the presence of four isomers of $C_{70}F^\bullet$ in which the fluorine is attached to the $A$, $B$, $C$, and $D$ carbon atoms of $C_{70}$ (Chapter 4). An analogous addition pattern has been found for the $C_{70}H^\bullet$ species [76]. In both cases, the observed lines in the EPR spectra have been assigned to individual radicals with the help of predictions based on QCFF/PI, MNDO, and AM1 spin density calculations.

## 9.6 Alkyl- and Arylfullerenes

With the help of Lewis acids as catalysts, benzene and toluene undergo addition reactions with fullerenes [105]. As many as 12 molecules of benzene can be added to a single $C_{60}$ cage. The additions are believed to proceed through the formation of hydrofullerenium cations that attack aromatic hydrocarbons in a process reminiscent of the Friedel–Crafts synthesis. Another interesting reaction is the highly selective electrochemical production of the $1,2\text{-}C_{60}(CH_3)_2/1,4\text{-}C_{60}(CH_3)_2$ mixture [106].

Alkyl radicals readily react with $C_{60}$, forming adducts $C_{60}R_n$, where $n$ varies between 1 and 34 [107]. In the case of $R=C_6H_5CH_2$, the $C_{60}R_3^\bullet$ and $C_{60}R_5^\bullet$ species have been identified, respectively, as stable allylic and cyclopentadienyl radicals, implying a single five-membered ring as the addition site. The radical monoadducts $C_{60}R^\bullet$ possess unpaired electrons confined mostly to the two six-membered rings that share a common edge including the point of addition [108]. The radicals exist in equilibria with the $RC_{60}\cdot C_{60}R$ dimers. The standard enthalpies of dimerization span the range from 17.0 ($R=CBr_3$) to 35.5 kcal/mol [$R=(CH_3)_2CH$] [109]. The dimerization is believed to involve a 4,4'-bond between two $C_{60}$ molecules. EPR studies suggest that rotation of the $C(CH_3)_3$ moiety in $C_{60}[C(CH_3)_3]^\bullet$ is hindered by a barrier that is *ca.* 8.2 kcal/mol high [110].

Reduction of $C_{60}[C(CH_3)_3]^{\bullet}$ produces the $C_{60}H[C(CH_3)_3]$ species in which the hydrogen and $t$-butyl addents are in the 1,4- position [111]. The 1,4- isomer is thermodynamically unstable and rearranges into the 1,2- isomer. A mixture of the 1,2- and 1,4- isomers is also obtained by protonation of the $C_{60}[C(CH_3)_3]^-$ anion, which can be prepared by the addition of $Li[C(CH_3)_3]$ to $C_{60}$ [84, 111]. Because this anion is well stabilized by resonance, $C_{60}H[C(CH_3)_3]$ is one of the most acidic hydrocarbons known ($pK_a = 5.7\pm0.1$) and its C–H bond is relatively weak, as reflected in its experimental dissociation energy of $71\pm2$ kcal/mol.

AM1 calculations predict the unobserved 1,6- isomer of $C_{60}H[C(CH_3)_3]$ to lie 18.1 kcal/mol above the 1,2- adduct [84]. For the $C_{60}H(CH_3)$ species, the analogous energy difference amounts to 18.5 kcal/mol, only slightly less than the relative energy of 18.8 kcal/mol computed for the corresponding isomers of $C_{60}H_2$ (Section 9.4). As in the case of $C_{70}H_2$, the AM1 method favors the 7,8- structures for the $C_{70}H(CH_3)$ and $C_{70}H[C(CH_3)_3]$ molecules, although the calculated energy differences between the 1,9- and the 7,8- isomers are as small as 0.66 kcal/mol [82]. At present, no experimental data are available to either confirm or disprove these predictions.


## 9.7  Fullerene Oxides

Several structures are possible for fullerene oxides. This is so because, for each addition pattern, the oxide can possess either an open C–O–C bridge or a closed three-membered ring of an epoxide. Moreover, the oxygen atom can either protrude from the fullerene cage (an exohedral position) or be located inside of it (an endohedral position).

Photooxidation of $C_{60}$ in benzene yields only one isomer of $C_{60}O$ [112]. On the basis of the observed NMR spectrum, this isomer has been assigned a closed-ring epoxide $C_{2v}$ structure arising from a 1,2-addition. The same isomer is afforded by the reaction of $C_{60}$ with dimethyldioxirane [113]. Mono- and polyoxides of $C_{60}$ are also produced in benzene–oxygen flames [114], by electrolysis of $C_{60}$ in water [115], and by oxidation of $C_{60}$ with $m$-chloroperbenzoic acid [9]. The $C_{70}O$ oxide has been isolated, but its structure has only been conjectured [116].

In variance with the experimental data, MNDO calculations favor the 1,6-addition of oxygen to $C_{60}$ [117, 118]. The 1,6-$C_{60}O$ molecule has an open-bridge structure (with the C–C bond length of 2.15 Å), which corresponds to a local minimum on the potential energy hypersurface. In contrast, the product of 1,2-addition is an epoxide, as reflected in the optimized C–C bond length of 1.60 Å). At the MNDO level of theory, the 1,2-epoxide of $C_{60}$ lies 6 kcal/mol above the 1,6- adduct. HF/3-21G [117, 118] and AM1 [119] calculations predict even larger values of the energy difference, equaling 9.3 and 12.5 kcal/mol, respectively. Similarly, the $C_{2h}$ 1,6,59,60- isomer of the $C_{60}O_2$ dioxide is favored by 24.6 kcal/mol over its $D_{2h}$ 1,2,55,60- counterpart in AM1 calculations [120].

There are two conceivable reasons for this discrepancy between experimental observations and theoretical predictions. One possibility is that the observed isomer of $C_{60}O$ is a product of kinetic, rather than thermodynamic, control. The other possibility is that the approximations used in the aforementioned calculations are inadequate in the case of $C_{60}O$. The fact that the relative energy is reduced to 2.2 kcal/mol at the HF/6-31G*(5d)//HF/3-21G level of theory [121] seems to support the latter explanation. The reduction is caused by the augmentation of the basis set with polarization functions that improves description of the strained epoxide. Since the interconversion between the two isomers of $C_{60}O$ is not an isodesmic reaction, further improvement in the estimates of their relative stability can be expected upon inclusion of electron correlation effects. In fact, LDA/DZP calculations favor the 1,6- isomer by only 1.1 kcal/mol, and the addition of gradient corrections to the LDA functional results in the 1,2- isomer being preferred by 0.5 kcal/mol [121].

Other, less obvious structures have been considered for $C_{60}O$. The zwitterionic $C_{60}^{+}-O^{-}$ species [122] with the oxygen atom attached to only *one* carbon through a bond 1.26 Å long has been found 59 and 57 kcal/mol less stable than the 1,2-$C_{60}O$ epoxide at the MNDO and HF/6-31G*(5d)//MNDO levels of theory, respectively [121]. Endohedral 1,2- and 1,6- structures have been predicted to lie 115–138 kcal/mol above their exohedral counterparts [118].

## Table 9.5. Relative energies of the exohedral isomers of $C_{70}O$ [118, 123]

| Isomer[a] | Symmetry | Oxygen bridge | Relative energy [kcal/mol] | |
|---|---|---|---|---|
| | | | MNDO | HF/3-21G//MNDO |
| 1,2- | $C_s$ | Open | 8.8 | 15.9 |
| 1,9- | $C_s$ | Epoxide | 15.7 | 26.3 |
| 6,7- | $C_1$ | Open | 12.2 | 19.4 |
| 7,8- | $C_s$ | Epoxide | 16.3 | 25.0 |
| 7,22- | $C_1$ | Open | 15.8 | 24.4 |
| 21,22- | $C_s$ | Open | 14.5 | 19.5 |
| 22,23- | $C_1$ | Epoxide | 27.8 | 41.1 |
| 23,24- | $C_{2v}$ | Open | 0.0 | 0.0 |

[a] See Fig. 9.4.

There are eight distinct C–C bonds in the $C_{70}$ fullerene (Chapter 4), giving rise to eight positional isomers of $C_{70}O$. MNDO calculations predict the 23,24- isomer with $C_{2v}$ symmetry to be more stable than the other seven structures by a wide margin (Table 9.5) [118, 123]. The ordering of stabilities persists at the HF/3-21G//MNDO level of theory. The energy differences between the most stable isomer and the other structures are large enough to make the reversal of stabilities upon going to higher levels of theory quite unlikely. Interestingly, as in the case of $C_{60}O$, the local structures of the oxides (open-bridge versus epoxide) are directly related to the character of the C–C bond in $C_{70}$ across which the addition has occurred. In particular, addition across "single" bonds that are shared by one five- and one six-membered ring (Chapter 4) results in open-bridge oxides. With the exception of 23,24-$C_{70}O$, addition across "double" bonds shared by two hexagons leads to epoxides. The exceptional behavior of the most stable isomer, which is an open-bridge oxide, can be accounted for by considering the Hückel bond orders in $C_{70}$ [77]. Although the 23,24- bond is formally "double", its order falls within the range characteristic of "single" bonds in $C_{70}$.

In summary, the structures of $C_{60}O$ and $C_{70}O$ are quite different. In addition, the patterns of the oxygen addition are distinct from those predicted and observed for hydrogens, halogens, and alkyl groups. The use of high-level electronic structure calculations is essential for obtaining reliable relative energies of fullerene oxides.

## 9.8  Fulleroids

The products of carbene additions to fullerenes are called *fulleroids* [124]. The fulleroids obtained from $C_{60}$, such as $C_{60}C(C_6H_5)_2$, have been originally speculated to possess open-bridge structures ensuing from the 1,2-addition [124], but further experimental studies have demonstrated that the reaction products are mixtures of the 1,2- adducts with rings composed of three carbon atoms and the 1,6- adducts with open bridges, the latter believed to be less stable [125]. The parent $C_{60}CH_2$ fulleroid has been synthesized by the addition of $CH_2N_2$ to $C_{60}$, followed by thermolysis of the resulting pyrazoline (which is a 1,2- adduct), and shown to possess the 1,6- open-bridge structure [126]. Photolysis of the same pyrazoline has been found to produce a 3:4 mixture of 1,2-$C_{60}CH_2$ and 1,6-$C_{60}CH_2$ [127].

In agreement with the experimental observation of both structures, *ab initio* electronic structure calculations employing a gradient-corrected exchange-correlation density functional predict the 1,2- isomer of $C_{60}CH_2$ to be more stable than its 1,6- counterpart by only 2.5 kcal/mol [121]. The computed energy differences are larger at the HF/3-21G, HF/6-31G*(5d)//HF/3-21G, and LDA/DZP levels of theory, amounting to 6.0, 9.3, and 4.5 kcal/mol, respectively.

AM1 calculations on the $1,2\text{-}C_{60}CH_2$ [128] and $1,2\text{-}C_{60}C(C_6H_5)_2$ [129] fulleroids have also been published. As expected, the closed-bridge structures with $C_{2v}$ symmetries correspond to local minima on the respective potential energy hypersurfaces. For the latter molecule, a discrepancy between the calculated (1.565 Å) and the observed (1.84 Å) lengths of the C–C bond across which the $C(C_6H_5)_2$ moiety is attached has been noted. However, the observation that the experimental value falls between bond lengths of *ca.* 1.6 and 2.2 Å predicted for the 1,2- and 1,6- structures of $C_{60}CH_2$ [121] seems to indicate that the crystals of $C_{60}C(p\text{-}BrC_6H_4)_2$ used in X-ray diffraction measurements [130] could have in fact contained both isomers (see also footnotes [29] and [30] in [130]).

## 9.9  Complexes of Fullerenes with Metals

Osmium derivatives, in which the Os atoms are linked to the carbon cages through oxygens, have played an important role in the characterization of fullerene structures and the understanding of their reactivities. NMR [131] and X-ray diffraction measurements [132] of the $1,2\text{-}C_{60}(OsO_4)(4\text{-}t\text{-}BuC_5H_4N)_2$ complex provided the first experimental proof of the soccerball structure of $C_{60}$ (Chapter 3). A similar complex with $C_{76}$ has been used to resolve that fullerene into enantiomers [133]. Asymmetric biosmylation of $C_{60}$ has also been studied [134]. Osmylation of $C_{70}$ has been shown to occur preferentially at the 1,9- and 7,8- positions [135].

In the formation of molecules with carbon–metal bonds, fullerenes exhibit reactivities characteristic of electron-poor alkenes rather than those of aromatic systems. In particular, the complexes of $C_{60}$ with $Pd[P(C_6H_5)_3]_2$ and $Pt[P(C_6H_5)_3]_2$ [136, 137] have the structures in which the metals are di-coordinated to the "double" bond of $C_{60}$ shared by two hexagons [138]. In other words, these complexes are products of the 1,2- addition to $C_{60}$ and are of the same type as the well-known $(\eta^2\text{-}C_2H_4)Pt[P(C_6H_5)_3]_2$ species. The same coordination is seen in the $(\eta^2\text{-}C_{60})Ir(CO)(\eta^5\text{-}C_9H_7)$ [139] and $C_{60}\{Ir(CO)Cl[P(CH_3)_2C_6H_5]_2\}_2$ complexes [140]. Since complexes of $C_{60}$ analogous to ferrocene or $Cr(C_6H_6)_2$ have never been isolated, the exohedral $C_{60}Fe^+$ [141, 142], $C_{60}Co^+$, $C_{60}Ni^+$, $C_{60}Cu^+$, $C_{60}Rh^+$, $C_{60}La^+$ [141], and $(C_{60})_2Ni^+$ species [143] observed in the gas phase most probably involve di-coordinated metal atoms as well.

Electronic structure calculations on the $(\eta^2\text{-}C_{60})Pt(PH_3)_2$ molecule demonstrate that bonding in the complexes between $C_{60}$ (and presumably other fullerenes) and $Pt(PH_3)_3$ is dominated by strong back-donation from the transition metal to the fullerene cage [144]. This back-donation is facilitated by the positive electron affinity of $C_{60}$ (Chapter 3). At the HF/3-21G level of theory (with effective core potentials and the (5s5p5d)/[3s3p3d] basis set on the Pt atom), the C–C bond across which the coordination takes place is found to be significantly weakened by back-donation, as reflected in the calculated bond lengths of 1.367 and 1.495 Å in the pristine $C_{60}$ and in the complex, respectively. The latter value is in good agreement with the experimentally determined bond length of 1.502 Å in $(\eta^2\text{-}C_{60})Pt[P(C_6H_5)_3]_2$ [138]. The extent of back-donation is also revealed by the calculated charge on the fullerene cage, which amounts to *ca.* $-1.0$. An analysis using localized orbitals shows the presence of two Pt–C $\sigma$ bonds, justifying the description of the PtCC group as a metallacyclopropane. Interestingly, although addition of the $Pt(PH_3)_2$ fragment is predicted to lower the electron affinity of $C_{60}$ by as much as 0.44 eV (which compares reasonably well with the experimental value of 0.34 eV [145]), the perturbation to the fullerene cage, as measured by the changes in bond lengths and atomic charges, is mostly localized. This prediction implies that multiple additions, leading to the $C_{60}[Pt(PH_3)_2]_n$ species ($n = 2-6$) should be facile. Indeed, such additions have been observed experimentally [146].

An entirely different type of bonding is present in the hypothetical $Li_{12}C_{60}$ molecule, in which the Li atoms are positioned above the centers of the five-membered rings of $C_{60}$, preserving the overall $I_h$ symmetry. LDA calculations show that the Li atoms become $Li^+$ ions in this species, donating 12 electrons into the empty LUMO and LUMO+1 orbitals of the $C_{60}$ cage [147]. The energy minimum for this unusual molecule, which still awaits experimental observation, is attained at the distances between the $Li^+$ cations and the pentagonal faces of $C_{60}$ equal to 1.54 Å.

Geometries of the $KC_{60}^\bullet$, $K_2C_{60}$, and $K_3C_{60}^\bullet$ molecules, which constitute the building blocks of solid-state fullerides (Chapter 10), have been optimized within the Hartree–Fock level of approximation [148]. In the cases of $KC_{60}^\bullet$ and $K_2C_{60}$, the potassium atoms prefer the positions over hexagonal faces of $C_{60}$ to those over pentagons. All three molecules are highly ionic, with the potassium atoms bearing charges close to 1.0. Using semiempirical Born–Haber cycles, the energies of the reactions

$$K_nC_{60} \rightarrow K_{n-1}C_{60} + K \tag{9.2}$$

have been estimated at 25.4, 30.0, and 27.7 kcal/mol for $n = 1, 2$, and 3, respectively.

## References

1. A. Hirsch, *The Chemistry of the Fullerenes: An Overview*, Angew. Chem. Int. Ed. Engl. **32**, 1138 (1993).

2. M. Prato, Q. Chan Li, F. Wudl, and V. Lucchini, *Addition of Azides to $C_{60}$: Synthesis of Azafulleroids*, J. Am. Chem. Soc. **115**, 1148 (1993).

3. S. H. Hoke II, J. Molstad, D. Dilettato, M. J. Jay, D. Carlson, B. Kahr, and R. G. Cooks, *Reaction of Fullerenes and Benzyne*, J. Org. Chem. **57**, 5069 (1992).

4. K.-D. Kampe, N. Egger, and M. Vogel, *Diamino and Tetraamino Derivatives of Buckminsterfullerene $C_{60}$*, Angew. Chem. Int. Ed. Engl. **32**, 1174 (1993).

5. L. Isaacs, A. Wehrsig, and F. Diederich, *Improved Purification of $C_{60}$ and Formation of $\sigma$- and $\pi$-Homoaromatic Methano-Bridged Fullerenes by Reaction with Alkyl Diazoacetates*, Helv. Chim. Acta **76**, 1231 (1993).

6. X. Zhang, A. Romero, and C. S. Foote, *Photochemical [2 + 2] Cycloaddition of N,N-Diethylpropynylamine to $C_{60}$*, J. Am. Chem. Soc. **115**, 11024 (1993).

7. L. Y. Chiang, R. B. Upasani, and J. W. Swirczewski, *Versatile Nitronium Chemistry for $C_{60}$ Fullerene Functionalization*, J. Am. Chem. Soc. **114**, 10154 (1992).

8. Y. Rubin, S. Khan, D. I. Freedberg, and C. Yeretzian, *Synthesis and X-ray Structure of a Diels-Alder Adduct of $C_{60}$*, J. Am. Chem. Soc. **115**, 344 (1993).

9. T. Nogami, M. Tsuda, T. Ishida, S. Kurono, and M. Ohashi, *Addition Reactions of Benzyne, Dienes, Dichlorocarbene, and Oxygen to $C_{60}$*, Full. Sci. Tech. **1**, 275 (1993).

10. P. Belik, A. Gügel, J. Spickermann, and K. Müllen, *Reaction of Buckminsterfullerene with ortho-Quinodimethane: A New Access to Stable $C_{60}$ Derivatives*, Angew. Chem. Int. Ed. Engl. **32**, 78 (1993).

11. A. Vasella, P. Uhlmann, C. A. A. Waldraff, F. Diederich, and C. Thilgen, *Fullerene Sugars: Preparation of Enantiomerically Pure, Spiro-Linked C-Glycosides of $C_{60}$*, Angew. Chem. Int. Ed. Engl. **31**, 1388 (1992).

12. S. Petrie, G. Javahery, and D. K. Bohme, *Attaching Handles to $C_{60}^{2+}$: The Double-Derivatization of $C_{60}^{2+}$*, J. Am. Chem. Soc. **115**, 1445 (1993).

13. T. Guo, C. Jin, and R. E. Smalley, *Doping Bucky: Formation and Properties of Boron-Doped Buckminsterfullerene*, J. Phys. Chem. **95**, 4948 (1991).

14. T. Pradeep, V. Vijayakrishnan, A. K. Santra, and C. N. R. Rao, *Interaction of Nitrogen with Fullerenes: Nitrogen Derivatives of $C_{60}$ and $C_{70}$*, J. Phys. Chem. **95**, 10564 (1991).

15. W. Andreoni, F. Gygi, and M. Parrinello, *Impurity States in Doped Fullerenes: $C_{59}B$ and $C_{59}N$*, Chem. Phys. Lett. **190**, 159 (1992).

16. N. Kurita, K. Kobayashi, H. Kumahora, K. Tago, and K. Ozawa, *Molecular Structures, Binding Energies and Electronic Properties of Dopyballs $C_{59}X$ (X = B, N and S)*, Chem. Phys. Lett. **198**, 95 (1992).

17. X. Xia, D. A. Jelski, J. R. Bowser, and T. F. George, *MNDO Study of Boron-Nitrogen Analogues of Buckminsterfullerene*, J. Am. Chem. Soc. **114**, 6493 (1992).

18. N. Kurita, K. Kobayashi, H. Kumahora, and K. Tago, *Bonding and Electronic Properties of Substituted Fullerenes $C_{58}B_2$, $C_{58}N_2$, and $C_{58}BN$*, Full. Sci. Tech. **1**, 319 (1993).

19. S. J. La Placa, P. A. Roland, and J. J. Wynne, *Boron Clusters ($B_n$, $n = 2-52$) Produced by Laser Ablation of Hexagonal Boron Nitride*, Chem. Phys. Lett. **190**, 163 (1992).

20. J. R. Bowser, D. A. Jelski, and T. F. George, *Stability and Structure of $C_{12}B_{24}N_{24}$: A Hybrid Analogue of Buckminsterfullerene*, Inorg. Chem. **31**, 154 (1992).

21. K. Kobayashi and N. Kurita, *Bonding and Electronic Properties of A Multicomponent Fullerene $C_{12}B_{24}N_{24}$ by a Non-Local-Density-Functional Calculation*, Phys. Rev. Lett. **70**, 3542 (1993).

22. J. C. Phillips, *Stability, Kinetics, and Magic Numbers of $Si_n^+$ (n = 7-45) Clusters*, J. Chem. Phys. **83**, 3330 (1985).

23. Y. Liu, Q.-L. Zhang, F. K. Tittel, R. F. Curl, and R. E. Smalley, *Photodetachment and Photofragmentation Studies of Semiconductor Cluster Anions*, J. Chem. Phys. **85**, 7434 (1986).

24. M. F. Jarrold, Y. Ijiri, and U. Ray, *Interaction of Silicon Cluster Ions with Ammonia: Annealing, Equilibria, High Temperature Kinetics, and Saturation Studies*, J. Chem. Phys. **94**, 3607 (1991).

25. M. F. Jarrold and J. E. Bower, *Mobilities of Silicon Cluster Ions: The Reactivity of Silicon Sausages and Spheres*, J. Chem. Phys. **96**, 9180 (1992).

26. M. F. Jarrold and V. A. Constant, *Silicon Cluster Ions: Evidence for a Structural Transition*, Phys. Rev. Lett. **67**, 2994 (1991).

27. J. C. Phillips, *Morphology of Medium-Size Silicon Clusters*, J. Chem. Phys. **88**, 2090 (1988).

28. D. A. Jelski, Z. C. Wu, and T. F. George, *Large Silicon Clusters: Confirmation of Phillips' Conjecture*, Chem. Phys. Lett. **150**, 447 (1988).

29. D. A. Jelski, Z. C. Wu, and T. F. George, *An Inquiry into the Structure of the $Si_{60}$ Cluster: Analysis of Fragmentation Data*, J. Cluster Sci. **1**, 143 (1990).

30. C. Zybill, *$Si_{60}$, an Analogue of $C_{60}$?*, Angew. Chem. Int. Ed. Engl. **31**, 173 (1992).

31. S. Nagase and K. Kobayashi, *On the Fullerene Structures of the $Si_{60}$ Cluster*, Full. Sci. Tech. **1**, 299 (1993).

32. S. Nagase and K. Kobayashi, *$Si_{60}$ and $Si_{60}X$ (X = Ne, $F^-$, and $Na^+$)*, Chem. Phys. Lett. **187**, 291 (1991).

33. F. S. Khan and J. Q. Broughton, *Relaxation of Icosahedral-Cage Silicon Clusters via Tight-Binding Molecular Dynamics*, Phys. Rev. B **43**, 11754 (1991).

34. M. Menon and K. R. Subbaswamy, *Structure of $Si_{60}$. Cage versus Network Structures*, Chem. Phys. Lett. **219**, 219 (1994).

35. A. A. Bliznyuk, M. Shen, and H. F. Schaefer III, *The Dodecahedral $N_{20}$ Molecule. Some Theoretical Predictions*, Chem. Phys. Lett. **198**, 249 (1992).

36. F. Jensen and H. Toftlund, *Structure and Stability of $C_{24}$ and $B_{12}N_{12}$ Isomers*, Chem. Phys. Lett. **201**, 89 (1993).

37. B. C. Guo, K. P. Kerns, and A. W. Castleman, Jr., *$Ti_8C_{12}^+$-Metallo-Carbohedrenes: A New Class of Molecular Clusters?*, Science **255**, 1411 (1992).

38. S. Wei, B. C. Guo, J. Purnell, S. Buzza, and A. W. Castleman, Jr., *Metallocarbohedrenes as a Class of Stable Neutral Clusters: Formation Mechanism of $M_8C_{12}$ (M = Ti and V)*, J. Phys. Chem. **96**, 4166 (1992).

39. B. C. Guo, S. Wei, J. Purnell, S. Buzza, and A. W. Castleman, Jr., *Metallo-Carbohedrenes [$M_8C_{12}^+$ (M = V, Zr, Hf, and Ti)]: A Class of Stable Molecular Cluster Ions*, Science **256**, 515 (1992).

40. S. Wei, B. C. Guo, J. Purnell, S. Buzza, and A. W. Castleman, Jr., *Metallo-Carbohedrenes: Formation of Multicage Structures*, Science **256**, 818 (1992).

41. A. Ceulemans and P. W. Fowler, *Bonding in $Ti_8C_{12}$ and the Substitutional Jahn–Teller Effect*, J. Chem. Soc. Faraday Trans. **88**, 2797 (1992).

42. L. Pauling, *Molecular Structure of $Ti_8C_{12}$ and Related Complexes*, Proc. Natl. Acad. Sci. USA **89**, 8175 (1992).

43. I. Dance, *Geometric and Electronic Structures of [$Ti_8C_{12}$]: Analogies with $C_{60}$*, J. Chem. Soc. Chem. Commun. 1779 (1992).

44. M.-M. Rohmer, P. de Vaal, and M. Bénard, *Jahn–Teller Distortion Predicted for Metallocarbohedrenes: An ab Initio SCF Geometry Optimization of the Lowest Singlet and Triplet States of $Ti_8C_{12}$ in the $T_h$ and $D_{2h}$ Point Groups*, J. Am. Chem. Soc. **114**, 9696 (1992).

45. R. W. Grimes and J. D. Gale, *Predicted Structures for $Ti_8C_{12}$ and $Si_8C_{12}$ Dodecahedron Molecules*, J. Chem. Soc. Chem. Commun. 1222 (1992).

46. M. Methfessel, M. van Schilfgaarde, and M. Scheffler, *Electronic Structure and Bonding in the Metallocarbohedrene $Ti_8C_{12}$*, Phys. Rev. Lett. **70**, 29 (1993).

47. B. V. Reddy, S. N. Khanna, and P. Jena, *Electronic, Magnetic, and Geometric Structure of Metallo-Carbohedrenes*, Science **258**, 1640 (1992).

48. L. Lou, T. Guo, P. Nordlander, and R. E. Smalley, *Electronic Structure of the Hollow-Cage $M_8X_{12}$ Clusters*, J. Chem. Phys. **99**, 5301 (1993).

49. P. J. Hay, *Theoretical Studies of $M_8C_{12}$ Species*, J. Phys. Chem. **97**, 3081 (1993).

50. M.-M. Rohmer, M. Bénard, C. Henriet, C. Bo, and J.-M. Poblet, *$Ti_8C_{12}$: A Polytopal Molecule with 36 Ti–C Bonds*, J. Chem. Soc. Chem. Commun. 1182 (1993).

51. M.-M. Rohmer, personal communication (1994).

52. B. V. Reddy and S. N. Khanna, *Formation and Stability of Dodecahedral and fcc Structures in Metal-Carbon Clusters*, Chem. Phys. Lett. **209**, 104 (1993).

53. Z. Lin and M. B. Hall, *Theoretical Studies on the Stability of $M_8 C_{12}$ Clusters*, J. Am. Chem. Soc. **115**, 11165 (1993).

54. A. Hamwi, C. Fabre, P. Chaurand, S. Della-Negra, C. Ciot, D. Djurado, J. Dupuis, and A. Rassat, *Preparation and Characterization of Fluorinated Fullerenes*, Full. Sci. Tech. **1**, 499 (1993).

55. A. A. Tuinman, P. Mukherjee, J. L. Adcock, R. L. Hettich, and R. N. Compton, *Characterization and Stability of Highly Fluorinated Fullerenes*, J. Phys. Chem. **96**, 7584 (1992).

56. H. Selig, C. Lifshitz, T. Peres, J. E. Fischer, A. R. McGhie, W. J. Romanow, J. P. McCauley, Jr., and A. B. Smith III, *Fluorinated Fullerenes*, J. Am. Chem. Soc. **113**, 5475 (1991).

57. K. Kniaź, J. E. Fischer, H. Selig, G. B. M. Vaughan, W. J. Romanow, D. M. Cox, S. K. Chowdhury, J. P. McCauley, R. M. Strongin, and A. B. Smith III, *Fluorinated Fullerenes: Synthesis, Structure, and Properties*, J. Am. Chem. Soc. **115**, 6060 (1993).

58. J. H. Holloway, E. G. Hope, R. Taylor, G. J. Langley, A. G. Avent, T. J. Dennis, J. P. Hare, H. W. Kroto, and D. R. M. Walton, *Fluorination of Buckminsterfullerene*, J. Chem. Soc. Chem. Commun. 966 (1991).

59. M. I. Attalla, A. M. Vassallo, B. N. Tattam, and J. V. Hanna, *Preparation of Hydrofullerenes by Hydrogen Radical Induced Hydrogenation*, J. Phys. Chem. **97**, 6329 (1993).

60. J. Cioslowski, *Electronic Structures of the Icosahedral $C_{60} H_{60}$ and $C_{60} F_{60}$ Molecules*, Chem. Phys. Lett. **181**, 68 (1991).

61. J. Cioslowski, unpublished.

62. G. E. Scuseria, *Ab Initio Theoretical Predictions of the Equilibrium Geometries of $C_{60}$, $C_{60} H_{60}$ and $C_{60} F_{60}$*, Chem. Phys. Lett. **176**, 423 (1991).

63. B. I. Dunlap, D. W. Brenner, J. W. Mintmire, R. C. Mowrey, and C. T. White, *Geometric and Electronic Structures of $C_{60} H_{60}$, $C_{60} F_{60}$, and $C_{60} H_{36}$*, J. Phys. Chem. **95**, 5763 (1991).

64. D. Bakowies and W. Thiel, *Theoretical Study of Buckminsterfullerene Derivatives $C_{60} X_n$ (X=H,F; $n = 2$, 36, 60)*, Chem. Phys. Lett. **193**, 236 (1992).

65. N. Matsuzawa, T. Fukunaga, and D. A. Dixon, *Electronic Structures of $1,2$- and $1,4$-$C_{60} X_{2n}$ Derivatives with $n = 1$, 2, 4, 6, 8, 10, 12, 18, 24, and 30*, J. Phys. Chem. **96**, 10747 (1992).

66. J. Cioslowski, *An Efficient Evaluation of Atomic Properties Using a Vectorized Numerical Integration with Dynamic Thresholding*, Chem. Phys. Lett. **194**, 73 (1992).

67. P. W. Fowler, H. W. Kroto, R. Taylor, and D. R. M. Walton, *Hypothetical Twisted Structure for $C_{60} F_{60}$*, J. Chem. Soc. Faraday Trans. **87**, 2685 (1991).

68. G. E. Scuseria and G. K. Odom, *Exo-Fluorinated $C_{60} F_{60}$ Has $I_h$ Symmetry*, Chem. Phys. Lett. **195**, 531 (1992).

69. M. Saunders, *Buckminsterfullerane: The Inside Story*, Science **253**, 330 (1991).

70. G. Thimm and W. E. Klee, *Enumerating Buckminsterfullerane Isomers*, Science **255**, 92 (1992).

71. T. Guo and G. E. Scuseria, *Ab Initio Calculations of Tetrahedral Hydrogenated Buckminsterfullerene*, Chem. Phys. Lett. **191**, 527 (1992).

72. B. I. Dunlap, J. W. Mintmire, D. H. Robertson, D. W. Brenner, R. C. Mowrey, and C. T. White, *Photoelectron Spectra of $C_{60}H_{36}$ and $C_{60}H_{60}$*, Mat. Res. Soc. Symp. Proc. **247**, 351 (1992).

73. A. Webster, *Comparison of a Calculated Spectrum of $C_{60}H_{60}$ with the Unidentified Astronomical Infrared Emission Features*, Nature **352**, 412 (1991).

74. C. C. Henderson and P. A. Cahill, *Semi-Empirical Calculations of the Isomeric $C_{60}$ Dihydrides*, Chem. Phys. Lett. **198**, 570 (1992).

75. C. C. Henderson, C. M. Rohlfing, and P. A. Cahill, *Theoretical Studies of Selected $C_{60}H_2$ and $C_{70}H_2$ Isomers*, Chem. Phys. Lett. **213**, 383 (1993).

76. J. R. Morton, F. Negri, and K. F. Preston, *The EPR Spectra of Hydrides of $C_{70}$. An Experimental and Theoretical Investigation*, Chem. Phys. Lett. **218**, 467 (1994).

77. R. Taylor, *$C_{60}$, $C_{70}$, $C_{76}$, $C_{78}$ and $C_{84}$: Numbering, $\pi$-Bond Order Calculations and Addition Pattern Considerations*, J. Chem. Soc. Perkin Trans. 2 813 (1993).

78. C. C. Henderson and P. A. Cahill, *$C_{60}H_2$: Synthesis of the Simplest $C_{60}$ Hydrocarbon Derivative*, Science **259**, 1885 (1993).

79. C. C. Henderson, C. M. Rohlfing, R. A. Assink, and P. A. Cahill, *$C_{60}H_4$: Kinetics and Thermodynamics of Multiple Addition to $C_{60}$*, Angew. Chem. Int. Ed. Engl. **33**, 786 (1994).

80. C. C. Henderson, C. M. Rohlfing, K. T. Gillen, and P. A. Cahill, *Synthesis, Isolation, and Equilibration of 1,9- and 7,8-$C_{70}H_2$*, Science **264**, 397 (1994).

81. A. G. Avent, A. D. Darwish, D. K. Heimbach, H. W. Kroto, M. F. Meidine, J. P. Parsons, C. Remars, R. Roers, O. Ohashi, R. Taylor, and D. R. M. Walton, *Formation of Hydrides of Fullerene-$C_{60}$ and Fullerene-$C_{70}$*, J. Chem. Soc. Perkin Trans. 2 15 (1994).

82. H. R. Karfunkel and A. Hirsch, *Product Predictions for Nucleophillic Additions to $C_{70}$*, Angew. Chem. Int. Ed. Engl. **31**, 1468 (1992).

83. D. A. Dixon, N. Matsuzawa, T. Fukunaga, and F. N. Tebbe, *Patterns for Addition to $C_{60}$*, J. Phys. Chem. **96**, 6107 (1992).

84. A. Hirsch, A. Soi, and H. R. Karfunkel, *Titration of $C_{60}$: A Method for the Synthesis of Organofullerenes*, Angew. Chem. Int. Ed. Engl. **31**, 766 (1992).

85. A. L. Balch, personal communication (1994).

86. N. Matsuzawa, D. A. Dixon, and P. J. Krusic, *Semiempirical Calcula-*

*tions of* $C_{60}$ *Derivatives: Addition to Double Bonds Radiating from a Five-Membered Ring*, J. Phys. Chem. **96**, 8317 (1992).

87. S. Ballenweg, R. Gleiter, and W. Krätschmer, *Hydrogenation of Buckminsterfullerene* $C_{60}$ *via Hydrozirconation: A New Way to Organofullerenes*, Tetrahedron Lett. **34**, 3737 (1993).

88. C. Jin, R. Hettich, R. Compton, D. Joyce, J. Blencoe, and T. Burch, *Solid-Phase Hydrogenation of Fullerenes*, J. Phys. Chem. **98**, 4215 (1994).

89. R. E. Haufler, J. Conceicao, L. P. F. Chibante, Y. Chai, N. E. Byrne, S. Flanagan, M. M. Haley, S. C. O'Brien, C. Pan, Z. Xiao, W. E. Billups, M. A. Ciufolini, R. H. Hauge, J. L. Margrave, L. J. Wilson, R. F. Curl, and R. E. Smalley, *Efficient Production of* $C_{60}$ *(Buckminsterfullerene)*, $C_{60}H_{36}$, *and the Solvated Buckide Ion*, J. Phys. Chem. **94**, 8634 (1990).

90. K. Shigematsu, K. Abe, M. Mitani, and K. Tanaka, *Catalytic Hydrogenation of Fullerenes in the Presence of Metal Catalysts in Toluene Solution*, Full. Sci. Tech. **1**, 309 (1993).

91. C. Rüchardt, M. Gerst, J. Ebenhoch, H.-D. Beckhaus, E. E. B. Campbell, R. Tellgmann, H. Schwarz, T. Weiske, and S. Pitter, *Transfer Hydrogenation and Deuteration of Buckminsterfullerene* $C_{60}$ *by 9,10-Dihydroanthracene and 9,9',10,10'[$D_4$]Dihydroanthracene*, Angew. Chem. Int. Ed. Engl. **32**, 584 (1993).

92. L. E. Hall, D. R. McKenzie, M. I. Attalla, A. M. Vassallo, R. L. Davis, J. B. Dunlop, and D. J. H. Cockayne, *The Structure of* $C_{60}H_{36}$, J. Phys. Chem. **97**, 5741 (1993).

93. B. I. Dunlap, D. W. Brenner, R. C. Mowrey, J. W. Mintmire, D. H. Robertson, and C. T. White, *Possible Isomers and Electronic Structure of* $C_{60}H_{36}$, Mat. Res. Soc. Symp. Proc. **206**, 687 (1991).

94. L. D. Book and G. E. Scuseria, *Isomers of* $C_{60}H_{36}$ *and* $C_{70}H_{36}$, J. Phys. Chem. **98**, 4283 (1994).

95. B. W. Clare and D. L. Kepert, *Structures and Stabilities of Hydrofullerenes. Completion of a Tetrahedral Fused Quadruple Crown Structure and a Double Crown Structure at* $C_{60}H_{36}$, J. Mol. Struct. (Theochem) **304**, 181 (1994).

96. B. W. Clare and D. L. Kepert, *Structures and Stabilities of Hydrofullerenes. Completion of Crown Structures at* $C_{60}H_{18}$ *and* $C_{60}H_{24}$, J. Mol. Struct. (Theochem) **303**, 1 (1994).

97. A. A. Gakh, A. A. Tuinman, J. L. Adcock, R. A. Sachleben, and R. N. Compton, *Selective Synthesis and Structure Determination of* $C_{60}F_{48}$, J. Am. Chem. Soc. **116**, 819 (1994).

98. G. A. Olah, I. Bucsi, C. Lambert, R. Aniszfeld, N. J. Trivedi, D. K. Sensharma, and G. K. S. Prakash, *Chlorination and Bromination of Fullerenes. Nucleophilic Methoxylation of Polychlorofullerenes and Their Aluminum Trichloride Catalyzed Friedel-Crafts Reaction with Aromatics to Polyarylfullerenes*, J. Am. Chem. Soc. **113**, 9385 (1991).

99. P. R. Birkett, P. B. Hitchcock, H. W. Kroto, R. Taylor, and D. R. M. Walton, *Preparation and Characterization of $C_{60}Br_6$ and $C_{60}Br_8$*, Nature **357**, 479 (1992).

100. F. N. Tebbe, R. L. Harlow, D. B. Chase, D. L. Thorn, G. C. Campbell, Jr., J. C. Calabrese, N. Herron, R. J. Young, Jr., and E. Wasserman, *Synthesis and Single-Crystal X-ray Structure of a Highly Symmetrical $C_{60}$ Derivative, $C_{60}Br_{24}$*, Science **256**, 822 (1992).

101. M. Kobayashi, Y. Akahama, H. Kawamura, H. Shinohara, H. Sato, and Y. Saito, *X-ray Diffraction Study of Iodine-Doped $C_{60}$*, Solid State Commun. **81**, 93 (1992).

102. T. Zenner and H. Zabel, *Synthesis, Characterization, and Stability of $C_{60}I_2$*, J. Phys. Chem. **97**, 8690 (1993).

103. Q. Zhu, D. E. Cox, J. E. Fischer, K. Kniaz, A. R. McGhie, and O. Zhou, *Intercalation of Solid $C_{60}$ with Iodine*, Nature **355**, 712 (1992).

104. J. R. Morton, K. F. Preston, and F. Negri, *The EPR Spectra of $FC_{60}$ and $FC_{70}$*, Chem. Phys. Lett. **221**, 59 (1994).

105. G. A. Olah, I. Bucsi, C. Lambert, R. Aniszfeld, N. J. Trivedi, D. K. Sensharma, and G. K. S. Prakash, *Polyarenefullerenes, $C_{60}(H\text{-}Ar)_n$, Obtained by Acid-Catalyzed Fullerenation of Aromatics*, J. Am. Chem. Soc. **113**, 9387 (1991).

106. C. Caron, R. Subramanian, F. D'Souza, J. Kim, W. Kutner, M. T. Jones, and K. M. Kadish, *Selective Electrosynthesis of $(CH_3)_2C_{60}$: A Novel Method for the Controlled Functionalization of Fullerenes*, J. Am. Chem. Soc. **115**, 8505 (1993).

107. P. J. Krusic, E. Wasserman, P. N. Keizer, J. R. Morton, and K. F. Preston, *Radical Reactions of $C_{60}$*, Science **254**, 1183 (1991).

108. J. R. Morton, K. F. Preston, P. J. Krusic, S. A. Hill, and E. Wasserman, *ESR Studies of the Reaction of Alkyl Radicals with $C_{60}$*, J. Phys. Chem. **96**, 3576 (1992).

109. J. R. Morton, K. F. Preston, P. J. Krusic, S. A. Hill, and E. Wasserman, *The Dimerization of $RC_{60}$ Radicals*, J. Am. Chem. Soc. **114**, 5454 (1992).

110. P. J. Krusic, D. C. Roe, E. Johnston, J. R. Morton, and K. F. Preston, *EPR Study of Hindered Internal Rotation in Alkyl-$C_{60}$ Radicals*, J. Phys. Chem. **97**, 1736 (1993).

111. P. J. Fagan, P. J. Krusic, D. H. Evans, S. A. Lerke, and E. Johnston, *Synthesis, Chemistry, and Properties of a Monoalkylated Buckminsterfullerene Derivative, $t\text{-}BuC_{60}$ Anion*, J. Am. Chem. Soc. **114**, 9697 (1992).

112. K. M. Creegan, J. L. Robbins, W. K. Robbins, J. M. Millar, R. D. Sherwood, P. J. Tindall, D. M. Cox, A. B. Smith, III, J. P. McCauley, Jr., D. R. Jones, and R. T. Gallagher, *Synthesis and Characterization of $C_{60}O$, the First Fullerene Epoxide*, J. Am. Chem. Soc. **114**, 1103 (1992).

113. Y. Elemes, S. K. Silverman, C. Sheu, M. Kao, C. S. Foote, M. M. Alvarez, and R. L. Whetten, *Reaction of $C_{60}$ with Dimethyldioxirane — Forma-*

*tion of an Epoxide and a 1,3-Dioxolane Derivative*, Angew. Chem. Int. Ed. Engl. **31**, 351 (1992).

114. J. B. Howard, J. T. McKinnon, M. E. Johnson, Y. Makarovsky, and A. L. Lafleur, *Production of $C_{60}$ and $C_{70}$ Fullerenes in Benzene–Oxygen Flames*, J. Phys. Chem. **96**, 6657 (1992).

115. W. A. Kalsbeck and H. H. Thorp, *Electrochemical Reduction of Fullerenes in the Presence of $O_2$ and $H_2O$: Polyoxygen Adducts and Fragmentation of the $C_{60}$ Framework*, J. Electroanal. Chem. **314**, 363 (1991).

116. F. Diederich, R. Ettl, Y. Rubin, R. L. Whetten, R. Beck, M. Alvarez, S. Anz, D. Sensharma, F. Wudl, K. C. Khemani, and A. Koch, *The Higher Fullerenes: Isolation and Characterization of $C_{76}$, $C_{84}$, $C_{90}$, $C_{94}$, and $C_{70}O$, an Oxide of $D_{5h}$-$C_{70}$*, Science **252**, 548 (1991).

117. K. Raghavachari, *Structure of $C_{60}O$: Unexpected Ground State Geometry*, Chem. Phys. Lett. **195**, 221 (1992).

118. K. Raghavachari, *Fullerene Derivatives: Structures and Stabilities of $C_{60}O$ and $C_{70}O$*, Int. J. Mod. Phys. B **6**, 3821 (1992).

119. Z. Slanina, F. Uhlík, J.-P. François, and L. Adamowicz, *Two $C_{60}O$ Structures and Their Relative Stabilities: AM1 Computations*, Full. Sci. Tech. **1**, 537 (1993).

120. Z. Slanina, F. Uhlík, S.-L. Lee, and L. Adamowicz, *AM1 Computations of $C_{60}O_2$*, Full. Sci. Tech. **2**, 73 (1994).

121. K. Raghavachari and C. Sosa, *Fullerene Derivatives. Comparative Theoretical Study of $C_{60}O$ and $C_{60}CH_2$*, Chem. Phys. Lett. **209**, 223 (1993).

122. M. Menon and K. R. Subbaswamy, *Optimized Structures of $C_{60}O$ and $C_{60}O_2$ Calculated by a Damped Molecular Dynamics Optimization Scheme*, Chem. Phys. Lett. **201**, 321 (1993).

123. K. Raghavachari and C. M. Rohlfing, *Ground State of $C_{70}O$: Oxygen Bridging at the Equatorial Belt*, Chem. Phys. Lett. **197**, 495 (1992).

124. T. Suzuki, Q. Li, K. C. Khemani, F. Wudl, and Ö. Almarsson, *Systematic Inflation of Buckminsterfullerene $C_{60}$: Synthesis of Diphenyl Fulleroids $C_{61}$ to $C_{66}$*, Science **254**, 1186 (1991).

125. M. Prato, V. Lucchini, M. Maggini, E. Stimpfl, G. Scorrano, M. Eiermann, T. Suzuki, and F. Wudl, *Energetic Preference in 5,6 and 6,6 Ring Junction Adducts of $C_{60}$: Fulleroids and Methanofullerenes*, J. Am. Chem. Soc. **115**, 8479 (1993).

126. T. Suzuki, Q. Li, K. C. Khemani, and F. Wudl, *Dihydrofulleroid $H_2C_{61}$: Synthesis and Properties of the Parent Fulleroid*, J. Am. Chem. Soc. **114**, 7301 (1992).

127. A. B. Smith III, R. M. Strongin, L. Brard, G. T. Furst, and W. J. Romanow, *1,2-Methanobuckminsterfullerene ($C_{61}H_2$), the Parent Fullerene Cyclopropane: Synthesis and Structure*, J. Am. Chem. Soc. **115**, 5829 (1993).

128. Z. Slanina, F. Uhlík, S.-L. Lee, and L. Adamowicz, $C_{61}H_2$ *Fulleroid: AM1 Computational Study*, Full. Sci. Tech. **2**, 13 (1994).

129. Z. Slanina, F. Uhlík, J.-P. François, and L. Adamowicz, *Diphenylmethane Fulleroid $C_{73}H_{10}$: AM1 Computational Study*, Full. Sci. Tech. **1**, 189 (1993).

130. F. Wudl, *The Chemical Properties of Buckminsterfullerene ($C_{60}$) and the Birth and Infancy of Fulleroids*, Acc. Chem. Res. **25**, 157 (1992).

131. J. M. Hawkins, S. Loren, A. Meyer, and R. Nunlist, *2D Nuclear Magnetic Resonance Analysis of Osmylated $C_{60}$*, J. Am. Chem. Sóc. **113**, 7770 (1991).

132. J. M. Hawkins, A. Meyer, T. A. Lewis, S. Loren, and F. J. Hollander, *Crystal Structure of Osmylated $C_{60}$: Confirmation of the Soccer Ball Framework*, Science **252**, 312 (1991).

133. J. M. Hawkins and A. Meyer, *Optically Active Carbon: Kinetic Resolution of $C_{76}$ by Asymmetric Osmylation*, Science **260**, 1918 (1993).

134. J. M. Hawkins, A. Meyer, and M. Nambu, *Asymmetric Bisosmylation of $C_{60}$: Novel Chiral $\pi$-Systems*, J. Am. Chem. Soc. **115**, 9844 (1993).

135. J. M. Hawkins, A. Meyer, and M. A. Solow, *Osmylation of $C_{70}$: Reactivity versus Local Curvature of the Fullerene Spheroid*, J. Am. Chem. Soc. **115**, 7499 (1993).

136. B. Chase and P. J. Fagan, *Substituted $C_{60}$ Molecules: A Study in Symmetry Reduction*, J. Am. Chem. Soc. **114**, 2252 (1992).

137. S. A. Lerke, B. A. Parkinson, D. H. Evans, and P. J. Fagan, *Electrochemical Studies on Metal Derivatives of Buckminsterfullerene ($C_{60}$)*, J. Am. Chem. Soc. **114**, 7807 (1992).

138. P. J. Fagan, J. C. Calabrese, and B. Malone, *The Chemical Nature of Buckminsterfullerene ($C_{60}$) and the Characterization of a Platinum Derivative*, Science **252**, 1160 (1991).

139. R. S. Koefod, C. Xu, W. Lu, J. R. Shapley, M. G. Hill, and K. R. Mann, *An Electrochemical and Spectroelectrochemical Study of an Iridium-Buckminsterfullerene Complex. Evidence for $C_{60}$-Localized Reductions*, J. Phys. Chem. **96**, 2928 (1992).

140. A. L. Balch, J. W. Lee, B. C. Noll, and M. M. Olmstead, *A Double Addition Product of $C_{60}$: $C_{60}\{Ir(CO)Cl(PMe_2Ph)_2\}_2$. Individual Crystallization of Two Conformational Isomers*, J. Am. Chem. Soc. **114**, 10984 (1992).

141. Y. Huang and B. S. Freiser, *Externally Bound Metal Ion Complexes of Buckminsterfullerene, $MC_{60}^+$, in the Gas Phase*, J. Am. Chem. Soc. **113**, 9418 (1991).

142. L. M. Roth, Y. Huang, J. T. Schwedler, C. J. Cassady, D. Ben-Amotz, B. Kahr, and B. S. Freiser, *Evidence for an Externally Bound $Fe^+-$Buckminsterfullerene Complex, $FeC_{60}^+$, in the Gas Phase*, J. Am. Chem. Soc. **113**, 6298 (1991).

143. Y. Huang and B. S. Freiser, *Synthesis of Bis(buckminsterfullerene)nickel Cation, $Ni(C_{60})_2^+$ in the Gas Phase*, J. Am. Chem. Soc. **113**, 8186 (1991).

144. N. Koga and K. Morokuma, *Ab Initio MO Calculation of ($\eta^2$-$C_{60}$) $Pt(PH_3)_2$. Electronic Structure and Interaction Between $C_{60}$ and Pt*, Chem. Phys. Lett. **202**, 330 (1993).

145. P. J. Fagan, J. C. Calabrese, and B. Malone, *Metal Complexes of Buckminsterfullerene ($C_{60}$)*, Acc. Chem. Res. **25**, 134 (1992).

146. P. J. Fagan, J. C. Calabrese, and B. Malone, *A Multiply-Substituted Buckminsterfullerene ($C_{60}$) with an Octahedral Array of Platinum Atoms*, J. Am. Chem. Soc. **113**, 9408 (1991).

147. J. Kohanoff, W. Andreoni, and M. Parrinello, *A Possible New Highly Stable Fulleride Cluster: $Li_{12}C_{60}$*, Chem. Phys. Lett. **198**, 472 (1992).

148. P. Weis, R. D. Beck, G. Bräuchle, and M. M. Kappes, *Properties of Size and Composition Selected Gas Phase Alkali Fulleride Clusters*, J. Chem. Phys. **100**, 5684 (1994).

# Chapter 10

# Solid-State Properties of Fullerenes and Their Derivatives

Fullerenes and their derivatives form both molecular and ionic crystals. The combination of factors, such as the spherical or nearly spherical shape, high polarizability, relatively low ionization potential and electron affinity, as well as the extraordinary stiffness of individual molecules, gives rise to several interesting properties of solid fullerenes [1]. Phenomena such as the phase transition in solid $C_{60}$ [2], the existence of a multitude of alkali metal fullerides [3–5], the superconductivity in the $A_3C_{60}$ intercalation compounds [6], and the low-temperature ferromagnetism of the $TDAE^+C_{60}^-$ salt [7] are of both theoretical and experimental importance. Quantum-mechanical calculations are indispensable in predicting and understanding these phenomena.

## 10.1  Solid-State Phases of $C_{60}$

Solid $C_{60}$ is a molecular crystal in which the fullerene cages are held together by cohesive van der Waals interactions. Two solid-state phases of $C_{60}$ are well characterized. The high-temperature phase possesses a face-centered cubic (fcc) structure with a primitive cell containing four $C_{60}$ molecules [2]. These four molecules are crystallographically equivalent due to their very rapid, continuous rotation. NMR measurements [8, 9] show that, despite the fact that the $C_{60}$ cages are faceted rather than smooth, the rotation is almost unhindered [10], as reflected by the apparent activation energy of only 1.4 kcal/mol [11]. Neutron scattering experiments indicate the lack of correlation between rotational motions of individual molecules [12].

Upon cooling, the high-temperature rotator phase undergoes a transition to a low-temperature ratchet phase in which the rotation of the $C_{60}$ molecules is no longer continuous [8, 11]. Instead, the fullerene cages jump rotationally between symmetry-equivalent minima. The height of the barrier between these minima has been estimated at 4.2 (NMR) [11], 5.8 (NMR) [8], and 6.0 kcal/mol (thermal conductivity) [13]. As the ratcheting motion is about 30 times slower than the rotation, the four $C_{60}$ molecules of the fcc cell become crystallographically unequivalent and a simple cubic (sc) structure (space group $Pa\bar{3}$) ensues [2]. According to neutron powder diffraction measurements at 5 K [14], the energy minimum is attained in the sc phase when the $C_{60}$ molecules are rotated in a way such that the "double" bonds of each cage (Chapter 3) face the five-membered rings of its neighbors. However, analysis of the X-ray data collected at 110 K [15] leads to the conclusion that in fact two distinct populations of $C_{60}$ molecules are present in the sc crystals at higher temperatures [16]. Molecules of the major population are oriented in accordance with the neutron diffraction data, whereas the $C_{60}$ cages of the minor population are rotated around their threefold axes by *ca.* 180°. The former orientation is lower in energy than the latter one by an estimated 0.25 kcal/mol. The coexistence of two populations in the sc crystals of $C_{60}$ is consistent with the observations of a continuous phase transition around 90 K [17, 18] and a low-temperature superstructure [19]. On the other hand, the presence of permanent dipole moments in low-temperature samples of $C_{60}$ still awaits explanation [20].

Depending on the purity of sample and the experimental method, the transition between the low- and the high-temperature phases is observed at 249 [2, 21], 252 [22], 256 [23], 260 [24–26], and 269 K [27] at ambient pressure. In accordance with its first-order character, the transition is accompanied by discontinuities in the lattice constant [26, 28], specific heat [2, 23, 25], thermal conductivity [13], bulk modulus [39], and vibrational frequencies [22, 24]. As expected, the transition temperature increases with pressure, the rate amounting to 104 [21] and 162 K/GPa [40]. Interestingly, there have been reports on the coexistence of the high-temperature fcc phase with either an hcp [41, 42] or a bct phase [42] in samples of sublimed $C_{60}$.

Not surprisingly, accurate estimates of solid-state properties of $C_{60}$, such as the lattice constant and the cohesive energy (Table 10.1), are furnished by calculations employing the Lennard–Jones carbon–carbon potential with parameters taken from graphite. Smearing nuclei over the entire surface of the $C_{60}$ sphere and considering the resulting sphere–sphere interactions does not appreciably worsen the accuracy of computed properties [43, 44]. However, neither the low-temperature phase nor the transition temperature are predicted correctly by calculations in which a pure Lennard–Jones potential is used [29, 31]. This deficiency can be eliminated by assigning fractional charges of 0.27 and −0.54 to each of the "single" and "double" bonds of $C_{60}$ (Chapter 3), respectively, and then adding the resulting Coulombic interaction between the bond midpoints to the Lennard–Jones potential [45]. Such a parameterization, although lacking clear physical basis, yields the correct sc

## Table 10.1. Properties of the fcc phase of solid $C_{60}$

| Method | Lattice constant [Å] | Cohesive energy [kcal/mol of $C_{60}$] | Bulk modulus [GPa] |
|---|---|---|---|
| Lennard–Jones [29, 30] | 14.14 | 34.0 | n/a |
| Lennard–Jones [31] | 14.13 | 43.9 | n/a |
| LDA [32] | 15.05 | n/a | 18.3 |
| LDA [33] | 14.04 | 36.9 | 16.5 |
| X-ray diffr. [28] | 14.154 | n/a | n/a |
| X-ray diffr. [34] | 14.11 | n/a | 14.5 |
| X-ray diffr. [35] | 14.198 | n/a | 18.1 |
| Enth. of subl. (at 707 K) [36] | n/a | 40.1 | n/a |
| Enth. of subl. (at 700 K) [37] | n/a | 43.3 | n/a |
| Enth. of subl. (at 843 K) [38] | n/a | 46.1 | n/a |

low-temperature phase with a cohesive energy of 45.9 kcal/mol of $C_{60}$ (at 0 K), *ca.* 90% of which comes from the van der Waals and 10% from the Coulomb interactions [46]. The corresponding bulk modulus and lattice constant are 19.3 GPa and 14.05 Å, respectively. The predicted transition temperature of 270 K and its pressure coefficient of 115 K/GPa are also in good agreement with the experimental values. Somewhat worse agreement (the cohesive energy of *ca.* 44 kcal/mol of $C_{60}$ and the transition temperature of 215 K) is obtained when the bond charges are reduced to 0.175 and $-0.350$ [47].

The center-to-center distances between the $C_{60}$ molecules in the fcc phase amount to *ca.* 10 Å, or *ca.* 3 Å more than the sum of their cage radii. This means that at low pressures the main contribution to the bulk modulus of solid $C_{60}$ comes from the van der Waals interactions. This is indeed the case, as reflected by the measured bulk modulus that is much smaller than that of either hexagonal graphite (37 GPa) or diamond (556 GPa) [34]. The experimental values of the bulk modulus are in good agreement with those obtained from LDA calculations [32, 33].

There have been some speculations on the bulk modulus of solid $C_{60}$ under high pressures. In principle, once the distances between the $C_{60}$ cages are greatly reduced, deformation of individual molecules is unavoidable [48]. Invoking this deformation could make solid $C_{60}$ stiffer than diamond, as the bulk modulus arising from a direct, uniform compression of the $C_{60}$ cage has been estimated at 1300 (AM1) [49] and 843 GPa (molecular mechanics) [50]. However, high-pressure experiments show that before compression sufficient to trigger such a deformation is reached, the $C_{60}$ molecules fuse through $\sigma$-type bonds and three-dimensional, polymeric structures are formed (Chapter 6).

## 10.2  Electronic Structure of Solid $C_{60}$

Solid $C_{60}$ is a semiconductor at ambient temperatures. Electric conductivity studies yield a band gap $\Delta E$ of 1.6 eV [51], in reasonable agreement with the measured optical gaps of 1.74 [52] and 1.8 eV [53], but substantially smaller than the gaps of 2.3 and 2.3±0.1 eV deduced from ellipsometric measurements [54] and the (inverse) photoelectron spectra (PES/IPES) [55], respectively. LDA calculations, which are known to underestimate band gaps, predict for the fcc phase of solid $C_{60}$ values of $\Delta E$ equal to 1.18 [33], 1.34 [56], 1.35 [32, 57], 1.5 [58], and 1.6 eV [59]. A band gap of 1.58 eV is obtained within the tight-binding Hamiltonian approximation [60]. The gap is direct and occurs at the $X$ point of the Brillouin zone.

As expected from the weak interactions between individual molecules, the bands of solid $C_{60}$ are quite narrow, with the deeper $\sigma$-type bands exhibiting almost no dispersion [58]. The highest occupied valence band is 0.55 [32, 56, 57], 0.58 [33], and 0.59 eV [60] wide. The lowest unoccupied conduction band has the calculated width of 0.40 [60], 0.46 [32], and 0.54 eV [56]. The effective electron mass equals *ca.* 1.3 au for the conduction band [58].

Electronic structure methods are successful in predicting the static dielectric constant $\epsilon_0$ of solid $C_{60}$. The experimental values of $4.4 \pm 0.2$ [61] and 4.7 [53] are well reproduced by LDA calculations that estimate $\epsilon_0$ at 4.27 [32] and 4.4 [56]. An increase in $\epsilon_0$ to 5.63 at a pressure of 3 GPa is also predicted [32, 57].

The band gap of solid $C_{60}$ narrows with pressure. According to electric conductivity studies, $\Delta E$ decreases by *ca.* 0.4 eV when a pressure of 10 GPa is applied [51]. Optical absorption measurements yield a slope of $-0.14$ eV/GPa for the pressure dependence of $\Delta E$ [52]. This value is commensurate with the LDA predictions, according to which the band gap equals 1.35, 0.99, and 0.69 eV at pressures of 0, 3, and 13 GPa [32], respectively. The same calculations indicate that at pressures higher than 1.7 GPa the highest occupied valence band merges with its lower-energy counterpart. Similarly, the lowest unoccupied conduction band is predicted to merge with the next higher band.

The Hubbard Hamiltonian approximation provides means for a qualitative understanding of the electronic structure of solid $C_{60}$. The magnitude of the on-site Coulomb interaction $U$ has been estimated at 1.6±0.2 eV from Auger spectra [55]. Values of 0.8–1.3 eV for $U$ and 0.3–0.5 eV for the magnitude of the nearest-neighbor Coulomb interaction $V$ have been suggested [62].

MNDO calculations on the $C_{60}^{2-}$ and $C_{60}^{3-}$ anions have been employed in conjunction with the multiplet theory to estimate $U$ and $V$ [63]. Upon inclusion of screening, the values of 1.3 and 0.3 eV have been obtained for $U$ and $V$, respectively. LDA calculations on model lattices of $C_{60}$ have yielded $U=3.32$ eV [64]. The fact that both $U$ and $V$ are comparable with the band widths means that electron correlation effects are important in solid $C_{60}$ [63].

The electronic structure of the sc phase of solid $C_{60}$ has been a subject of only a few theoretical studies [60, 65]. Calculations employing the tight-binding Hamiltonian approximation predict this phase to possess an indirect (from $\Gamma$ to $R$) gap.


## 10.3  Electronic Structures of Solid $A_3C_{60}$ Fullerides

Reactions between alkali metals and the $C_{60}$ fullerene produce solid fullerides (alkali metal salts of fullerenes) that possess well-defined stoichiometries. Among those, the $A_3C_{60}$ fullerides have been studied most extensively. As revealed by X-ray diffraction experiments, crystals of the $A_3C_{60}$ species possess fcc lattices of the well-known cryolite structure (space group $Fm\bar{3}m$) in which the alkali metal cations occupy both the tetrahedral and octahedral interstitial sites between the $C_{60}^{3-}$ ions [Fig. 10.1($a$)] [3]. At room temperature, the fullerene cages are disordered between two distinct orientations, which are related by a 90° rotation and separated by a barrier of *ca.* 11 kcal/mol [75].

Several simple and mixed fcc $A_3C_{60}$ fullerides have been characterized experimentally (Table 10.2). It is believed that the lattice constants $a_0$ of these species are controlled by the size of the alkali metal cation that occupies the smaller tetrahedral site with a radius of *ca.* 1.1 Å (calculated for freely rotating $C_{60}$ cages) [66]. Solid fullerides involving the $Li^+$ and $Na^+$ cations possess lattice constants that are smaller than that of the fcc phase of solid $C_{60}$. This lattice contraction is easily explained by the electrostatic attraction between the negatively charged fullerene cages and the positively charged counterions. On the other hand, doping $C_{60}$ with heavier alkali metals results in lattice expansion necessary for the accommodation of the larger $K^+$, $Rb^+$, and $Cs^+$ cations. In the extreme case of $Cs_3C_{60}$, the $Cs^+$ cation with a radius of *ca.* 1.7 Å is too large to fit in the tetrahedral site [70], and a bcc rather than an fcc structure ensues [76]. However, solid $CsC_{60}$ possesses an fcc lattice, as the octahedral sites with a radius of *ca.* 2.1 Å are spacious enough to house the $Cs^+$ cations.

Electronic structure calculations find solid $K_3C_{60}$ to be an ionic crystal composed of $C_{60}^{3-}$ anions and $K^+$ counterions. The donation of three electrons to each fullerene cage results in a half-filled band derived from the triply degenerate $t_{1u}$ LUMO of $C_{60}$. With the width of 0.41 [77], 0.44 [78], 0.50 [79], and 0.60 eV [80] calculated within the LDA formalism, the band is quite narrow.

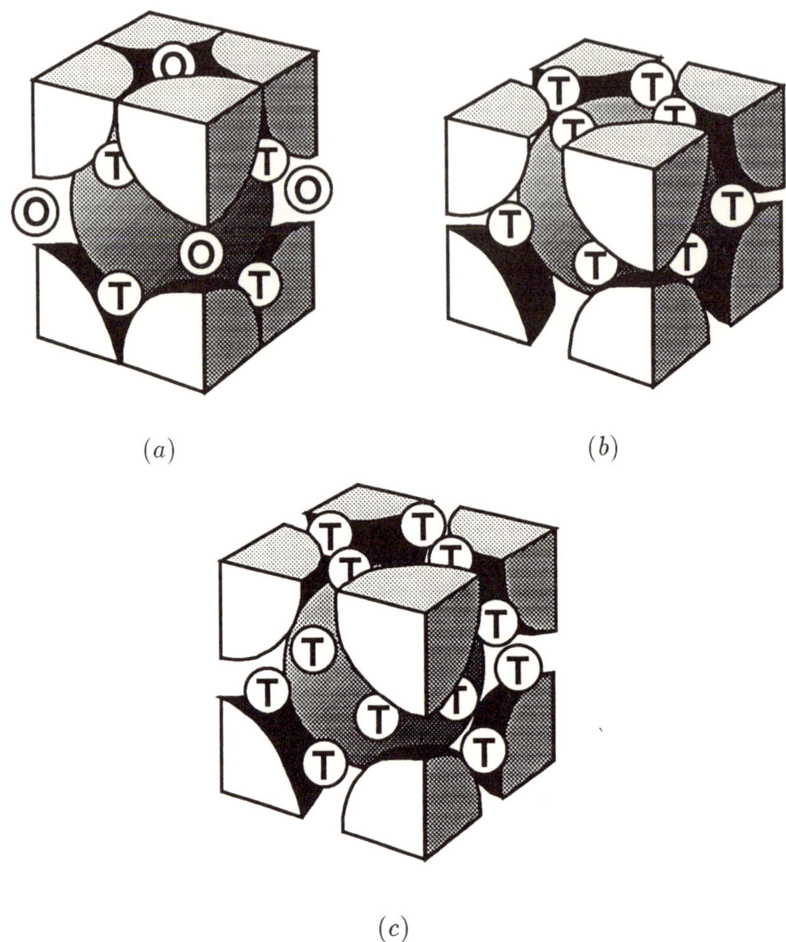



Fig. 10.1. Crystal structures of alkali metal fullerides: (a) $A_3 C_{60}$ (fcc), (b) $A_4 C_{60}$ (bct), (c) $A_6 C_{60}$ (bcc) [4]. The octahedral and tetrahedral interstitial sites are marked by $O$ and $T$, respectively.

Its density of states exhibits a bimodal distribution, with the Fermi level $E_F$ located near the maximum of the lower-energy peak [80]. The occupied portion of the band is only *ca.* 0.18 eV wide [81]. The narrowness of this band, together with the fortuitous position of the Fermi level, gives rise to a large density of states $N_F$ at $E_F$. LDA calculations estimate $N_F$ at 12 [82], 13.2 [80], 17 [81], and 17.96 eV$^{-1}$ per molecule of $K_3 C_{60}$ [78], in reasonable agreement with the experimental values of 17 [83] and 20 eV$^{-1}$ per molecule of $K_3 C_{60}$ [84] derived

## Table 10.2. Lattice constants and temperatures of transition to the superconducting state of the fcc $A_3C_{60}$ fullerides

| Fulleride | $a_0$ [Å] | | $T_c$ [K] | | |
|---|---|---|---|---|---|
| $Li_2RbC_{60}$ | 13.896 [66] | | | | |
| $Li_2CsC_{60}$ | 13.998 [66] | | | | |
| $Na_3C_{60}$ | 14.191 [67] | | | | |
| $Na_2KC_{60}$ | 14.025 [68] | 2.5 [68] | | | |
| $Na_2RbC_{60}$ | 14.028 [68] | 2.5 [68] | | | |
| $Na_2CsC_{60}$ | 14.134 [68] | 12 [68] | | | |
| $K_3C_{60}$ | 14.240 [3] | 18 [6] | 19.3 [70, 71] | 19.8 [72] | |
| $K_2RbC_{60}$ | 14.299 [70] | 21.8 [70] | | | |
| $KRb_2C_{60}^a$ | 14.350 [70] | 25.4 [70] | | | |
| $K_2CsC_{60}$ | 14.292 [66] | | | | |
| $Rb_3C_{60}$ | 14.425 [73] | 29 [74] | 29.4 [70] | 30 [71] | |
| $Rb_2CsC_{60}$ | 14.493 [70] | 31 [74] | 31.3 [70] | | |
| $RbCs_2C_{60}$ | 14.555 [66] | 33 [74] | | | |

[a] Average of two different samples.

from NMR measurements. The theoretically predicted band structure of $K_3C_{60}$ is consistent with that inferred from spectroscopic measurements [85], although the observed density of states at $E_F$ is much smaller.

The presence of ionic interactions in $K_3C_{60}$ stiffens the lattice and gives rise to quite a large bulk modulus (experimental: 28 GPa [86], LDA: 23 GPa [77]). The lattice constant is experimentally found to be greater than that of $C_{60}$; a phenomenon that is reproduced by some [77] but not all [82] LDA calculations. The reaction

$$C_{60}(\text{solid}) + 3\,K(\text{solid}) \rightarrow K_3C_{60}(\text{solid}) \,, \tag{10.1}$$

is predicted to be exothermic, with the enthalpy amounting to $-115$ kcal/mol [77], in reasonable agreement with the estimate of $-145$ kcal/mol obtained from a semiempirical Born–Haber cycle [87]. The latter figure can be compared with those of $-100$ and $-147$ kcal/mol quoted for the enthalpies of formation of solid $Na_3C_{60}$ and $Rb_3C_{60}$, respectively.

The densities of states at the Fermi levels of solid fullerides depend strongly on their lattice constants. Expanding the lattice brings about diminished interactions between the $C_{60}^{3-}$ ions, resulting in band narrowing, which in turn increases $N_F$. LDA calculations on simple and mixed $A_3C_{60}$ fullerides confirm this prediction [78]. With the experimental lattice constants [70], the widths of the half-filled bands in $RbK_2C_{60}$, $Rb_2KC_{60}$, $Rb_3C_{60}$, and $Rb_2CsC_{60}$ have been estimated at 0.43, 0.38, 0.36, and 0.35 eV, respectively. The corresponding values of 19.35, 20.10, 21.13, and 22.54 $eV^{-1}$ per molecule of $A_3C_{60}$ have been found for $N_F$. The predicted increase in $N_F$ upon replacing potassium by rubidium as the dopant is corroborated by NMR measurements [83, 84]. However, the observed relative change in $N_F$ (*ca.* 29% [83] and *ca.* 40% [84]) is larger than the predicted one (*ca.* 18%). As the interactions between the $C_{60}^{3-}$ ions grow in strength upon compression of the $A_3C_{60}$ crystals, positive pressure effect on the band width is expected [77].

## 10.4 Superconductivity in Solid $A_3C_{60}$ Fullerides

Solid fcc $A_3C_{60}$ fullerides, which are good conductors of electricity at ambient temperatures [69], become superconductors upon cooling [6]. The temperatures $T_c$ at which the transition to the superconducting state occur span the range of 12–30 K (Table 10.2) and depend strongly on the intercalating alkali metals. Superconductivity is not observed in $A_3C_{60}$ fullerides involving the $Li^+$ cation. Among fullerides containing the $Na^+$ cation, only $Na_2CsC_{60}$ is superconducting above 10 K [67].

The following experimental facts, which have been the subject of a recent review [88], shed some light on the nature of the superconducting state. First of all, external pressure has a pronounced effect on $T_c$. The experimentally determined pressure coefficient is negative and equals $-7.8$ K/GPa for $K_3C_{60}$ [89]. For $Rb_3C_{60}$, $(\partial T_c/\partial p)_{p=0}$ has been measured at $-9.7$ [90] and $-10$ K/GPa [89]. Even more importantly, studies of both pressure [86, 90] and composition dependence [70] establish an approximately universal relationship between the transition temperatures of the $A_3C_{60}$ fullerides (with the exception of those containing the $Na^+$ cation [67, 68]) and their lattice constants. From these measurements one concludes that lattice contraction decreases $T_c$ by as much as $45\pm1$ [86] and *ca.* 50 K/Å [70].

Second, isotope effects on the transition temperature are observed in the $K_3C_{60}$ and $Rb_3C_{60}$ superconductors. In the former case, replacing $C_{60}$ by $^{13}C_{60}$ results in the depression of $T_c$ by $0.45\pm0.1$ K [91], whereas in the latter case a depression of $0.7\pm0.1$ K is found [92]. When used as input to the equation

$$T_c = a\,M^{-\alpha},\tag{10.2}$$

where $M$ is the atomic mass and $a$ is a proportionality constant, the observed decrease in $T_c$ translates into the exponent $\alpha$ equal to $0.30\pm0.06$ for $K_3C_{60}$

[91]. For $Rb_3C_{60}$, the respective exponent has been estimated at $0.30\pm0.05$ [92] and $0.37\pm0.05$ [93]. Third, the experimentally determined dependence of the transition temperatures on the applied magnetic field yields the upper critical fields at $T = 0$ K of 28 and 38 T for $K_3C_{60}$ and $Rb_3C_{60}$, respectively [94]. These values correspond to the Ginzburg–Landau coherence lengths of only 34 and 30 Å, indicating that both $K_3C_{60}$ and $Rb_3C_{60}$ belong to the class of extreme type-II superconductors [88].

The observed universal relationship between $T_c$ on $a_0$ can be explained in a qualitative manner within the weak-coupling limit of the BCS theory, at which the transition temperature is given by

$$T_c = \frac{h\nu}{k} \exp\left(-\frac{1}{\lambda}\right), \qquad \lambda = VN_F, \tag{10.3}$$

where $\nu$ is the characteristic phonon frequency and $V$ is the electron–phonon coupling parameter [70, 86, 89, 90]. As the lattice constant decreases, the density of states at the Fermi level drops (Section 10.3) and $T_c$ falls. The insensitivity of $T_c$ to the nature of the alkali metal cation implies that both $\nu$ and $V$ are intercalant-independent, meaning that they are unique properties of the $C_{60}^{3-}$ ions.

The simple formula (10.3) is incapable of accounting for the smallness of the exponent $\alpha$ in Eq. (10.2). As the vibrational frequencies of the fullerene cages scale inversely with the square root of the carbon mass, Eq. (10.3) predicts $\alpha = 0.5$, which is in variance with the experimental observations. This discrepancy can be eliminated by considering the more accurate McMillan equation [95]

$$T_c = \frac{h <\nu>}{1.2k} \exp\left[\frac{-1.04\,(1+\lambda)}{\lambda - \mu^*\,(1+0.62\lambda)}\right], \tag{10.4}$$

where $<\nu>$ is the logarithmically averaged phonon frequency and $\mu^*$ is the phonon-renormalized Coulomb interaction parameter. As the carbon mass enters not only $<\nu>$ but also $\mu^*$, the deviation of $\alpha$ from 0.5 can be explained [93].

There is much controversy concerning vibrational modes responsible for the electron pairing that gives rise to superconductivity in the $A_3C_{60}$ fullerides. Theories invoking intermolecular vibrational modes that involve the alkali metal cations [96] can be readily ruled out on the basis of the observed insensitivity of $\nu$ to the nature of the intercalant, leaving high-frequency intracage vibrations [97–100], low-frequency intercage librations, or a mixture of both [101] as the possible candidates. NMR measurements appear to favor the second possibility [83]. However, tight-binding Hamiltonian calculations estimate the electron–phonon coupling parameter $V$ at 0.04 eV and the average vibrational frequency $<\nu>$ at 800–1000 cm$^{-1}$ [98]. When these values are used in conjunction with the calculated densities of states at the Fermi level and with $\mu^*$ equal to 0.19, Eq. (10.4) produces transition temperatures that are in reasonable agreement with the experimental data [60]. Other calculations yield somewhat lower estimates for $<\nu>$. In particular, with the values of $N_F$ obtained within the LDA formalism used as input, frequencies $<\nu>$ as low as 138 [102] and 417 cm$^{-1}$ [59] have been reported to furnish a good fit to the experimental transition temperatures. LDA calculations reproduce experimental values of $T_c$ for the $K_n Rb_{3-n} C_{60}$ series with the RMS error of only 1.7 K when the set of parameters $V$=0.057 eV, $\mu^*$=0.09, and $<\nu>$=127 cm$^{-1}$ is employed [78]. Very good agreement between theoretical predictions and experimental data is also attained by solving the Eliashberg equations [103] directly rather than using the approximate Eq. (10.4) [104].

A dissenting explanation for superconductivity in the $A_3 C_{60}$ fullerides is offered by a theory in which the electron–electron interactions are solely responsible for the pairing [105, 106]. Contrary to naive expectations, this approach, in which electron correlation effects play a central role, is capable of accounting for the observed isotope effects by invoking anharmonicities of the cage vibrations.

## 10.5 Electronic Structures of Solid $A_6 C_{60}$ Fullerides

The $A_6 C_{60}$ fullerides (with the exception of $Na_6 C_{60}$, which has an fcc structure [67]) possess bcc lattices in which each fullerene cage is surrounded by 24 alkali metal atoms [5]. The rapid rotation of fullerene cages that takes place in solid $C_{60}$ at ambient temperatures (Section 10.1) is completely suppressed in these species. X-ray diffraction measurements show that in $K_6 C_{60}$, $Cs_6 C_{60}$ [5], and $Rb_6 C_{60}$ [73] the $C_{60}^{6-}$ ions are fully ordered and the alkali metal cations occupy distorted tetrahedral interstitial sites [Fig. 10.1(c)]. The lack of rotational motion is caused by strong interactions between the $C_{60}^{6-}$ ions and the electric field produced by the counterions [107]. These interactions induce substantial distortions of the fullerene cages, reducing their symmetry from $I_h$ to $T_h$ [108].

Several electronic structure calculations have been carried out for the solid $K_6 C_{60}$ fulleride. All of these calculations indicate charge transfer of about six electrons from the alkali metal atoms to each of the fullerene cages. The presence of the $C_{60}^{6-}$ ions is consistent with the experimental optical absorption spectrum of $K_6 C_{60}$ [109]. In agreement with the results of electric conductivity measurements [110], LDA calculations predict $K_6 C_{60}$ to be a semiconductor with a narrow, indirect (from $\Gamma$ to $N$) gap of 0.21 [79] and 0.48 eV [107]. Both the valence and the conduction bands derive almost entirely from the molecular orbitals of the fullerene ions. In particular, the valence band originates from the triply degenerate $t_{1u}$ LUMOs of $C_{60}$ (Chapter 3), which are completely filled in $K_6 C_{60}$. The valence band is predicted by LDA calculations to be 0.46 [79, 81] and 0.5 eV [107] wide. The static dielectric constant of $K_6 C_{60}$ is estimated at 19.2 [81]. The computed enthalpy of the reaction

$$C_{60}(\text{solid}) + 6\ K(\text{solid}) \rightarrow K_6C_{60}(\text{solid})\ , \qquad (10.5)$$

amounts to $-235$ kcal/mol [77], which is close to the estimate of $-259$ kcal/mol afforded by a semiempirical Born–Haber cycle [87].

## 10.6  Electronic Structures of Other Solid Fullerides

In addition to the $A_3C_{60}$ and $A_6C_{60}$ fullerides, solids with the composition of $A_4C_{60}$ are known. The $K_4C_{60}$, $Rb_4C_{60}$, and $Cs_4C_{60}$ crystals possess bct lattices [4, 73, 111]. The $C_{60}^{4-}$ ions in the $A_4C_{60}$ fullerides either rotate rapidly or are disordered at room temperature. The alkali metal cations occupy tetrahedral interstitial sites of the bct lattice [Fig. 10.1($c$)]. The $A_4C_{60}$ phases do not exhibit superconductivity at low temperatures. ESR spectra of solid $K_4C_{60}$ indicate that it is not metallic, implying that its $C_{60}^{4-}$ ions are deformed by the Jahn–Teller effect [112].

Several distinct phases have been observed upon doping of $C_{60}$ with alkaline earth metals. Among those, solid $Ca_3C_{60}$ has been the subject of both experimental and theoretical studies. Its photoemission spectra indicate that the band originating from the $t_{1u}$ LUMO of $C_{60}$ is completely occupied, implying the electronic structure of $(Ca^{2+})_3C_{60}^{6-}$ [113, 114]. LDA calculations, in which the fcc structure of the alkali metal $A_3C_{60}$ fullerides is assumed for $Ca_3C_{60}$, predict it to be a narrow-gap semiconductor with the conduction band derived from both the $t_{1g}$ LUMO+1 of $C_{60}$ and the $4s$ orbitals of Ca [113, 115]. The enthalpy of the reaction

$$C_{60}(\text{solid}) + 3\ Ca(\text{solid}) \rightarrow Ca_3C_{60}(\text{solid})\ , \qquad (10.6)$$

is calculated at $-97$ kcal/mol, in significant disagreement with the estimate of $-172$ kcal/mol obtained from the Born–Haber cycle [87].

Further doping of $C_{60}$ with Ca produces a $Ca_5C_{60}$ phase (an sc lattice, $a_0 = 14.01$ Å) [116]. Solid $Ca_5C_{60}$ is metallic at room temperature [114] and becomes a superconductor upon cooling to 8.4 K [116]. Images obtained with tunneling microscopy suggest that in $Ca_5C_{60}$ four $Ca^{2+}$ ions occupy the octahedral interstitial sites, whereas the fifth cation resides in one of the tetrahedral sites [117]. LDA calculations on $Ca_5C_{60}$ uncover a partially filled band originating from both the $t_{1g}$ LUMO+1 of $C_{60}$ and $4s$ orbitals of Ca [113]. The predicted enthalpy of the reaction

$$C_{60}(\text{solid}) + 5\ Ca(\text{solid}) \rightarrow Ca_5C_{60}(\text{solid})\ , \qquad (10.7)$$

equals $-208$ kcal/mol.

Exhaustive doping of $C_{60}$ with Ca results in a $Ca_6C_{60}$ phase. As one would expect from the complete filling of both $t_{1u}$ and $t_{1g}$ bands derived from the LUMO and the LUMO+1 of $C_{60}$, photoemission spectra indicate that solid $Ca_6C_{60}$ is not a metal [114]. However, LDA calculations, carried out for bcc lattices of $Sr_6C_{60}$ and $Ba_6C_{60}$, predict these species to be semimetallic thanks to a partial overlap between strongly hybridized valence and conduction bands [118]. These predictions contradict resistivity measurements of solid $Ca_xC_{60}$ and $Sr_xC_{60}$, which reveal resistivity minima around $x = 5$ rather than $x = 6$ [119]. On the other hand, superconductivity of $Ba_6C_{60}$ observed below 7 K confirms its metallic character [120]. One should also mention that calculations based on the Born–Haber cycle predict that among the $A_6C_{60}$ alkaline earth fullerides only $Ca_6C_{60}$, $Sr_6C_{60}$, and $Ba_6C_{60}$ are stable with respect to the decomposition into $C_{60}$ and the corresponding metal [87].

There have been both experimental reports and theoretical calculations on other fullerides. Potassium-doped $C_{76}$ has been characterized experimentally as a narrow-gap semiconductor [121]. A $KC_{60}$ phase has been observed [122]. Fullerides with lanthanides and actinides have been the subject of a theoretical study [123] and the $Y_xC_{60}$ species have been investigated experimentally [124].

## 10.7  Solid-State Phases of $C_{70}$

The lower symmetry of the $C_{70}$ fullerene makes its solid-state phase diagram more complex than that of solid $C_{60}$. Between two [125, 126] and three [127] phases of $C_{70}$ have been reported to coexist at ambient temperatures, the fcc phase being stable and the hcp phase(s) being metastable. All of these high-temperature phases are characterized by a total orientational disorder of the constituting $C_{70}$ cages. Upon cooling, phase transitions at 276 and 337 K are observed [125]. The first transition produces a deformed hcp phase in which the $C_{70}$ molecules rotate only around their long axes [126, 128]. The rotational motions are completely frozen in the low-temperature monoclinic phase [129]. Finally, a rhombohedral phase of $C_{70}$ can be obtained by applying pressure [130] or irradiating solid $C_{70}$ with high-energy ions [131].

Molecular dynamics studies employing the Lennard–Jones carbon-carbon potential confirm [132] the experimental observation that freezing of the rotational motion around the short and long axes of $C_{70}$ occurs at different temperatures. However, in variance with experiment, either a triclinic [132] or an hcp [31] phase is predicted to be the most stable one at 0 K. On the other hand, application of the Lennard–Jones potential augmented with Coulombic interactions (Section 10.1) results not only in reasonably accurate predictions for the transition temperatures, but also in correct symmetries of the low-, middle-, and high-temperature phases [133].

## 10.8  Solid-State Properties of Other Species Related to Fullerenes

$C_{60}$ and other fullerenes readily form inclusion compounds with many organic and inorganic molecules. Interactions between fullerene cages and guest molecules in solid-state adducts such as $(C_{60})_2 \cdot C_6H_5Cl$, $(C_{60})_2 \cdot (S_8)_4 \cdot C_6H_5Cl$ [134], $C_{70} \cdot (S_8)_6$ [135], $C_{76} \cdot (S_8)_6$ [136], and $C_{60} \cdot S_8 \cdot CS_2$ [137] are very weak. The same is true about the solids with the $C_{60} \cdot I_2$ [138] and $C_{60} \cdot (I_2)_2$ [139] compositions that are formed upon doping of $C_{60}$ with elemental iodine [140]. In contrast to the fullerides (Sections 10.3–10.6), charge transfer between $C_{60}$ and $I_2$ is negligible in these solids [139, 141].

Reaction of $C_{60}$ with tetrakis(dimethylamino)ethylene ($C_2[N(CH_3)_2]_4$, TDAE), a well-known organic $\pi$-electron donor, yields a solid $TDAE^+C_{60}^-$ salt [7, 142]. The $TDAE^+C_{60}^-$ fulleride, which forms monoclinic crystals with highly anisotropic unit cells [7], becomes a soft ferromagnet below *ca.* 16 K. Interestingly, the analogous solid-state compounds of $C_{70}$ [143], $C_{84}$, $C_{90}$, and $C_{96}$ [144] are not ferromagnetic even at 4.5 K. One may expect the need for theoretical explanation of this distinct behavior of higher fullerenes to prompt detailed band-structure calculations on the TDAE fullerides in the near future.

The present knowledge of phase transitions in crystals of fullerene derivatives is rather limited. The $C_{60}O$ species has been reported to possess an fcc lattice ($a = 14.185$ Å) above 278 K [145]. The low-temperature phase has an sc lattice ($a = 14.062$ Å at 19 K), which is similar to that of $C_{60}$ (Section 10.1). The phase transition appears to be of the first order.

Solid-state properties of the hypothetical $C_{59}B$ heterofullerene (Chapter 9) have been studied with the LDA method [146]. The calculations predict $C_{59}B$ to form an fcc lattice with the constant of 13.49 Å and the cohesive energy of 41.5 kcal/mol of $C_{59}B$, which means that the solid-state properties of $C_{59}B$ are close to those of $C_{60}$. The computed total energy of solid $C_{59}B$ is only weakly dependent on the orientation of the individual molecules. On the other hand, the calculated densities of states exhibit strong dependence on the positions of the boron atoms relative to the crystallographic axes. Solid $C_{59}B$ is predicted to be a metal with a half-filled band. This band appears above the valence band and is similar to those of acceptor impurities in semiconductors.

## References

1. J. E. Fischer, P. A. Heiney, and A. B. Smith III, *Solid-State Chemistry of Fullerene-Based Materials*, Acc. Chem. Res. **25**, 112 (1992).
2. P. A. Heiney, J. E. Fischer, A. R. McGhie, W. J. Romanow, A. M. Denenstein, J. P. McCauley, Jr., A. B. Smith III, and D. E. Cox, *Orientational Ordering Transition in Solid $C_{60}$*, Phys. Rev. Lett. **66**, 2911 (1991).

3. P. W. Stephens, L. Mihaly, P. L. Lee, R. L. Whetten, S.-M. Huang, R. Kaner, F. Diederich, and K. Holczer, *Structure of Single-Phase Super-conducting $K_3 C_{60}$*, Nature **351**, 632 (1991).

4. R. M. Fleming, M. J. Rosseinsky, A. P. Ramirez, D. W. Murphy, J. C. Tully, R. C. Haddon, T. Siegrist, R. Tycko, S. H. Glarum, P. Marsh, G. Dabbagh, S. M. Zahurak, A. V. Makhija, and C. Hampton, *Preparation and Structure of the Alkali-Metal Fulleride $A_4 C_{60}$*, Nature **352**, 701 (1991).

5. O. Zhou, J. E. Fischer, N. Coustel, S. Kycia, O. Zhu, A. R. McGhie, W. J. Romanow, J. P. McCauley, Jr., A. B. Smith III, and D. E. Cox, *Structure and Bonding in Alkali-Metal-Doped $C_{60}$*, Nature **351**, 462 (1991).

6. A. F. Hebard, M. J. Rosseinsky, R. C. Haddon, D. W. Murphy, S. H. Glarum, T. T. M. Palstra, A. P. Ramirez, and A. R. Kortan, *Superconductivity at $18\,K$ in Potassium-Doped $C_{60}$*, Nature **350**, 600 (1991).

7. P. W. Stephens, D. Cox, J. W. Lauher, L. Mihaly, J. B. Wiley, P.-M. Allemand, A. Hirsch, K. Holczer, Q. Li, J. D. Thompson, and F. Wudl, *Lattice Structure of the Fullerene Ferromagnet $TDAE$–$C_{60}$*, Nature **355**, 331 (1992).

8. R. Tycko, G. Dabbagh, R. M. Fleming, R. C. Haddon, A. V. Makhija, and S. M. Zahurak, *Molecular Dynamics and the Phase Transition in Solid $C_{60}$*, Phys. Rev. Lett. **67**, 1886 (1991).

9. R. Tycko, R. C. Haddon, G. Dabbagh, S. H. Glarum, D. C. Douglass, and A. M. Mujsce, *Solid-State Magnetic Resonance Spectroscopy of Fullerenes*, J. Phys. Chem. **95**, 518 (1991).

10. P. C. Chow, X. Jiang, G. Reiter, P. Wochner, S. C. Moss, J. D. Axe, J. C. Hanson, R. K. McMullan, R. L. Meng, and C. W. Chu, *Synchrotron X-ray Study of Orientational Order in Single Crystal $C_{60}$ at Room Temperature*, Phys. Rev. Lett. **69**, 2943 (1992).

11. R. D. Johnson, C. S. Yannoni, H. C. Dorn, J. R. Salem, and D. S. Bethune, *$C_{60}$ Rotation in the Solid State: Dynamics of a Faceted Spherical Top*, Science **255**, 1235 (1992).

12. D. A. Neumann, J. R. D. Copley, R. L. Cappelletti, W. A. Kamitakahara, R. M. Lindstrom, K. M. Creegan, D. M. Cox, W. J. Romanow, N. Coustel, J. P. McCauley, Jr., N. C. Maliszewskyj, J. E. Fischer, and A. B. Smith III, *Coherent Quasielastic Neutron Scattering Study of the Rotational Dynamics of $C_{60}$ in the Orientationally Disordered Phase*, Phys. Rev. Lett. **67**, 3808 (1991).

13. R. C. Yu, N. Tea, M. B. Salamon, D. Lorents, and R. Malhotra, *Thermal Conductivity of Single Crystal $C_{60}$*, Phys. Rev. Lett. **68**, 2050 (1992).

14. W. I. F. David, R. M. Ibberson, J. C. Matthewman, K. Prassides, T. J. S. Dennis, J. P. Hare, H. W. Kroto, R. Taylor, and D. R. M. Walton, *Crystal Structure and Bonding of Ordered $C_{60}$*, Nature **353**, 147 (1991).

15. S. Liu, Y. Lu, M. M. Kappes, and J. A. Ibers, *The Structure of the $C_{60}$ Molecule: X-ray Crystal Structure Determination of a Twin at $110\ K$*, Science **254**, 408 (1991).

16. H.-B. Bürgi, E. Blanc, D. Schwarzenbach, S. Liu, Y. Lu, M. M. Kappes, and J. A. Ibers, *The Structure of $C_{60}$: Orientational Disorder in the Low-Temperature Modification of $C_{60}$*, Angew. Chem. Int. Ed. Engl. **31**, 640 (1992).

17. V. S. Babu and M. S. Seehra, *Temperature Dependence of the Infrared Spectra of $C_{60}$: Orientational Transition and Freezing*, Chem. Phys. Lett. **196**, 569 (1992).

18. W. I. F. David, R. M. Ibberson, T. J. S. Dennis, J. P. Hare, and K. Prassides, *Structural Phase Transitions in the Fullerene $C_{60}$*, Europhys. Lett. **18**, 219 (1992).

19. G. Van Tendeloo, A. Amelinckx, M. A. Verheijen, P. H. M. van Loosdrecht, and G. Meijer, *New Orientationally Ordered Low-Temperature Superstructure in High-Purity $C_{60}$*, Phys. Rev. Lett. **69**, 1065 (1992).

20. G. B. Alers, B. Golding, A. R. Kortan, R. C. Haddon, and F. A. Thiel, *Existence of an Orientational Electric Dipolar Response in $C_{60}$ Single Crystals*, Science **257**, 511 (1992).

21. G. A. Samara, J. E. Schirber, B. Morosin, L. V. Hansen, D. Loy, and A. P. Sylwester, *Pressure Dependence of the Orientational Ordering in Solid $C_{60}$*, Phys. Rev. Lett. **67**, 3136 (1991).

22. P. H. M. van Loosdrecht, P. J. M. van Bentum, and G. Meijer, *Rotational Ordering Transition in Single-Crystal $C_{60}$ Studied by Raman Spectroscopy*, Phys. Rev. Lett. **68**, 1176 (1992).

23. Y. Jin, J. Cheng, M. Varma-Nair, G. Liang, Y. Fu, B. Wunderlich, X.-D. Xiang, R. Mostovoy, and A. K. Zettl, *Thermodynamic Characterization of $C_{60}$ by Differential Scanning Calorimetry*, J. Phys. Chem. **96**, 5151 (1992).

24. B. Chase, N. Herron, and E. Holler, *Vibrational Spectroscopy of $C_{60}$ and $C_{70}$ Temperature-Dependent Studies*, J. Phys. Chem. **96**, 4262 (1992).

25. M. Chung, Y. Wang, J. W. Brill, X.-D. Xiang, R. Mostovoy, J. G. Hou, and A. Zettl, *AC Calorimetry of $C_{60}$ Single Crystals*, Phys. Rev. B **45**, 13831 (1992).

26. H. Kasatani, H. Terauchi, Y. Hamanaka, and S. Nakashima, *X-ray-Diffraction Study of the Phase Transition in a $C_{60}$ Single Crystal*, Phys. Rev. B **47**, 4022 (1993).

27. S. Hoen, N. G. Chopra, X.-D. Xiang, R. Mostovoy, J. Hou, W. A. Vareka, and A. Zettl, *Elastic Properties of a van der Waals Solid: $C_{60}$*, Phys. Rev. B **46**, 12737 (1992).

28. P. A. Heiney, G. B. M. Vaughan, J. E. Fischer, N. Coustel, D. E. Cox, J. R. D. Copley, D. A. Neumann, W. A. Kamitakahara, K. M. Creegan, D. M. Cox, J. P. McCauley, Jr., and A. B. Smith III, *Discontinuous Volume Change at the Orientational-Ordering Transition in Solid $C_{60}$*, Phys. Rev. B **45**, 4544 (1992).

29. A. Cheng and M. L. Klein, *Molecular-Dynamics Investigation of Orientational Freezing in Solid $C_{60}$*, Phys. Rev. B **45**, 1889 (1992).

30. A. Cheng and M. L. Klein, *Molecular Dynamics Simulations of Solid Buckminsterfullerenes*, J. Phys. Chem. **95**, 6750 (1991).

31. Y. Guo, N. Karasawa, and W. A. Goddard III, *Prediction of Fullerene Packing in $C_{60}$ and $C_{70}$ Crystals*, Nature **351**, 464 (1991).

32. Y.-N. Xu, M.-Z. Huang, and W. Y. Ching, *Theoretical Determination of the Pressure Dependence of the Electronic and the Optical Properties of fcc $C_{60}$*, Phys. Rev. B **46**, 4241 (1992).

33. N. Troullier and J. L. Martins, *Structural and Electronic Properties of $C_{60}$*, Phys. Rev. B **46**, 1754 (1992).

34. J. E. Fischer, P. A. Heiney, A. R. McGhie, W. J. Romanow, A. M. Denenstein, J. P. McCauley, Jr., and A. B. Smith III, *Compressibility of Solid $C_{60}$*, Science **252**, 1288 (1991).

35. S. J. Duclos, K. Brister, R. C. Haddon, A. R. Kortan, and F. A. Thiel, *Effects of Pressure and Stress on $C_{60}$ Fullerite to 20 GPa*, Nature **351**, 380 (1991).

36. C. Pan, M. P. Sampson, Y. Chai, R. H. Hauge, and J. L. Margrave, *Heats of Sublimation from a Polycrystalline Mixture of $C_{60}$ and $C_{70}$*, J. Phys. Chem. **95**, 2944 (1991).

37. C. K. Mathews, M. Sai Baba, T. S. Lakshmi Narasimhan, R. Balasubramanian, N. Sivaraman, T. G. Srinivasan, and P. R. Vasudeva Rao, *Vaporization Studies on Buckminsterfullerene*, J. Phys. Chem. **96**, 3566 (1992).

38. C. Pan, M. S. Chandrasekharaiah, D. Agan, R. H. Hauge, and J. L. Margrave, *Determination of Sublimation Pressures of a $C_{60}/C_{70}$ Solid Solution*, J. Phys. Chem. **96**, 6752 (1992).

39. X. D. Shi, A. R. Kortan, J. M. Williams, A. M. Kini, B. M. Savall, and P. M. Chaikin, *Sound Velocity and Attenuation in Single-Crystal $C_{60}$*, Phys. Rev. Lett. **68**, 827 (1992).

40. G. A. Samara, L. V. Hansen, R. A. Assink, B. Morosin, J. E. Schirber, and D. Loy, *Effects of Pressure and Ambient Species on the Orientational Ordering in Solid $C_{60}$*, Phys. Rev. B **47**, 4756 (1993).

41. Z. G. Li and P. J. Fagan, *Coexistence of fcc and hcp Phases of $C_{60}$. Microstructural Characterization of $C_{60}$ and Metal Complexes of $C_{60}$*, Chem. Phys. Lett. **194**, 461 (1992).

42. B. W. van de Waal, *Comment on "Coexistence of fcc and hcp Phases of $C_{60}$"*, Chem. Phys. Lett. **202**, 341 (1993).

43. L. A. Girifalco, *Interaction Potential for $C_{60}$ Molecules*, J. Phys. Chem. **95**, 5370 (1991).

44. L. A. Girifalco, *Molecular Properties of $C_{60}$ in the Gas and Solid Phases*, J. Phys. Chem. **96**, 858 (1992).

45. J. P. Lu, X.-P. Li, and R. M. Martin, *Ground State and Phase Transitions in Solid $C_{60}$*, Phys. Rev. Lett. **68**, 1551 (1992).

46. X.-P. Li, J. P. Lu, and R. M. Martin, *Ground-State Structural and Dynamical Properties of Solid $C_{60}$ from an Empirical Intermolecular Potential*, Phys. Rev. B **46**, 4301 (1992).

47. M. Sprik, A. Cheng, and M. L. Klein, *Modeling the Orientational Ordering Transition in Solid $C_{60}$*, J. Phys. Chem. **96**, 2027 (1992).

48. Y. Wang, D. Tománek, and G. F. Bertsch, *Stiffness of a Solid Composed of $C_{60}$ Clusters*, Phys. Rev. B **44**, 6562 (1991).

49. P. Haaland, R. Pachter, M. Pachter, and W. Adams, *Anisotropic Stress and Molecular Moduli of $C_{60}$*, Chem. Phys. Lett. **199**, 379 (1992).

50. R. S. Ruoff and A. L. Ruoff, *Is $C_{60}$ Stiffer than Diamond?*, Nature **350**, 663 (1991).

51. Y. Saito, H. Shinohara, M. Kato, H. Nagashima, M. Ohkohchi, and Y. Ando, *Electric Conductivity and Band Gap of Solid $C_{60}$ under High Pressure*, Chem. Phys. Lett. **189**, 236 (1992).

52. D. W. Snoke, K. Syassen, and A. Mittelbach, *Optical Absorption Spectrum of $C_{60}$ at High Pressure*, Phys. Rev. B **47**, 4146 (1993).

53. P. L. Hansen and P. J. Fallon, *An EELS Study of Fullerite — $C_{60}/C_{70}$*, Chem. Phys. Lett. **181**, 367 (1991).

54. S. L. Ren, Y. Wang, A. M. Rao, E. McRae, J. M. Holden, T. Hager, K. Wang, W.-T. Lee, H. F. Ni, J. Selegue, and P. C. Eklund, *Ellipsometric Determination of the Optical Constants of $C_{60}$ (Buckminsterfullerene) Films*, Appl. Phys. Lett. **59**, 2678 (1991).

55. R. W. Lof, M. A. van Veenendaal, B. Koopmans, H. T. Jonkman, and G. A. Sawatzky, *Band Gap, Excitons, and Coulomb Interaction in Solid $C_{60}$*, Phys. Rev. Lett. **68**, 3924 (1992).

56. W. Y. Ching, M.-Z. Huang, Y.-N. Xu, W. G. Harter, and F. T. Chan, *First-Principles Calculation of Optical Properties of $C_{60}$ in the fcc Lattice*, Phys. Rev. Lett. **67**, 2045 (1991).

57. W. Y. Ching, M.-Z. Huang, Y.-N. Xu, and F. Gan, *Electronic Structure and Optical Properties of Crystalline $C_{60}$*, Mod. Phys. Lett. B **6**, 309 (1992).

58. S. Saito and A. Oshiyama, *Cohesive Mechanism and Energy Bands of Solid $C_{60}$*, Phys. Rev. Lett. **66**, 2637 (1991).

59. S. Satpathy, V. P. Antropov, O. K. Andersen, O. Jepsen, O. Gunnarsson, and A. I. Liechtenstein, *Conduction-Band Structure of Alkali-Metal-Doped $C_{60}$*, Phys. Rev. B **46**, 1773 (1992).

60. N. Tit and V. Kumar, *A Tight Binding Model for the Electronic Structure of FCC and SC $C_{60}$*, Int. J. Mod. Phys. B **6**, 3959 (1993).

61. A. F. Hebard, R. C. Haddon, R. M. Fleming, and A. R. Kortan, *Deposition and Characterization of Fullerene Films*, Appl. Phys. Lett. **59**, 2109 (1991).

62. V. P. Antropov, O. Gunnarsson, and O. Jepsen, *Coulomb Integrals and Model Hamiltonians for $C_{60}$*, Phys. Rev. B **46**, 13647 (1992).

63. R. L. Martin and J. P. Ritchie, *Coulomb and Exchange Interactions in $C_{60}^{n-}$*, Phys. Rev. B **48**, 4845 (1993).

64. M. R. Pederson and A. A. Quong, *Polarizabilities, Charge States, and Vibrational Modes of Isolated Fullerene Molecules*, Phys. Rev. B **46**, 13584 (1992).

65. M. P. Gelfand and J. P. Lu, *Orientational Disorder and Electronic States in $C_{60}$ and $A_3C_{60}$, where A Is an Alkali Metal*, Phys. Rev. Lett. **68**, 1050 (1992).

66. K. Tanigaki, I. Hirosawa, J. Mizuki, and T. W. Ebbesen, *Lattice Parameters of Alkali-Metal-Doped $C_{60}$ Fullerides*, Chem. Phys. Lett. **213**, 395 (1993).

67. M. J. Rosseinsky, D. W. Murphy, R. M. Fleming, R. Tycko, A. P. Ramirez, T. Siegrist, G. Dabbagh, and S. E. Barrett, *Structural and Electronic Properties of Sodium-Intercalated $C_{60}$*, Nature **356**, 416 (1992).

68. K. Tanigaki, I. Hirosawa, T. W. Ebbesen, J. Mizuki, Y. Shimakawa, Y. Kubo, J. S. Tsai, and S. Kuroshima, *Superconductivity in Sodium and Lithium-Containing Alkali-Metal Fullerides*, Nature **356**, 419 (1992).

69. R. C. Haddon, A. F. Hebard, M. J. Rosseinsky, D. W. Murphy, S. J. Duclos, K. B. Lyons, B. Miller, J. M. Rosamilia, R. M. Fleming, A. R. Kortan, S. H. Glarum, A. V. Makhija, A. J. Muller, R. H. Eick, S. M. Zahurak, R. Tycko, G. Dabbagh, and F. A. Thiel, *Conducting Films of $C_{60}$ and $C_{70}$ by Alkali-Metal Doping*, Nature **350**, 320 (1991).

70. R. M. Fleming, A. P. Ramirez, M. J. Rosseinsky, D. W. Murphy, R. C. Haddon, S. M. Zahurak, and A. V. Makhija, *Relation of Structure and Superconducting Transition Temperatures in $A_3C_{60}$*, Nature **352**, 787 (1991).

71. K. Holczer, O. Klein, S.-M. Huang, R. B. Kaner, K.-J. Fu, R. L. Whetten, and F. Diederich, *Alkali-Fulleride Superconductors: Synthesis, Composition, and Diamagnetic Shielding*, Science **252**, 1154 (1991).

72. X.-D. Xiang, J. G. Hou, G. Briceño, W. A. Vareka, R. Mostovoy, A. Zettl, V. H. Crespi, and M. L. Cohen, *Synthesis and Electronic Transport of Single Crystal $K_3C_{60}$*, Science **256**, 1190 (1992).

73. P. W. Stephens, L. Mihaly, J. B. Wiley, S.-M. Huang, R. B. Kaner, F. Diederich, R. L. Whetten, and K. Holczer, *Structure of $Rb{:}C_{60}$ Compounds*, Phys. Rev. B **45**, 543 (1992).

74. K. Tanigaki, T. W. Ebbesen, S. Saito, J. Mizuki, J. S. Tsai, Y. Kubo, and S. Kuroshima, *Superconductivity at 33 K in $Cs_x Rb_y C_{60}$*, Nature **352**, 222 (1991).

75. S. E. Barrett and R. Tycko, *Molecular Orientational Dynamics in $K_3C_{60}$ Probed by Two-Dimensional Nuclear Magnetic Resonance*, Phys. Rev. Lett. **69**, 3754 (1992).

76. A. Messaoudi, J. Conard, R. Setton, and F. Béguin, *New Intercalation Compounds of $C_{60}$ with Cesium*, Chem. Phys. Lett. **202**, 506 (1993).

77. J. L. Martins and N. Troullier, *Structural and Electronic Properties of $K_n C_{60}$*, Phys. Rev. B **46**, 1766 (1992).

78. M.-Z. Huang, Y.-N. Xu, and W. Y. Ching, *Electronic Structures of $K_3C_{60}$, $RbK_2C_{60}$, $Rb_2KC_{60}$, $Rb_3C_{60}$, $Rb_2CsC_{60}$, and $Cs_3C_{60}$ Crystals*, Phys. Rev. B **46**, 6572 (1992).

79. M.-Z. Huang, Y.-N. Xu, and W. Y. Ching, *Band Structures, Density of*

*States and Fermi Surfaces of* $K_x C_{60}$, $x = 1, 2, 3, 6$, J. Chem. Phys. **96**, 1648 (1992).

80. S. C. Erwin and W. E. Pickett, *Theoretical Fermi-Surface Properties and Superconducting Parameters for* $K_3 C_{60}$, Science **254**, 842 (1991).

81. Y.-N. Xu, M.-Z. Huang, and W. Y. Ching, *Optical Properties of Superconducting* $K_3 C_{60}$ *and Insulating* $K_6 C_{60}$, Phys. Rev. B **44**, 13171 (1991).

82. S. Saito and A. Oshiyama, *Ionic Metal* $K_x C_{60}$: *Cohesion and Energy Bands*, Phys. Rev. B **44**, 11536 (1991).

83. R. Tycko, G. Dabbagh, M. J. Rosseinsky, D. W. Murphy, A. P. Ramirez, and R. M. Fleming, *Electronic Properties of Normal and Superconducting Alkali Fullerides Probed by* $^{13}C$ *Nuclear Magnetic Resonance*, Phys. Rev. Lett. **68**, 1912 (1992).

84. R. Tycko, G. Dabbagh, M. J. Rosseinsky, D. W. Murphy, R. M. Fleming, A. P. Ramirez, and J. C. Tully, $^{13}C$ *NMR Spectroscopy of* $K_x C_{60}$: *Phase Separation, Molecular Dynamics, and Metallic Properties*, Science **253**, 884 (1991).

85. C. T. Chen, L. H. Tjeng, P. Rudolf, G. Meigs, J. E. Rowe, J. Chen, J. P. McCauley, Jr., A. B. Smith III, A. R. McGhie, W. J. Romanow, and E. W. Plummer, *Electronic States and Phases of* $K_x C_{60}$ *from Photoemission and X-ray Absorption Spectroscopy*, Nature **352**, 603 (1991).

86. O. Zhou, G. B. M. Vaughan, Q. Zhu, J. E. Fischer, P. A. Heiney, N. Coustel, J. P. McCauley, Jr., and A. B. Smith III, *Compressibility of* $M_3 C_{60}$ *Fullerene Superconductors: Relation Between* $T_c$ *and Lattice Parameter*, Science **255**, 833 (1992).

87. Y. Wang, D. Tománek, G. F. Bertsch, and R. S. Ruoff, *Stability of* $C_{60}$ *Fullerite Intercalation Compounds*, Phys. Rev. B **47**, 6711 (1993).

88. K. Holczer, *Superconducting and Normal State Properties of the* $A_3 C_{60}$ *Compounds*, Int. J. Mod. Phys. B **6**, 3967 (1992).

89. G. Sparn, J. D. Thompson, S.-M. Huang, R. B. Kaner, F. Diederich, R. L. Whetten, G. Grüner, and K. Holczer, *Pressure Dependence of Superconductivity in Single-Phase* $K_3 C_{60}$, Science **252**, 1829 (1991).

90. G. Sparn, J. D. Thompson, R. L. Whetten, S.-M. Huang, R. B. Kaner, F. Diederich, G. Grüner, and K. Holczer, *Pressure and Field Dependence of Superconductivity in* $Rb_3 C_{60}$, Phys. Rev. Lett. **68**, 1228 (1992).

91. C.-C. Chen and C. M. Lieber, *Synthesis of Pure* $^{13}C_{60}$ *and Determination of the Isotope Effect for Fullerene Superconductors*, J. Am. Chem. Soc. **114**, 3141 (1992).

92. C.-C. Chen and C. M. Lieber, *Isotope Effect and Superconductivity in Metal-Doped* $C_{60}$, Science **259**, 655 (1993).

93. A. P. Ramirez, A. R. Kortan, M. J. Rosseinsky, S. J. Duclos, A. M. Mujsce, R. C. Haddon, D. W. Murphy, A. V. Makhija, S. M. Zahurak, and K. B. Lyons, *Isotope Effect in Superconducting* $Rb_3 C_{60}$, Phys. Rev. Lett. **68**, 1058 (1992).

94. C. E. Johnson, H. W. Jiang, K. Holczer, R. B. Kaner, R. L. Whetten,

and F. Diederich, *Upper-Critical-Field–Temperature Phase Diagram of Alkali-Metal-Intercalated $C_{60}$ Superconductors*, Phys. Rev. B **46**, 5880 (1992).

95. W. L. McMillan, *Transition Temperature of Strong-Coupled Superconductors*, Phys. Rev. **167**, 331 (1968).

96. F. C. Zhang, M. Ogata, and T. M. Rice, *Attractive Interaction and Superconductivity for $K_3 C_{60}$*, Phys. Rev. Lett. **67**, 3452 (1991).

97. C. M. Varma, J. Zaanen, and K. Raghavachari, *Superconductivity in the Fullerenes*, Science **254**, 989 (1991).

98. M. Schlüter, M. Lannoo, M. Needels, G. A. Baraff, and D. Tománek, *Electron-Phonon Coupling and Superconductivity in Alkali-Intercalated $C_{60}$ Solid*, Phys. Rev. Lett. **68**, 526 (1992).

99. S. Chakravarty, S. Khlebnikov, and S. Kivelson, *Comment on "Electron-Phonon Coupling and Superconductivity in Alkali-Intercalated $C_{60}$ Solid"*, Phys. Rev. Lett. **69**, 212 (1992).

100. M. A. Schlüter, M. Lannoo, M. F. Needels, G. A. Baraff, and D. Tománek, *Reply to Comment on "Electron-Phonon Coupling and Superconductivity in Alkali-Intercalated $C_{60}$ Solid"*, Phys. Rev. Lett. **69**, 213 (1992).

101. I. I. Mazin, O. V. Dolgov, A. Golubov, and S. V. Shulga, *Strong-Coupling Effects in Alkali-Metal-Doped $C_{60}$*, Phys. Rev. B **47**, 538 (1993).

102. I. I. Mazin, S. N. Rashkeev, V. P. Antropov, O. Jepsen, A. I. Liechtenstein, and O. K. Andersen, *Quantitative Theory of Superconductivity in Doped $C_{60}$*, Phys. Rev. B **45**, 5114 (1992).

103. G. M. Éliashberg, *Interactions Between Electrons and Lattice Vibrations in a Superconductor*, Sov. Phys. JETP **11**, 696 (1960).

104. H. Zheng and K.-H. Bennemann, *Calculation of the Superconducting Critical Temperature in Doped Fullerenes*, Phys. Rev. B **46**, 11993 (1992).

105. S. Chakravarty, M. P. Gelfand, and S. Kivelson, *Electronic Correlation Effects and Superconductivity in Doped Fullerenes*, Science **254**, 970 (1991).

106. S. Chakravarty, S. A. Kivelson, M. I. Salkola, and S. Tewari, *Isotope Effect in Superconducting Fullerenes*, Science **256**, 1306 (1992).

107. S. C. Erwin and M. R. Pederson, *Electronic Structure of Crystalline $K_6 C_{60}$*, Phys. Rev. Lett. **67**, 1610 (1991).

108. W. Andreoni, F. Gygi, and M. Parrinello, *Doping-Induced Distortions and Bonding in $K_6 C_{60}$ and $Rb_6 C_{60}$*, Phys. Rev. Lett. **68**, 823 (1992).

109. K.-J. Fu, W. L. Karney, O. L. Chapman, S.-M. Huang, R. B. Kaner, F. Diederich, K. Holczer, and R. L. Whetten, *Giant Vibrational Resonances in $A_6 C_{60}$ Compounds*, Phys. Rev. B **46**, 1937 (1992).

110. G. P. Kochanski, A. F. Hebard, R. C. Haddon, and A. T. Fiory, *Electrical Resistivity and Stoichiometry of $K_x C_{60}$ Films*, Science **255**, 184 (1992).

111. R. M. Fleming, M. J. Rosseinsky, A. P. Ramirez, D. W. Murphy, J. C. Tully, R. C. Haddon, T. Siegrist, R. Tycko, S. H. Glarum, P. Marsh, G.

Dabbagh, S. M. Zahurak, A. V. Makhija, and C. Hampton, *Preparation and Structure of the Alkali-Metal Fulleride, $A_4C_{60}$*, Nature **353**, 868 (1991).

112. M. Kosaka, K. Tanigaki, I. Hirosawa, Y. Shimakawa, S. Kuroshima, T. W. Ebbesen, J. Mizuki, and Y. Kubo, *ESR Studies of K-Doped $C_{60}$*, Chem. Phys. Lett. **203**, 429 (1993).

113. Y. Chen, D. M. Poirier, M. B. Jost, C. Gu, T. R. Ohno, J. L. Martins, J. H. Weaver, L. P. F. Chibante, and R. E. Smalley, *Electronic Structure of $Ca_x C_{60}$ Fullerides*, Phys. Rev. B **46**, 7961 (1992).

114. G. K. Wertheim, D. N. E. Buchanan, and J. E. Rowe, *Charge Donation by Calcium into the $t_{1g}$ Band of $C_{60}$*, Science **258**, 1638 (1992).

115. S. Saito and A. Oshiyama, *Electronic Structure of Calcium-Doped $C_{60}$*, Solid State Commun. **83**, 107 (1992).

116. A. R. Kortan, N. Kopylov, S. Glarum, E. M. Gyorgy, A. P. Ramirez, R. M. Fleming, F. A. Thiel, and R. C. Haddon, *Superconductivity at $8.4\,K$ in Calcium-Doped $C_{60}$*, Nature **355**, 529 (1992).

117. Y. Z. Li, J. C. Patrin, M. Chander, J. H. Weaver, L. P. F. Chibante, and R. E. Smalley, *Real-Space Imaging of $Ca_x C_{60}$ Using Scanning Tunneling Microscopy*, Phys. Rev. B **46**, 12914 (1992).

118. S. Saito and A. Oshiyama, *$Sr_6 C_{60}$ and $Ba_6 C_{60}$: Semimetallic Fullerides*, Phys. Rev. Lett. **71**, 121 (1993).

119. R. C. Haddon, G. P. Kochanski, A. F. Hebard, A. T. Fiory, and R. C. Morris, *Electrical Resistivity and Stoichiometry of $Ca_x C_{60}$ and $Sr_x C_{60}$ Films*, Science **258**, 1636 (1992).

120. A. R. Kortan, N. Kopylov, S. Glarum, E. M. Gyorgy, A. P. Ramirez, R. M. Fleming, O. Zhou, F. A. Thiel, P. L. Trevor, and R. C. Haddon, *Superconductivity in Barium Fulleride*, Nature **360**, 566 (1992).

121. S. Hino, K. Matsumoto, S. Hasegawa, H. Inokuchi, T. Morikawa, T. Takahashi, K. Seki, K. Kikuchi, S. Suzuki, I. Ikemoto, and Y. Achiba, *Ultraviolet Photoelectron Spectra of $C_{76}$ and $K_x C_{76}$*, Chem. Phys. Lett. **197**, 38 (1992).

122. D. M. Poirier and J. H. Weaver, *$KC_{60}$ Fulleride Phase formation: An X-ray Photoemission Study*, Phys. Rev. B **47**, 10959 (1993).

123. R. S. Ruoff, Y. Wang, and D. Tománek, *Lanthanide- and Actinide-Based Fullerite Compounds: Potential $A_x C_{60}$ Superconductors?*, Chem. Phys. Lett. **203**, 438 (1993).

124. T. R. Ohno, G. H. Kroll, J. H. Weaver, L. P. F. Chibante, and R. E. Smalley, *Yb and Yb-K Fulleride Formation, Bonding, and Electrical Character*, Phys. Rev. B **46**, 10437 (1992).

125. G. B. M. Vaughan, P. A. Heiney, J. E. Fischer, D. E. Luzzi, D. A. Ricketts-Foot, A. R. McGhie, Y.-W. Hui, A. L. Smith, D. E. Cox, W. J. Romanow, B. H. Allen, N. Coustel, J. P. McCauley, Jr., and A. B. Smith III, *Orientational Disorder in Solvent-Free Solid $C_{70}$*, Science **254**, 1350 (1991).

126. G. Van Tendeloo, S. Amelinckx, J. L. De Boer, S. van Smaalen, M. A. Verheijen, H. Meekes, and G. Meijer, *Structural Phase Transitions in $C_{70}$*, Europhys. Lett. **21**, 329 (1993).

127. M. A. Green, M. Kurmoo, P. Day, and K. Kikuchi, *Structural Phase Transformations in $C_{70}$*, J. Chem. Soc. Chem. Commun. **92**, 1676 (1992).

128. M. A. Verheijen, H. Meekes, G. Meijer, P. Bennema, J. L. de Boer, S. van Smaalen, G. V. van Tendeloo, S. Amelinckx, S. Muto, and J. van Landuyt, *The Structure of Different Phases of Pure $C_{70}$ Crystals*, Chem. Phys. **166**, 287 (1992).

129. G. B. M. Vaughan, P. A. Heiney, D. E. Cox, J. E. Fischer, A. R. McGhie, A. L. Smith, R. M. Strongin, M. A. Cichy, and A. B. Smith III, *Structural Phase Transitions and Orientational Ordering in $C_{70}$*, Chem. Phys. **178**, 599 (1993).

130. H. Kawamura, M. Kobayashi, Y. Akahama, H. Shinohara, H. Sato, and Y. Saito, *Orientational Ordering in Solid $C_{70}$ under High Pressure*, Solid State Commun. **83**, 563 (1992).

131. K. Misof, P. Fratzl, and G. Vogl, *Experimental Evidence for Rhombohedral Phase of $C_{70}$ after Irradiation*, Europhys. Lett. **22**, 585 (1993).

132. A. Cheng and M. L. Klein, *Solid $C_{70}$: A Molecular-Dynamics Study of the Structure and Orientational Ordering*, Phys. Rev. B **46**, 4958 (1992).

133. M. Sprik, A. Cheng, and M. L. Klein, *Orientational Ordering in Solid $C_{70}$: Predictions from Computer Simulation*, Phys. Rev. Lett. **69**, 1660 (1992).

134. N. D. Kushch, I. Majchrzak, W. Ciesielski, and A. Graja, *Preparation and Spectral Properties of $C_{60}S_{16} \cdot 0.5 C_6H_5Cl$ and $C_{60} \cdot 0.5 C_6H_5Cl$ Compounds*, Chem. Phys. Lett. **215**, 137 (1993).

135. G. Roth and P. Adelmann, *The Crystal Structure of $C_{70}S_{48}$: The First a Priori Structure Determination of a $C_{70}$-Containing Compound*, J. Phys. I France **2**, 1541 (1992).

136. M. M. Kappes, personal communication (1994).

137. G. Roth, P. Adelmann, and R. Knitter, *Preparation and Crystal Structure of $C_{60}S_8 CS_2$, a New Fullerene-Containing Heteromolecular Solid*, Mater. Lett. **16**, 357 (1993).

138. T. Zenner and H. Zabel, *Synthesis, Characterization, and Stability of $C_{60}I_2$*, J. Phys. Chem. **97**, 8690 (1993).

139. Q. Zhu, D. E. Cox, J. E. Fischer, K. Kniaz, A. R. McGhie, and O. Zhou, *Intercalation of Solid $C_{60}$ with Iodine*, Nature **355**, 712 (1992).

140. M. Kobayashi, Y. Akahama, H. Kawamura, H. Shinohara, H. Sato, and Y. Saito, *X-ray Diffraction Study of Iodine-Doped $C_{60}$*, Solid State Commun. **81**, 93 (1992).

141. T. R. Ohno, G. H. Kroll, J. H. Weaver, L. P. F. Chibante, and R. E. Smalley, *Doping of $C_{60}$ with Iodine*, Nature **355**, 401 (1992).

142. P.-M. Allemand, K. C. Khemani, A. Koch, F. Wudl, K. Holczer, S. Donovan, G. Grüner, and J. D. Thompson, *Organic Molecular Soft Ferromagnetism in a Fullerene $C_{60}$*, Science **253**, 301 (1991).

143. K. Tanaka, A. A. Zakhidov, K. Yoshizawa, K. Okahara, T. Yamabe, K. Yakushi, K. Kikuchi, S. Suzuki, I. Ikemoto, and Y. Achiba, *Magnetic Properties of TDAE-$C_{60}$ and TDAE-$C_{70}$. A Comparative Study*, Phys. Lett A **164**, 221 (1992).

144. K. Tanaka, A. A. Zakhidov, K. Yoshizawa, K. Okahara, T. Yamabe, K. Kikuchi, S. Suzuki, I. Ikemoto, and Y. Achiba, *Magnetic Properties of Higher Fullerides TDAE-$C_{84}$, $-C_{90}$ and $-C_{96}$*, Solid State Commun. **85**, 69 (1993).

145. G. B. M. Vaughan, P. A. Heiney, D. E. Cox, A. R. McGhie, D. R. Jones, R. M. Strongin, M. A. Cichy, and A. B. Smith III, *The Orientational Phase Transition in Solid Buckminsterfullerene Epoxide ($C_{60}O$)*, Chem. Phys. **168**, 185 (1992).

146. Y. Miyamoto, N. Hamada, A. Oshiyama, and S. Saito, *Electronic Structures of Solid $BC_{59}$*, Phys. Rev. B **46**, 1749 (1992).

# Chapter 11

# Conclusions and Future Directions

The last five years of fullerene research have been punctuated by many unexpected discoveries amid rapid broad advances. The unprecedented success of fullerene research can be credited without doubt to its multidisciplinary nature. Among the disciplines involved, quantum chemistry has made a significant contribution to the understanding of fullerenes and their properties, as valuable insights have been gained and useful predictions have been gathered through the judicious application of a broad spectrum of electronic structure methods.

As mentioned in Chapter 3, during the early stages of experimental research on $C_{60}$, quantum-chemical calculations furnished crucial data on its IR and Raman spectra. Simple theoretical approaches, such as the Hückel approximation, have helped establish the viability of the icosahedral structure for the $C_{60}$ cluster. The calculated orbital energy levels have provided the explanation for the observed properties of $C_{60}$ anions. Theoretical predictions have also contributed to the resolution of the controversy surrounding the second vertical ionization potential of $C_{60}$.

In the case of the $C_{70}$ fullerene (Chapter 4), results of electronic structure calculations have confirmed the experimental value of its standard enthalpy of formation. Theoretical predictions of bond lengths have uncovered systematic errors in the experimental geometry of $C_{70}$ obtained from electron diffraction measurements. Calculations have also demonstrated that $C_{70}$ constitutes an island of stability among carbon clusters, explaining the experimental failure to isolate the $C_{68}$ and $C_{72}$ species.

The knowledge gained from theoretical studies of $C_{60}$ and $C_{70}$ has been highly useful in research on higher fullerenes. In particular, the isolated pentagon rule (IPR; Chapter 1) has been indispensable in predicting relative stabilities of fullerene cages. Application of this rule to the medium-size fullerenes (Chapter 5) has made it possible to preselect a small number of viable structures from the plethora of possible fullerene isomers, opening the avenue to reliable theoretical predictions of the most stable species. Although the simple Hückel approach has been found inadequate for this purpose, *ab initio* methods have furnished the lowest-energy isomers that consistently match those observed experimentally.

Extensive electronic structure calculations have been performed on various hypothetical allotropes of carbon (Chapter 6). Since these modifications of carbon are presently unknown, the future effort toward their synthesis will undoubtedly be influenced by the theoretical predictions. On the other hand, the research on carbon clusters with fewer than sixty atoms has already greatly benefited from quantum-chemical investigations of their relative stabilities, which are described in Chapter 7.

Properties of endohedral complexes (Chapter 8) had been predicted before macroscopic quantities of these truly remarkable species have become available. A complete understanding of bonding in endohedral complexes has been attained. The absence or presence of phenomena such as charge transfer and displacement of the guests from the cage centers in individual complexes has been explained. GIAO CPHF calculations have correctly predicted both the signs and the magnitudes of changes in the chemical shifts experienced by the guests upon encagement.

Significant progress in theoretical research on properties of fullerene derivatives has been achieved. As described in Chapter 9, the futility of experimental attempts to synthesize the icosahedral $C_{60}H_{60}$ and $C_{60}F_{60}$ species has been accounted for by the computed large steric repulsions between the addents. The regiochemistry of the fullerene hydrogenation has been successfully reproduced for $C_{60}$ and $C_{70}$ by high-quality *ab initio* calculations.

Theoretical investigations of solid-state fullerenes and their derivatives have abounded in the recent literature. The presence of almost complete charge transfer in the alkali metal fullerides has been predicted. Building on these predictions, several theories of superconductivity in the $A_3C_{60}$ fullerides have been proposed, providing a sound explanation for the experimentally observed pressure and composition dependences of the critical temperature. Reasonably accurate predictions of this and other properties of the superconducting phase have been reported.

As one may judge from this summary, the successes of electronic structure calculations on fullerenes are legion. However, unresolved questions and occasional failures remain to be addressed. While trying to assess the scope of future theoretical research on carbon clusters, one should be reminded that most of the calculations described in this book were deemed unfeasible a decade ago. Extrapolating from the rapid pace of improvement in the past, we can confidently predict that properties of fullerenes with fewer than 200 atoms will be accurately computed in a routine manner with high-quality *ab initio* techniques by the year 2000. Keeping this prediction in mind, one may compose a "wish list" of future developments that would significantly advance our knowledge of fullerenes.

First, taking into account the fast progress in both the exohedral and the endohedral chemistry of fullerenes, and the ensuing emergence of $^3$He NMR spectroscopy, one concludes that the development of methods capable of more accurate chemical shift calculations should take a priority. The GIAO CPHF formalism employed in the contemporary calculations on fullerenes yields reasonable shielding tensors, but its accuracy is insufficient for assigning the observed lines in the $^{13}$C NMR spectra to individual nuclei (Chapters 4 and 5). Test calculations indicate that errors in the computed chemical shifts are mostly due to inaccurate geometries. Since the Hartree–Fock approximation, which presently dominates the field of large-scale electronic structure calculations, exaggerates bond alternation in fullerenes (Chapters 3 and 5), inclusion of electron correlation effects is crucial for improvements in the computed bond lengths and, in turn, the NMR shifts. As the MP2 calculations will almost certainly remain too expensive except for systems with high molecular symmetries, one may expect approaches based on the density functional theory to emerge as the methods of choice for optimizations of fullerene geometries.

Second, although the currently used *ab initio* techniques are very successful in predicting energies of isodesmic reactions involving fullerenes (Chapters 4 and 5), they are practically useless for energy comparisons involving systems with unequal numbers of bonds, such as isomers of the $C_{20}$ cluster (Chapter 7). Rectification of this deficiency will require the development of either more sophisticated electron density functionals for the correlation energy or a new semiempirical method parameterized expressly for carbon clusters.

Third, with the number of computed structures rapidly rising, the importance of being able to determine whether the optimized geometries correspond to energy minima is becoming more evident. Currently, this ability is mostly limited to calculations involving semiempirical approaches. As exemplified by several of the $C_{82}$ fullerene isomers (Chapter 5), systems with small HOMO–LUMO gaps are often prone to symmetry lowering, which is unrelated to a Jahn–Teller distortion. It is therefore entirely possible that some of the symmetry-constrained *ab initio* optimizations, such as those for the highly charged cations of $C_{60}$ (Chapter 3), have not produced genuine energy minima. In light of these concerns, the need for large-scale *ab initio* calculations of vibrational frequencies is apparent.

Fourth, several cases remain where theoretical predictions are in variance with the experimental data. Among these, the predicted instability of the $C_{60}^{2-}$ anion with respect to electron detachment, contradicting its experimental observation in the gas phase (Chapter 3), is particularly puzzling. Another unexplained discrepancy involves the $C_{60}O$ species for which the 1,6-open-bridge structure is consistently favored over the 1,2- epoxide isomer by a broad spectrum of quantum-chemical methods (Chapter 9). The experimental second-order hyperpolarizabilities of $C_{60}$ and $C_{70}$, discussed in Chapters 3 and 4, are very poorly reproduced by theoretical calculations. One anticipates these disagreements between experiment and theory to be resolved with the advent of more accurate electronic structure approaches.

Fifth, in a few instances a consensus among theoretical predictions is lacking for properties of species not yet amenable to experimental research. For example, neither the ground-state structure nor the corresponding electronic configuration of the $Ti_8C_{12}$ metallocarbohedrene (Chapter 9) is presently known. The $Si_{60}$ cluster has been speculated to possess a cage structure analogous to $C_{60}$, but other calculations have found the arrangement of six stacked naphthalene-like rings more favorable. As mentioned in Chapter 3, both allowed and forbidden transitions have been ascribed to the first singlet excited state of $C_{60}$. Again, final resolution of these contradictions has to await more sophisticated calculations.

Sixth, many experimental findings that run contrary to "chemical intuition" remain to be investigated with theoretical methods. A good example of such findings is provided by the reported formation of endohedral complexes between Kr and Xe, and $C_{60}$ (Chapter 8). Although these noble gas atoms appear to be too large to fit inside the $C_{60}$ cage, a possibility exists that some form of direct guest–host bonding or attractive dispersion interactions furnishes enough stabilization to assure the existence of the $Kr@C_{60}$ and $Xe@C_{60}$ species. Electronic structure calculations will be capable of providing a decisive verdict in this case only when formalisms affording accurate *ab initio* estimates of the dispersion energies become available.

Seventh, despite the recent significant advances, phenomena pertinent to solid-state fullerenes and their derivatives (Chapter 10), such as the superconductivity in the $A_3C_{60}$ fullerides, the presence of the low-temperature ferromagnetism in the $TDAE^+C_{60}^-$ salt, and its absence in the analogous compounds of higher fullerenes, are still lacking detailed theoretical descriptions. With progress in band structure calculations, the presently used phenomenological theories will certainly give way to more rigorous approaches, making it possible to predict solid-state properties of fullerenes, fullerene salts, and various allotropes of carbon with greater accuracy.

In summary, the serendipity of empirical discovery and the deliberateness of theoretical calculations have proved to be a winning combination for fullerene research. The benefits have been mutual. On several occasions, the results of theoretical calculations have guided the experimental studies of fullerenes. At the same time, the new substances and their experimentally measured properties have created the need for large-scale electronic structure calculations, driving the development of new theoretical techniques.

# Index

Pages containing definitions of individual entries are listed in bold face.